Oxford Resources for IB
Diploma Programme

IB PREPARED

2023 EDITION
PHYSICS

David Homer
Maciej Pietka

OXFORD
UNIVERSITY PRESS

OXFORD
UNIVERSITY PRESS

Great Clarendon Street, Oxford, OX2 6DP, United Kingdom

Oxford University Press is a department of the University of Oxford. It furthers the University's objective of excellence in research, scholarship, and education by publishing worldwide. Oxford is a registered trade mark of Oxford University Press in the UK and in certain other countries.

© Oxford University Press 2024

The moral rights of the authors have been asserted

First published in 2024

All rights reserved. No part of this publication may be reproduced, stored in a retrieval system, transmitted, used for text and data mining, or used for training artificial intelligence, in any form or by any means, without the prior permission in writing of Oxford University Press, or as expressly permitted by law, by licence or under terms agreed with the appropriate reprographics rights organization. Enquiries concerning reproduction outside the scope of the above should be sent to the Rights Department, Oxford University Press, at the address above.

You must not circulate this work in any other form and you must impose this same condition on any acquirer

British Library Cataloguing in Publication Data
Data available

9781382058391

9781382058421 (ebook)

10 9 8 7 6 5 4 3 2 1

Paper used in the production of this book is a natural, recyclable product made from wood grown in sustainable forests.

The manufacturing process conforms to the environmental regulations of the country of origin.

Printed in China by Shanghai Offset Printing Products Ltd

Acknowledgements

The "In cooperation with IB" logo signifies the content in this textbook has been reviewed by the IB to ensure it fully aligns with current IB curriculum and offers high-quality guidance and support for IB teaching and learning.

The Publisher wishes to thank the International Baccalaureate Organization for permission to reproduce their intellectual property.

The Publisher also wishes to thank David Lyons for his review of the book content.

The publisher would like to thank the following for permissions to use copyright material:

Cover: Michael Gollotti / Getty Images

Artworks: QBS Learning

Photos: piii, **piv**, **p239**, **p244**, **p253**: CLAUDE NURIDSANY & MARIA PERENNOU/SCIENCE PHOTO LIBRARY; **p1:** fuyu liu / Shutterstock; **p17:** amlukyeee / Shutterstock; **p53:** Abstract51 / Shutterstock; **p97:** saurabh7275 / Shutterstock; **p119:** GIPHOTOSTOCK / SCIENCE PHOTO LIBRARY; **p136:** Brian Maudsley / Alamy Stock Photo; **p180:** NASA images / Shutterstock; **p227:** maxpro / Shutterstock.

Although we have made every effort to trace and contact all copyright holders before publication this has not been possible in all cases. If notified, the publisher will rectify any errors or omissions at the earliest opportunity.

Links to third party websites are provided by Oxford in good faith and for information only. Oxford disclaims any responsibility for the materials contained in any third party website referenced in this work.

Contents

Introduction ... iv

A Space, time and motion — 1
- A.1 Kinematics — 1
- A.2 Forces and momentum — 8
- A.3 Work, energy and power — 24
- A.4 Rigid body mechanics — 32
- A.5 Galilean and special relativity — 40

B The particulate nature of matter — 53
- B.1 Thermal energy transfers — 53
- B.2 Greenhouse effect — 63
- B.3 Gas laws — 69
- B.4 Thermodynamics — 77
- B.5 Current and circuits — 87

C Wave behaviour — 97
- C.1 Simple harmonic motion — 97
- C.2 Wave model — 107
- C.3 Wave phenomena — 112
- C.4 Standing waves and resonance — 122
- C.5 Doppler effect — 131

D Fields — 136
- D.1 Gravitational fields — 136
- D.2 Electric and magnetic fields — 148
- D.3 Motion in electromagnetic fields — 163
- D.4 Induction — 172

E Nuclear and quantum physics — 180
- E.1 Structure of the atom — 180
- E.2 Quantum physics — 188
- E.3 Radioactive decay — 196
- E.4 Fission — 211
- E.5 Fusion and stars — 217

Tools for physics — 227
Internal Assessment — 239
Practice Exam papers — 244
Index — 253

Answers
www.oxfordsecondary.com/ib-prepared-2e

Introduction

This book provides full coverage of the Diploma Programme physics course and offers support to students preparing for their examinations. The book will help you:
- revise the study material
- learn the essential terms and concepts
- strengthen your problem-solving skills
- improve your approach to IB examinations.

The book is packed with worked examples and exam tips that demonstrate best practices and warn against common errors.

All topics are illustrated by annotated student answers to questions from past examinations, which explain why marks may be scored or missed. Numerous examples show you how to tackle unfamiliar situations, interpret, and analyse experimental data, and suggest possible improvements to experimental procedures. Practice problems and a complete set of IB-style examination papers provide further opportunities to check your knowledge and skills, boost your confidence and monitor the progress of your studies.

Full solutions to all problems and examination papers are given online at www.oxfordsecondary.com/ib-prepared-2e

This book is not intended to replace your course materials, such as textbooks, laboratory manuals, past papers and mark schemes, the IB Diploma *Physics guide*, DP *Physics data booklet* and your own notes. To succeed in the assessment, you will need to use a broad range of resources, many of which are available online. The authors hope that this book will navigate you through this critical part of your studies, making your preparation for the exam less stressful and more efficient.

Overview of the book structure

The book is divided into several sections that cover core standard level (SL) and additional higher level (AHL) topics, the internal assessment, tools for physics and a complete set of practice examination papers.

- The largest section of the book follows the structure of the syllabus for the Diploma Programme physics course and covers everything that you need.
- The **internal assessment** section outlines the nature of the investigation that you will have to carry out. It explains how to select a suitable topic, collect and process experimental data, draw conclusions, and present your report in a suitable format to satisfy the marking criteria and achieve your best possible grade.
- The final section contains IB-style **practice examination papers 1 and 2**, written exclusively for this book. These papers will give you an opportunity to test yourself before the actual exam and at the same time provide additional practice problems for every topic.

The answers and solutions to all practice problems and examination papers are given online at www.oxfordsecondary.com/ib-prepared-2e

DP Physics assessment

All standard level (SL) and higher level (HL) students must complete the internal assessment and take two papers as part of their external assessment. Papers 1 and 2 are usually sat on different days.

The internal and external assessment marks are combined as shown in Table 1 to give your overall IB Diploma Programme physics grade, from 1 (lowest) to 7 (highest).

Assessment	Description	Duration	Standard Level		Higher Level	
			Marks	Weight	Marks	Weight
Internal	Experimental work with a report	10 hours	24	20%	24	20%
Paper 1a	Multiple-choice questions	SL: 1.5 hours	45	36%	60	36%
Paper 1b	Data-based questions	HL: 2 hours				
Paper 2	Short-answer and extended-response questions	SL: 1.5 hours HL: 2.5 hours	50	44%	90	44%

▲ Table 1 Assessment overview

Your overall IB Diploma score is calculated by combining grades for six subjects with up to three additional points from theory of knowledge and extended essay components.

Students undertaking standard level (SL) do not cover the same material as students studying higher level (HL). However, material common to both levels is covered to the same depth and is examined at the same standard.

The topics that are only examined at HL are identified in the margin.

Table 2 shows the topics that are completely excluded from the SL assessments and Table 3 lists the topics that include some content just for HL students.

Topic	Topics that are not part of the SL assessment
A.4	Rigid body mechanics
A.5	Galilean and special relativity
B.4	Thermodynamics
D.4	Induction
E.2	Quantum physics

▲ **Table 2** Topics completely excluded from the SL assessments

Topic	Areas that are not part of the SL assessment
A.2	Conservation of momentum for collisions in two dimensions
C.1	Phase angle
	Knowledge and use of kinematic and energy equations for simple harmonic motion
C.3	Single-slit diffraction patterns
	The modulation of an interference pattern by a single-slit diffraction pattern
	Diffraction gratings
C.5	Knowledge and use of Doppler-shift equations for observed frequency for sound and mechanical waves
D.1	Gravitational potential and gravitational potential energy
	Gravitational field strength as the negative of the derivative of gravitational potential with respect to distance
	Escape speed and orbital speed
	The effect of viscous drag on the motion of an orbiting body
D.2	Electric potential and electric potential energy
	Electric field strength as the negative of the derivative of electric potential with respect to distance
	Equipotential surfaces
E.1	The dependence of nuclear radius on nucleon number
	Deviations from Rutherford scattering and closest approach in head-on scattering
	The Bohr model for the hydrogen atom
	• discrete energy levels in the model
	• quantized energy and quantized angular momentum
E.3	Evidence for the strong nuclear force
	Nuclear stability
	• the role of the ratio $\frac{\text{neutron number}}{\text{proton number}}$
	• the significance of the binding energy per nucleon curve for $A > 60$
	Discrete nuclear levels as evidenced by alpha emission and gamma radiation spectra
	Decay constant and the probability of decay in unit time
	Solving radioactive decay problems where the time intervals concerned are not integer values of half-lives

▲ **Table 3** Some material within these topics is for HL students only

Command terms

Command terms are pre-defined words and phrases used in all Diploma Programme physics questions and problems. Each command term specifies the type and depth of the response expected from you in a particular question.

For example, the command terms *state*, *outline*, *explain* and *discuss* require answers with increasingly higher levels of detail, from a single word, short sentence, or numerical value ("state") to comprehensive analysis ("discuss"). A list of the command terms used in Diploma Programme physics examination questions is given in Table 4. However, it is possible that you will see other words used where they are more appropriate.

Understanding the exact meaning of the command terms is essential for your success in the examination. Explore this table and use it regularly as a reference when answering questions during your examination preparation.

Introduction

Command term	Definition	Typical question
Analyse	Break a complex idea down in order to bring out the essential elements or structure.	**Analyse** this expression to show that the system will perform simple harmonic motion.
Annotate	Add notes or labelling to a diagram or a graph.	**Annotate** your graph to show the point at which the ball reaches the top of its trajectory.
Calculate	Obtain a numerical answer to a problem showing the stages in the working.	**Calculate** the resistance of the filament lamp.
Deduce	Reach a conclusion from the information given.	**Deduce,** using the graph, the value of the Planck constant.
Describe	Give a detailed account of a piece of physics (e.g. an experiment or an idea).	**Describe** how pressure arises at the walls of a container filled with an ideal gas.
Determine	Obtain the only possible answer to a problem. You should show all your working in the event that there is an error somewhere in your solution.	**Determine** the gravitational field strength at that point.
Discuss	Give a review of an issue that includes a range of the factors, assumptions or hypotheses that apply. Support your answer with evidence where possible.	**Discuss** why the moderator in a nuclear reactor should be a poor neutron absorber.
Draw	Represent a piece of physics using a labelled diagram or graph. Graphs should be neatly drawn in pencil with data points correct and joined by a straight or curved line as appropriate.	**Draw** the best straight line for the data.
Estimate	Obtain an approximate value.	**Estimate** the change in gravitational potential energy of an apple as it falls from a tree.
Explain	Give a detailed account of a piece of physics.	**Explain** why there is a node at the closed end of a pipe that contains a stationary wave in the air of the pipe.
Identify	Provide an answer from a number of possibilities.	**Identify** the direction in which the force acts.
Outline	Give a summary or a brief account of a piece of physics.	**Outline** what is meant by the binding energy of a nucleus.
Predict	Give an expected result. This may involve a calculation to obtain the result (in which case, you must show your working) or it may involve you using qualitative reasoning to make a prediction.	**Predict** the subsequent motion of the alpha-decay products of an initially stationary nuclide.
Show (that)	Give all the steps in a calculation or a derivation. If the question involves a calculation, you should calculate the answer to at least one more significant figure than the "show that" value.	**Show that** the average output power of the car is about 30 kW.
Sketch	Use a diagram or graph to show the detailed relationship involved in the question. Include all the relevant features in your graph, including axes labels and values on the axes (unless specifically told not to). Pay particular attention to any asymptotic values.	**Sketch** a graph to show how I varies with V.
State	Give an answer without an explanation.	**State** what is meant by resistivity.
Suggest	Propose a solution, an explanation or a theory for a piece of physics.	**Suggest** why the orbital speed of a satellite changes when atmospheric drag acts on it.

▲ Table 4 The meaning and use of command terms

Preparation and exam strategies

Try to follow these simple rules during your preparation study and the exam itself.

1. **Get ready for study**. Have enough sleep, eat well, drink plenty of water and reduce your stress by positive thinking and physical exercise. A good night's sleep is particularly important before the exam day, as it can improve your score.

2. **Organize your study** environment. Find a comfortable place with adequate lighting, temperature and ventilation. Avoid distractions. Keep your papers and computer files organized. Bookmark useful online and offline material.

3. **Plan your studies**. Make a list of your tasks and arrange them by importance. Break up large tasks into smaller, more manageable parts. Create an agenda for your studying time and make sure that you can complete each task on time.

4. Use this book as your first point of **reference**. Work your way through the topics systematically and identify any gaps in your understanding and skills. Spend extra time on the areas where improvement is required. Check your textbook and online resources for more information.

5. **Read actively**. Focus on understanding rather than rote learning. Recite key points and definitions using your own words. Try to solve every worked example and practice problem before looking at the answer. Make notes for future reference.

6. **Get ready for the exams**. Practise answering exam-style questions under a time constraint. Learn how to use the *Physics data booklet* quickly and efficiently. Solve as many problems from past papers as you can. Take a trial exam using the papers at the end of this book.

7. **Optimize your exam approach**. Read all questions carefully, paying extra attention to command terms. Keep your answers as short and clear as possible. You should provide an answer and not simply re-write the question for the examiner. Double-check all numerical values and units and quote answers to sensible numbers of significant figures. Label graph axes with the quantity and unit, and annotate diagrams. Use exam tips from this book.

8. **Do not panic**. Take a positive attitude and concentrate on things you can improve. Set realistic goals and work systematically to achieve them. Reflect on your performance and learn from mistakes to improve your future results.

Key features of this book

Each chapter covers one topic from the DP *Physics guide*. It starts with 'You must know' and 'You should be able to' checklists. These outline the knowledge and skills that you should have after studying the topic. Some statements have been reworded or combined to make them more accessible and to simplify your navigation through the topic. These changes do not affect the coverage of key material, which is always explained within the chapter.

Chapters contain the features shown below and on the following page.

> ### Examples
> **Examples** offer solutions to typical problems and demonstrate common problem-solving techniques.

> **Theoretical concepts** and **key definitions** are discussed at a level sufficient for answering typical examination questions. Many concepts are illustrated by diagrams, tables or worked examples. Most definitions are explained in a panel like this one.

> ### Assessment tip
> This feature highlights essential terms and statements that have appeared in past mark schemes, warns against common errors and shows how to optimize your approach to particular questions.

Introduction

Links provide a reference to relevant material, within another part of this book or the IB *Physics data booklet*, that relates to the text in question

Nature of science

This relates a physics concept to the overarching principles of the scientific approach and the development of your own learning skills.

Approaches to learning

These ATL features give examples of how famous scientists have demonstrated the ATL skills of communication, self-management, research, thinking and social skills, and prompt you to think about how to develop your own strategies.

Sample student answers show typical student responses to IB-style questions (most of which are taken from past examination papers). Positive or negative feedback on the student's response is given in the blue and red pull-out boxes. An example is given below.

Sample student answer

A cable is wound onto a cylinder of diameter 1.2 m.

Calculate the angular velocity of the cylinder when the linear speed of the cable is 27 m s⁻¹.

State an appropriate unit for your answer. [2]

This answer could have achieved 2/2 marks:

▲ The calculation is correct and well explained. The unit is also correct.

$$\omega = \frac{v}{r} = \frac{27 \times 2\pi}{1.2 \times \pi} = \frac{27}{0.6} = 4.5 \text{ rad s}^{-1}$$

This answer could have achieved 0/2 marks:

▼ Although there is a reasonable start with a correct identification of the radius and the correct equation has been quoted, the solution becomes confused when *T* appears. The unit is incorrect for circular motion.

$$D = 1.2 \text{ m} \quad r = 0.6 \text{ m}$$
$$v = \omega r = \frac{2\pi}{T} \times r = \frac{2\pi}{11} \times 0.6 \text{ m}$$
$$= 0.343 \text{ m s}^{-1}$$

Questions not taken from past IB examinations will not have the exam paper icon.

Practice problems

Practice problems are given at the end of each chapter. These are IB-style questions that provide you with an opportunity to test yourself and improve your problem-solving skills. Some questions introduce factual or theoretical material from the syllabus that can be studied independently.

A Space, time and motion

A.1 Kinematics

You must know:
- ✔ the meaning of the terms distance, displacement, speed, velocity and acceleration
- ✔ the difference between instantaneous and average values for velocity, speed and acceleration
- ✔ the qualitative effect of fluid resistance on projectiles, including time of flight, trajectory, velocity, acceleration, range and terminal speed.

You should be able to:
- ✔ use the kinematic equations of motion (*suvat*) for situations involving uniform acceleration
- ✔ represent and interpret motion using distance–time, speed–time, displacement–time, velocity–time and acceleration–time graphs
- ✔ analyse quantitatively projectile motion in the absence of fluid resistance.

The motion of an object is described by the vector quantities displacement, velocity and acceleration. The scalar counterparts of displacement and velocity are distance and speed.

The length of the curved path in Figure 1 is the distance travelled. The displacement is the vector joining X and Y. Both the length of XY and its direction relative to north are required to specify the displacement completely.

Distance is a **scalar** quantity and is the length of a path between two points.

Displacement is the **vector** difference between the initial position and the final position.

Position means the point in space where an object is situated. This can be referred to numerically as the distance or displacement from an origin.

Average **speed** is the $\dfrac{\text{total distance travelled by an object}}{\text{total time taken}}$. It is a scalar quantity.

Average **velocity** is $\dfrac{\text{displacement}}{\text{time}}$. It is a vector quantity.

Acceleration is $\dfrac{\text{change in velocity}}{\text{time taken for change}}$. It is a vector quantity.

Typical units for these quantities are:
- distance and displacement: metre (m), kilometre (km), megametre (Mm), astronomical unit (AU), light year (ly)
- speed and velocity: $m\,s^{-1}$, $km\,s^{-1}$, $km\,h^{-1}$
- acceleration: $m\,s^{-2}$, $km\,s^{-2}$.

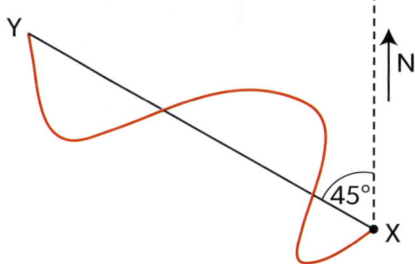

▲ Figure 1 The difference between distance (red curve) and displacement (straight black line)

> **Assessment tip**
>
> When a question asks for a vector quantity such as velocity or displacement, you must give the direction of the quantity **and** the magnitude to gain full marks.

> **Approaches to learning**
>
> You may find that linking graphs to equations and theory is a good way to reinforce your learning. Graph lines have gradients and there are areas associated with graphs. Use these ideas to enhance your understanding.

Motion can be represented using distance–time and speed–time graphs. Vector quantities can be shown as displacement–time, velocity–time and acceleration–time graphs. When vector graphs are used, the signs of the axes, gradients and areas indicate the direction, so take care with these quantities. Figure 2 shows how quantities can be derived from the motion graphs.

1

A Space, time and motion

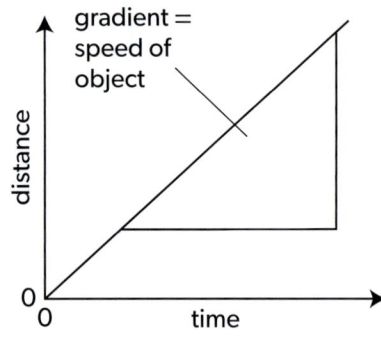 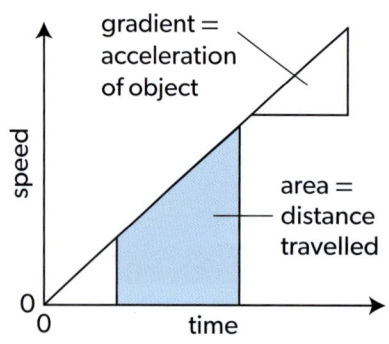 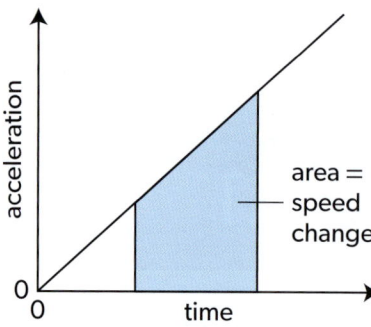

▲ Figure 2 The meaning of the gradient and area for motion graphs

Symbols used for motion in the IB Diploma Programme physics course:

s	displacement
u	initial velocity/speed
v	final velocity/speed
a	acceleration
t	change in time

These are connected by the four kinematic equations of motion for uniform acceleration, often known as the ***suvat* equations**:

$v = u + at$

$v^2 = u^2 + 2as$

$s = ut + \frac{1}{2}at^2$

$s = \frac{(v + u)t}{2}$

The *suvat* equations apply for **uniform acceleration only**. "Uniform" means constant here. The speed–time graph must be a straight line for the acceleration to be constant.

Nature of science

These kinematic equations for uniform acceleration can be used to determine the acceleration of free fall, g, for objects released near the Earth's surface. The distance fallen from rest and the time taken to fall are required.

The quantity g links to the concept of (gravitational) field strength which is a crucial quantity in field theory.

On page 234, there is more advice for constructing graphs, how to determine the gradient of a graph at a point and how to find the area under the graph.

Kinematic equations

The kinematic equations are a powerful way to solve problems involving motion in both one and two dimensions. They can only be used when the acceleration is constant.

For non-uniform accelerations (e.g. motion down a slope with a changing angle) you may have to equate the energies in a system.

Example 1 shows a common type of motion where an object is dropped from rest (or thrown vertically up or down) close to the Earth's surface. The object is then said to be in **freefall**.

Example 1

A ball is released, from rest, from a distance above the ground. The graph plots vertical speed of the ball v against time t.

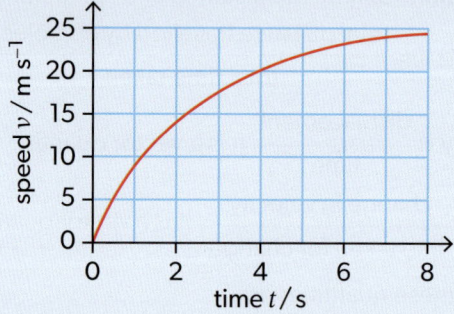

a) Estimate the distance fallen by the ball from $t = 0$ to $t = 8.0$ s.

b) Determine:

   (i) the instantaneous acceleration of the ball when $t = 4.0$ s

   (ii) the average acceleration of the ball from $t = 0$ to $t = 8.0$ s.

Solution

a) The distance fallen by the ball is equal to the area under the curve. To calculate this area, count the number of squares under the curve. Then multiply this number by the area of one square.

In this example, there are approximately 28 squares under the curve. Each square has an area of $1\,\text{s} \times 5\,\text{m s}^{-1}$, which is equivalent to a distance of $1 \times 5 = 5$ m. Total distance = $28 \times 5 = 140$ m.

b) (i) You can find the instantaneous acceleration of the ball when $t = 4.0\,\text{s}$ by drawing a tangent to the curve at $t = 4.0\,\text{s}$. The gradient of this tangent is the acceleration. Make sure you read off values as far apart as possible (at least half the length of the line).

Acceleration $= \dfrac{25 - 12}{6.4} = 2.03 \approx 2.0\,\text{m s}^{-2}$ (to 2 s.f.)

(ii) The average acceleration is given by $\dfrac{\text{change in velocity}}{\text{time taken}}$. The velocity of the ball increases from zero to about $24\,\text{m s}^{-1}$ in a time of $8.0\,\text{s}$.

Average acceleration $= \dfrac{24}{8.0} = 3.0\,\text{m s}^{-2}$

Nature of science

The kinematic equations allow physicists to model the future behaviour of a system that involves movement. This model has important limitations. The moving object is taken to be a point in space. Acceleration is taken to be uniform.

Example 2

A car changes its speed uniformly from $28\,\text{m s}^{-1}$ to $12\,\text{m s}^{-1}$ in $8.0\,\text{s}$.

a) Calculate the acceleration of the car.

b) Determine the distance travelled by the car during the 8.0 s of acceleration.

Solution

a) Use the *suvat* equations to calculate the acceleration. Start by writing down what you know, including symbols and units, and what you want to find out.

The equation you need is $v = u + at$.

u	initial speed	$28\,\text{m s}^{-1}$
v	final speed	$12\,\text{m s}^{-1}$
a	acceleration	$?\,\text{m s}^{-2}$
t	change in time	$8.0\,\text{s}$

Note that the units are consistent, if they were not (e.g. the speed was given in km h^{-1}), you would need to convert them. Take care to substitute correctly into the equation:

$12 = 28 + a \times 8.0$; therefore $a = -2.0\,\text{m s}^{-2}$.

The acceleration is negative. With this sign convention, this means the car is slowing down.

b) $s = ut + \dfrac{1}{2}at^2 = 28 \times 8 + \dfrac{1}{2} \times (-2.0) \times 8^2 = 224 - 64 = 160\,\text{m}$

Projectile motion

When the motion is in two dimensions, then it is known as projectile motion. The solutions still use the kinematic equations.

Follow these steps to solve questions about projectile motion.

- Resolve the initial velocity into horizontal and vertical components.
- Recognize that the acceleration due to free-fall acts only on the vertical component. The symbol for the acceleration due to free-fall is g.
- As g is constant, the *suvat* equations are used in the vertical direction.
- Equations for constant velocity must be used in the horizontal direction. This is because the horizontal component of velocity does not change (assuming negligible air resistance).
- The horizontal and vertical motions can be combined at the end of the problem by adding the speeds vectorially.

Acceleration in any uniform field leads to similar behaviour. You should link your understanding here to the movement of a charged particle in a uniform electric field. An electron travelling with an initial velocity that is at an angle to the electric field undergoes uniform acceleration and follows a parabolic path (Topic D.3).

In a uniform gravitational field (near the Earth's surface, for example) the motion of the projectile is parabolic when there is no air resistance.

A Space, time and motion

Approaches to learning

Sometimes it is easier to use energy conservation (Topic A.3) for problems of projectile motion. This is particularly the case for problems that involve the velocity of a projectile rather than its time of flight. In a gravitational field, and in the absence of friction, the gravitational potential energy is transferred to the kinetic energy of the projectile (Topic A.3).

Nature of science

Example 3 involves a modelling process that is not always obvious. The solution involves assuming a value for g that is an average value quoted in the *Physics data booklet*. The assumption is that this value is constant for the whole motion (a very reasonable assumption). The soccer ball is assumed to be a point in space. Some of these assumptions are backed up by observations of the real-world behaviour of a soccer ball. Some are assumptions that any soccer player can tell you are not true.

Assessment tip

Look for sentences such as "Assume that air resistance is negligible." This makes your job easier. You will not have to carry out calculations involving air resistance for projectile motion, but you may have to answer qualitative (non-mathematical) questions about it.

Example 3

Idris kicks a soccer ball over a wall that is a horizontal distance of 32 m away. Assume that air resistance is negligible.

a) The ball takes 1.6 s to reach a point vertically above the wall. Calculate the horizontal component of the velocity of the ball.

b) The ball is at its maximum height as it passes above the wall. State the vertical component of its velocity at maximum height.

c) Determine the magnitude of the initial velocity of the ball when it was kicked.

d) Deduce the angle to the horizontal at which the ball was kicked.

Solution

a) The ball travels 32 m horizontally in a time of 1.6 s.

The horizontal component of velocity is $\frac{32}{1.6} = 20 \text{ m s}^{-1}$ and is constant during the motion.

b) At its maximum height, the ball only moves horizontally, and so the vertical component is zero.

c) Begin by finding the initial vertical component of the velocity, so you can combine it with the horizontal component. Use a *suvat* equation.

u	initial speed	? m s^{-1}
v	final speed	0
a	acceleration	-9.8 m s^{-2}
t	change in time	1.6 s

Note that, in projectile motion, typically the positive vertical direction is defined as upwards, so the initial vertical component u of the velocity is positive, but the acceleration a is negative.

Use $v = u + at$. Substituting gives $0 = u + (-9.8) \times 1.6$.

Rearranging gives $u = 15.68 \text{ m s}^{-1}$.

You now have the information needed to calculate the vector velocity.

Magnitude $= \sqrt{20^2 + 15.68^2} = 25.4 \approx 25 \text{ m s}^{-1}$ (to 2 s.f.)

d) Angle to the horizontal $= \tan^{-1}\left(\frac{15.68}{20}\right) = 38°$

Assessment tip

Avoid early rounding of intermediate answers in multi-step calculations. Only round the final value to the appropriate number of significant figures.

A.1 Kinematics

Sample student answer

A tennis ball is hit with a racket from a point 1.5 m above the floor. The ceiling is 8.0 m above the floor. The initial velocity of the ball is 15 m s^{-1} at an angle of 50° above the horizontal. Assume that air resistance is negligible.

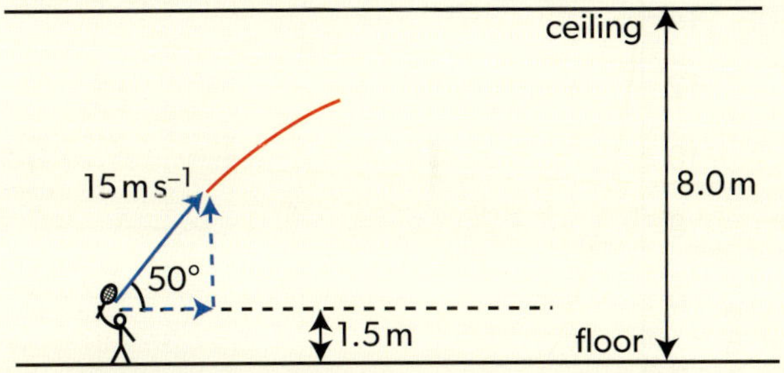

Determine whether the ball will hit the ceiling. [3]

This answer could have achieved 2/3 marks:

Vertical: $15 \times \sin 50° = 11.5$ m s^{-1}

$0 = (11.5)^2 - 2 \times 9.8 \times s$

$132.25 = 19.6 s$

$s = 6.75$ m

$6.75 + 1.5 = 8.25$ (m)

$8.25 > 8$ so the ball won't hit the ceiling.

This answer could have achieved 0/3 marks:

$A_v = A \cos \theta$

$= 15 \times \cos 50°$

$= 9.6$

Yes, 9.6 m > the 6.5 m required to hit the ceiling.

▲ The component of speed is correct.

▲ There is a stated comparison.

▼ The *suvat* equation used is not quoted and the substitution appears directly. Had there been an error here it would have been difficult to allow the error to be carried forward.

▲ The total height reached is calculated correctly.

▼ The total height (8.25 m) is compared with the height of the ceiling (8.0 m) but the wrong conclusion is drawn.

▼ The component of speed is incorrect. This is the horizontal component.

▼ The answer assumes that the calculated speed is equal to the distance travelled.

▼ There is no addition of the 1.5 m from the position where the ball begins.

Effects of air (fluid) resistance

When an object moves through a fluid (gas or liquid), a viscous drag force acts on the object when energy is transferred from the moving object to the fluid. The faster the speed, the greater the retarding force that acts.

In the case of an object accelerating vertically downwards, the retarding force upwards eventually equals the weight downwards and the object reaches a terminal speed. In practice:

- the time to fall a particular distance increases
- the acceleration varies from *g* to zero (when terminal speed is reached)
- the terminal speed depends on the dimensions (size) of the object.

Figure 3 shows the difference between projectile motion in a vacuum and in air (drag). With air resistance:

- the overall range is shorter and speeds are slower
- the maximum height is lower
- the vertical acceleration is not constant and there is horizontal deceleration
- the trajectory is no longer a parabola.

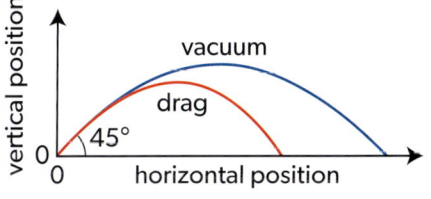

▲ Figure 3 The effect of drag on an object projected at 45° to the horizontal

Example 4

In badminton, a shuttlecock is hit from an initial height of 1.5 m. The graph shows how the vertical component v_y of the velocity of the shuttlecock varies with time t. The positive vertical direction is upwards. The shuttlecock lands on the ground at $t = 1.8$ s.

a) Estimate the initial vertical acceleration of the shuttlecock. Comment on your answer.

b) Describe how the magnitude of the vertical acceleration of the shuttlecock varies with t.

c) Estimate the maximum height reached by the shuttlecock.

A steel ball is shot from the same initial height and with the same initial velocity as the shuttlecock. Air resistance on the steel ball can be ignored.

d) Determine the maximum height reached by the steel ball.

e) Draw, on the same axes, a graph to show how v_y varies with t for the steel ball.

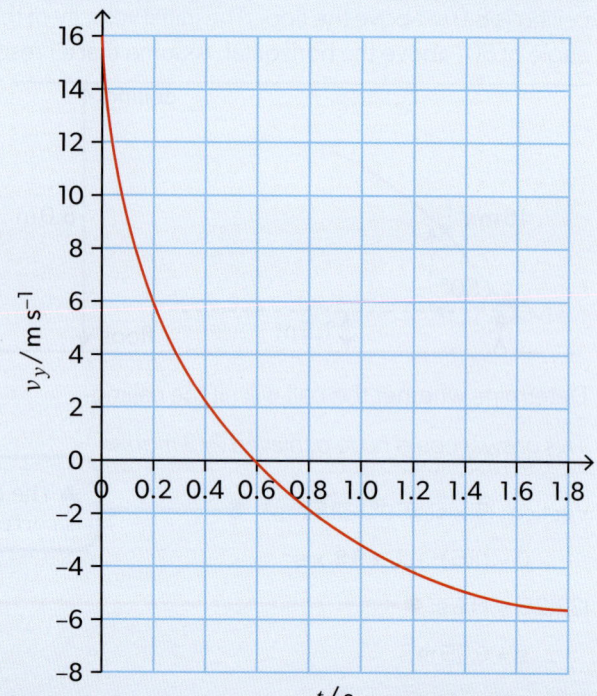

Solution

a) The tangent to the velocity–time graph drawn at $t = 0$ crosses the time axis at about $t = 0.15$ s. The initial vertical acceleration is equal to the gradient of this tangent, $\frac{0 - 16}{0.15 - 0} = -110$ m s^{-2}. The magnitude of this acceleration is more than $10g$, which means that the initial motion is very strongly affected by air resistance.

b) The magnitude of the vertical acceleration decreases with time, reaching 9.8 m s^{-2} at the maximum height (when $t = 0.6$ s) and decreasing further as the shuttlecock falls to the ground. The acceleration just before the impact with the ground (at $t = 1.8$ s) is still non-zero, which means that the shuttlecock has not yet reached its terminal speed.

c) The vertical displacement of the shuttlecock is equal to the area under the velocity–time graph from $t = 0$ to $t = 0.6$ s. The area covers about 7 grid squares, so the distance travelled in the vertical direction is approximately $7 \times 0.2 \times 2 = 2.8$ m. The maximum height is obtained by adding this value to the initial height.

Maximum height = 1.5 + 2.8 = 4.3 m

d) The vertical distance moved by the ball can be found from *suvat* equations.

The equation to use is $v^2 = u^2 + 2as$.

s	displacement	? m
u	initial speed	16 m s^{-1}
v	final speed	0
a	acceleration	-9.8 m s^{-2}

$0 = 16^2 - 2 \times 9.8 s$, so $s = 13.1$ m. The final height will be $1.5 + 13.1 = 14.6$ m, which is much greater than for the badminton shuttlecock!

e) The graph is a straight line with a negative gradient of -9.8 m s^{-2}. The graph crosses the time axis at about $t = 1.6$ s, so the steel ball reaches its maximum height more than a second later than the shuttlecock. The total time of flight to the ground will also be longer for the steel ball.

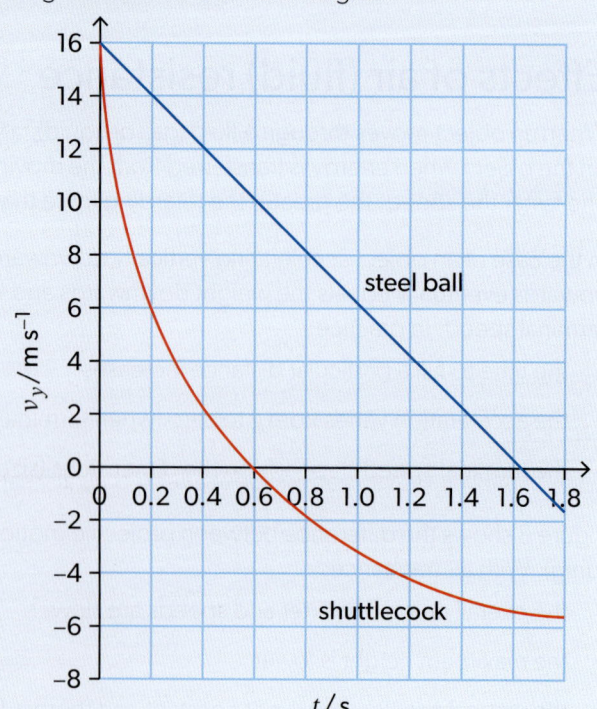

Practice problems

Problem 1
An object is thrown vertically upwards from the edge of a vertical sea cliff. The initial vertical speed of the object is 16 m s^{-1} and it is released 95 m above the surface of the sea.

Air resistance is negligible.

a) Calculate the maximum height reached by the object above the sea.

b) Determine the time taken for the object to reach the surface of the sea.

Problem 2
An object is at rest at time $t = 0$. The object then accelerates for 12.0 s at 1.25 m s^{-2}.

Determine, for time $t = 12.0$ s:

a) the speed of the object

b) the distance travelled by the object from its rest position.

Problem 3
A cart moves up a ramp with an initial velocity of 4.0 m s^{-1}. The graph shows the variation with time t of the velocity v of the cart.

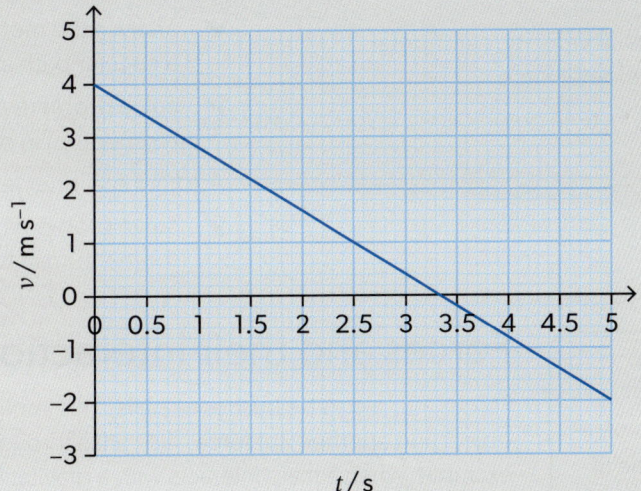

a) Calculate the acceleration of the cart.

b) Determine:

   (i) the maximum displacement of the cart from the starting position

   (ii) the displacement of the cart at time $t = 5.0$ s.

Problem 4
A projectile is fired horizontally with an initial velocity of 12 m s^{-1} from the top of a building that is 8.0 m tall.

a) Ignoring air resistance, calculate:

   (i) the time of flight of the projectile

   (ii) the horizontal range.

b) The motion of the projectile is affected by air resistance. Explain what effect it has on your answers in part (a).

A Space, time and motion

A.2 Forces and momentum

You must know:
- ✔ real objects can be represented as points in space
- ✔ forces are interactions between bodies
- ✔ the nature of field forces (gravitational force, electric force and magnetic force)
- ✔ what is meant by translational equilibrium
- ✔ Newton's three laws of motion
- ✔ what is meant by weight
- ✔ what is meant by linear momentum and impulse, and that an applied external impulse is equal to the change in momentum of a system
- ✔ what is meant by conservation of linear momentum, and how it applies to explosions and elastic and inelastic collisions
- ✔ what is meant by angular velocity, centripetal acceleration and centripetal force.

You should be able to:
- ✔ represent a force as a vector
- ✔ represent the forces acting on a body as a free-body diagram
- ✔ interpret the forces acting on a system in terms of Newton's first law of motion
- ✔ use Newton's second law of motion to relate the resultant force acting on an object to the object's acceleration
- ✔ identify action–reaction pairs of forces using Newton's third law of motion
- ✔ describe and solve problems involving contact forces (normal force, surface frictional force, elastic restoring force, viscous drag force and the buoyancy force)
- ✔ analyse the motion of an object moving around a circular path at a constant speed
- ✔ solve problems involving conservation of linear momentum in one dimension

Additional higher level:
- ✔ solve problems involving conservation of linear momentum in two dimensions.

This topic is about contact forces.

The field forces tested in the IB Diploma Programme physics course are:
- gravitational force (Topic D.1)
- electric force (Topics D.2, D.3 and B.5)
- magnetic force (Topics D.2, D.3 and D.4).

The **resultant force** on an object is the (vector) sum of all the forces that act on it. It is important to take into account the directions when determining a resultant force.

Forces and their interaction

Forces are said to be pushes or pulls. This implies a contact between one object and another. But forces also arise through field interactions; for example, when a mass interacts with a gravitational field.

A free-body diagram is used to represent the complete set of forces that act on an object. It allows you to analyse the effect of the forces on the object.
- The object is normally drawn as a single point (often representing the position of the centre of mass).
- The forces acting on the object are drawn to scale and labelled clearly.
- These force vectors can be summed (by calculation or by drawing) to give the resultant force that acts on the object.

Assessment tip

You may be asked to draw a free-body diagram to show the relative magnitudes and directions of all the forces acting on a single body. The body is usually represented by a dot. Take care that:
- the vector lengths represent the relative magnitudes
- the directions are correct
- every force is labelled unambiguously
- only forces acting on the object are shown.

In the case of a ball falling freely close to the Earth's surface, only a gravitational force acts on the object (Figure 1). The point is the centre of the sphere. The resultant force is downwards.

A.2 Forces and momentum

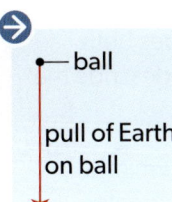

◀ Figure 1 The free-body diagram for a ball falling freely in the Earth's gravitational field

In a lift (elevator) accelerating upwards, an object experiences a force on it acting upwards from the lift and a gravitational force acting downwards (Figure 2). The sum of the two vector lengths (drawn to scale) has a net upwards force. This is the resultant force acting on the object.

◀ Figure 2 The free-body diagram for a ball in a lift that is accelerating upwards

> **Assessment tip**
>
> Remember that "accelerating upwards" can mean two things.
> - A lift starting from rest or moving upwards is accelerating upwards when it gets faster in the upwards direction.
> - A lift that is already moving downwards and accelerating upwards is getting slower and will eventually come to rest.

Contact forces

Normal force F_N

When an object rests on a surface in a gravitational field, a force acts at right angles to the surface. This force, called the normal force F_N, has the same magnitude as the apparent weight of the object when the surface is horizontal. The magnitude of the normal force changes when the object accelerates. For example, suppose an object is in a lift:

- when the lift is at rest or moving at a constant velocity, the normal force is mg (the mass of the lift is m)
- when the lift is accelerating upwards with acceleration a, then the normal force will be $mg + ma$
- when the lift is accelerating downwards, then the normal force is $mg - ma$.

Surface frictional force F_f

The magnitude of the frictional force acting between two surfaces depends on whether or not the surfaces are moving relative to each other.

- When there is no relative motion, the friction between the surfaces is called static friction.
- When there is relative movement between the surfaces, the friction is said to be dynamic.

In both cases, the magnitude of the frictional force F_f depends on the normal (perpendicular) force F_N that acts between the surfaces.

Mathematically, this is written as:

Static friction	**Dynamic friction**
$F_f \leq \mu_s F_N$	$F_f = \mu_d F_N$
where μ_s is the coefficient of static friction	where μ_d is the coefficient of dynamic friction

In general, the maximum magnitude of static friction is greater than that for dynamic friction, and so the size of the frictional force between two surfaces decreases as the surfaces begin to slide.

Different language is used for fluid friction (the term "fluid" includes both liquids and gases). The terms air resistance and drag are used to describe the friction on a solid as it moves through a fluid.

The amount of frictional drag between the object and the fluid in which it moves (called the medium) depends on their relative speed. The greater this relative speed, the greater the drag force. As an object accelerates, frictional drag increases and the magnitude of the drag eventually equals the accelerating force. The two forces oppose each other, so the resultant force is zero and there is no further change in velocity. The speed at which this occurs is called the terminal speed. It depends on the size and shape of the object and the "stickiness" of the medium. In an automobile, when the maximum force of the engine equals the drag force, the vehicle travels at its terminal speed. Objects that are falling freely in a planet's gravitational field also reach a terminal speed that depends on their size and shape.

Elastic restoring force F_H

When solid objects are deformed, the forces between atoms lead to a force that attempts to restore the object to its original size and shape. This is known as the elastic restoring force F_H (sometimes the "elastic force" for short).

An empirical equation is used to relate F_H to the deformation of the object. In the case of a spring, the extension x of the spring is assumed to be directly proportional to F_H:

$$F_H = -kx$$

where k is called the spring constant and has the SI units $N\,m^{-1}$. The extension x is the change between the original length and the final length, and F_H is the force needed to achieve this change. The equation was suggested by Robert Hooke following experiments in the 1660s.

Buoyancy F_b

When an object is submerged in a fluid and both are in a gravity field, then the object is subject to a buoyancy force F_b, which acts upwards on the object. This buoyancy is related to Archimedes' principle.

> **Archimedes' principle** states that, for an object wholly or partly in a fluid, an upward buoyancy force acts on the object that is equal to the weight of fluid displaced by the object. The buoyancy force is sometimes known as the **upthrust**.
>
> It is easy to show that $F_b = \rho V g$, where ρ is the density of the fluid and V is the volume of the fluid displaced (this is, of course, the same as the volume of the object when the object is fully submerged).
>
> The resultant force on a stationary object of density σ in a fluid of density ρ is $\rho V g$ (upwards) $-\sigma V g$ (downwards) $= (\rho - \sigma)V g$. When:
>
> - $\rho > \sigma$, the object floats on the surface
> - $\rho < \sigma$, the object sinks to the bottom
> - $\rho = \sigma$, the object has neutral buoyancy and neither rises nor sinks.

Viscous drag force F_d

Real fluids have viscosity that varies according to temperature. Some fluids are more viscous than others (e.g. at the same temperature, compare the flow of honey with water). George Stokes analysed the resistive (drag) force F_d acting on a sphere of radius r due to a viscous fluid while the sphere was falling through the fluid under conditions of streamline flow (flow that is not turbulent). When an object falls in a resistive medium, it will reach a terminal speed.

Nature of science

An empirical relationship is one that is determined by experiment, not by proof. There are other examples of empirical equations in the IB Diploma Programme physics course (e.g. the gas laws in Topic B.3). As an equation, $F_H = -kx$ is not particularly useful as most materials deviate from it at large extensions.

The energy stored by a stretched spring is discussed in Topic A.3.

Viscosity η is the resistance of a fluid to stress, where one plane of the liquid slides relative to another parallel plane. The unit of viscosity is the $Pa\,s$ or $N\,m^{-2}\,s$. In fundamental SI units, this is $kg\,m^{-1}\,s^{-1}$.

Stokes' law: when a sphere of radius r has a speed v, the viscous drag force on the sphere is $F_d = 6\pi\eta r v$.

Including the effects of buoyancy, when the density of the fluid is ρ and the density of the sphere is σ, the **terminal speed**
$$v_t = \frac{2r^2 g(\sigma - \rho)}{9\eta}.$$

Nature of science

Viscous drag in liquids and air resistance in gases both involve this empirical equation, which was suggested by the Irish scientist George Stokes.

Example 1

A sphere is falling at terminal speed through a fluid.

The following data are available.

diameter of sphere	= 3.0 mm
density of sphere	= 2500 kg m^{-3}
density of fluid	= 875 kg m^{-3}
terminal speed of sphere	= 160 mm s^{-1}

a) Draw a free-body diagram of the forces acting on the sphere.

b) Determine the viscosity of the fluid.

Solution

a) Drag force F_d and buoyancy F_b act upwards, and their combined magnitude is equal to the weight W of the sphere.

drag force,
$F_d = 6\pi\eta r v_t$

buoyancy,
$F_b = \frac{4}{3}\pi r^3 \rho g$

weight,
$W = \frac{4}{3}\pi r^3 \sigma g$

b) Rearranging the Stokes' law expression gives:

$$\eta = \frac{2r^2 g(\sigma - \rho)}{9v_t} = \frac{2 \times (1.5 \times 10^{-3})^2 \times 9.8 \times (2500 - 875)}{9 \times 0.16} = 0.050 \text{ Pa s}$$

> The kinematics of motion when viscous drag is involved are discussed in Topic A.1.

Newton's laws of motion

Newton's three laws of motion describe the interaction between an object and a force. Uniform velocity has both magnitude and direction, so both must be unchanging when the motion is uniform. For acceleration to occur, an object must be subject to a net external force.

When more than one external force acts on a body, you need to sum the vectors—these must be added using the rules for vector addition (Topic T.3). When the forces cancel out to give zero resultant force or net force, the body is in translational equilibrium. When the forces do not cancel out to give zero resultant force, the forces are unbalanced.

Newton's first law of motion (N1) states that an object will remain at rest or continue with uniform velocity unless a net external force acts on it.

Newton's second law of motion (N2), in its simplest form, relates the resultant force F acting on an object of mass m to the acceleration of the object a by the equation $F = ma$. Both force and acceleration are vectors, whereas mass is a scalar. The direction of the net force and the direction of the acceleration that it produces are identical.

N2 defines the SI unit of force, the **newton**. One newton (1 N) is the force that gives a mass of 1 kg an acceleration of 1 m s^{-2}. In fundamental units, this is kg m s^{-2}.

Newton's third law (N3) states that action and reaction are equal and opposite.

Other interpretations of N2 and N3 are considered later in this topic.

> **Assessment tip**
>
> This book uses N1, N2 and N3 to abbreviate Newton's first, second and third laws of motion. It is best not to use your own abbreviations in examinations unless you state clearly what they mean.
>
> When you apply any law in an answer, it is good practice to state the law as well—that way, the examiner will be sure that you understand how the physics arises from the law.

A Space, time and motion

> **Assessment tip**
>
> When a question specifies that a speed or a velocity is constant, then remember the implications of this for Newton's laws of motion. A constant velocity implies that neither the magnitude nor the direction is changing.

> **Example 2**
>
> A cyclist travels along a horizontal road at a constant velocity.
>
> Discuss the forces acting on the system of the cyclist and the bicycle.
>
> **Solution**
>
> N1 states that, because the velocity of the system is constant, there is no net external force. The forward force generated by the cyclist pedalling the bicycle is equal in magnitude to the frictional forces acting on the system in the opposite direction.
>
> The forces are also balanced in the vertical direction. The weight of the cyclist and the bicycle is equal in magnitude and has the opposite direction to the normal force from the road on the bicycle.

> **Example 3**
>
> A car of mass 9.0×10^2 kg travels in a straight line away from traffic lights after they turn green. The graph shows how the speed of the car varies with time.
>
> a) Calculate the acceleration of the car between $t = 2.0$ s and $t = 5.0$ s.
>
> b) Calculate the force used to accelerate the car.
>
>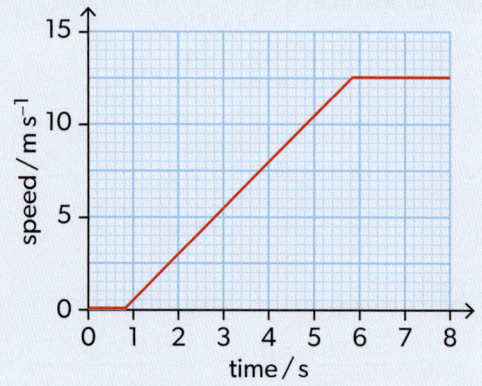
>
> **Solution**
>
> a) The change in speed is from 3.0 to 10.5 m s^{-1}. It takes 3.0 s.
>
> $$\text{Acceleration} = \frac{\text{change in speed}}{\text{time taken for change}} = \frac{7.5}{3.0} = 2.5 \text{ m s}^{-2}$$
>
> b) Force that accelerates the car = mass of car × acceleration
>
> $$= 900 \times 2.5 = 2.25 \text{ kN}$$
>
> Since the data used in the calculation were given to 2 significant figures, you should give the answer to 2 significant figures as well (2.3 kN).

> **Nature of science**
>
> Observations can sometimes be explained in more than one way. If you have seen a "Newton's cradle" you can explain its motion using either Newton's laws or energy ideas from Topic A.3.

> **Nature of science**
>
> Although the motion laws are credited to Newton, in fact, their development took place over many years by many scientists. Science was not then the shared endeavour that it is today. Nevertheless, Newton acknowledged his debt to earlier workers with his famous quotation where he described himself as standing on the shoulders of giants.

N3 implies that action–reaction pairs exist. These are pairs of forces that are:

- equal in magnitude and opposite in direction
- the same **type** of force (such as both gravitational, both electrostatic or both elastic restoring forces).

When a ball is released from rest above the Earth's surface, a gravitational force acts downwards on the ball due to the attraction of the ball by the Earth. At the same time, the ball attracts the Earth with an upwards gravitational force that is equal in magnitude to the weight of the ball. This is an action–reaction pair.

When the ball sits at rest on a table, an additional force pair now acts. This pair is electrostatic in origin. It is:

- the upwards force of the table on the ball
- the equal downwards force of the ball on the table.

These forces arise because the table and the ball are both deformed by the other object. The electrostatic forces arise from the attempt by each to return to its original shape.

Example 4

A car of mass 750 kg accelerates at 0.30 m s^{-2} in a straight line along a horizontal road. A resistive force of 550 N acts on the car.

a) Calculate the resultant force that acts on the car.

b) Calculate the force that the engine exerts on the car.

Solution

a) The accelerating force is $750 \times 0.3 = 225$ N. This is the net force acting. The answer is 230 N (to 2 s. f.).

b) The resistive force is 550 N. The engine must overcome this **and** provide the 225 N. So, the total engine force must be: $550 + 225 = 775 \approx 780$ N.

Note: the full number of significant figures is used until the **final** round-off and answer.

Momentum and impulse

Momentum is a conserved quantity. The momentum of a system is never lost or gained in a collision unless external forces act on it.

The law of conservation of momentum states that, when no external forces act on a system, the vector sum of the momenta before the collision is equal to the vector sum after the collision:

$$m_1 \times u_1 + m_2 \times u_2 + \cdots = m_1 \times v_1 + m_2 \times v_2 + \cdots$$

where $m_1, m_2, \ldots$ are the masses of the objects, $u_1, u_2, \ldots$ are the initial velocities and $v_1, v_2, \ldots$ are the final velocities of the objects.

The equation can also be expressed as $\sum \Delta(m \times v) = 0$, which means that the total change in momentum for a collision including every object that takes part must be zero. If you use this expression, you must use the signs of the velocities correctly.

When a girl throws a ball high into the air, she remains stationary but the force she exerts on the ball causes it to accelerate and gain vertical speed. The ball gains momentum because an external force has acted on it. When the system consists of the Earth plus the girl and ball, the ball still gains upward momentum. However, the girl also increases her downwards force on the ground as she throws. This gives momentum to the Earth, which recoils in the opposite direction to the ball's velocity. The momentum change of the ball and the momentum change of the Earth are equal and opposite.

> **Newton's second law of motion (N2)** can be written in terms of change in velocity:
>
> $F = m \times \left(\dfrac{\Delta v}{\Delta t}\right)$, where Δ means "change in". This can be rewritten (provided mass is constant) as: $F = \left(\dfrac{m \times \Delta v}{\Delta t}\right) = \dfrac{\Delta(\text{momentum})}{\Delta t}$
>
> $= \dfrac{\text{change in momentum}}{\text{change in time}}$
>
> This interpretation of N2 shows that the force acting on an object is its rate of change of momentum.
>
> When N2 is written as $F = ma$, then the mass is assumed to be constant.
> When N2 is written as $F = \dfrac{\Delta(\text{momentum})}{\Delta t}$, then the mass can be changing (e.g. when a rocket burns fuel as it accelerates).

Momentum = mass × velocity

Velocity is a vector quantity, mass is a scalar quantity, and so momentum is also a vector—always specify both its magnitude and direction.

The fundamental unit of momentum is kg m s^{-1}; this is equivalent to the newton second (N s).

Assessment tip

There are two parts to the conservation of momentum law:

- there is no change in momentum during a collision

- providing no **external** forces act, internal forces do not make any difference as they must act equally (N3) on all parts of the object.

Remember to quote both parts when writing about this law.

A Space, time and motion

Example 5

An apple is released from rest and falls towards the surface of Earth. Discuss how the conservation of momentum applies to the Earth–apple system.

Solution

The points to make are:
- The forces on the Earth and the apple are equal and opposite.
- No external force acts on this isolated system.
- Changes in the momentum of the Earth and the apple are equal and opposite.
- The momentum of the Earth–apple system stays the same and is conserved.

Assessment tip

When carrying out calculations, keep track of the direction in which objects move. Your work needs to be presented so that it is immediately clear what changes during a collision.

One way is to define a direction as positive (perhaps even drawing it on a diagram)—velocities in the opposite direction are then negative. When the answer is negative, it indicates that the direction has changed.

In Topic A.3 you meet the concept of kinetic energy E_k. This is linked to momentum through $\frac{1}{2}mv^2 = \frac{1}{2} \times \frac{1}{m}(mv)^2 = \frac{p^2}{2m}$, where p is the momentum of a mass m.

A collision in which kinetic energy is removed from the system is known as inelastic. When energy is conserved in a collision, it is said to be elastic. In another type of collision, kinetic energy can appear during the collision. Examples of this include:

- explosions
- the release of a compressed spring (Example 7)
- when a stationary gun is fired.

Example 6

A small steel ball is released from rest into a fluid. Ignore buoyancy effects in this question.

The graph shows how the speed v of the ball varies with time t.

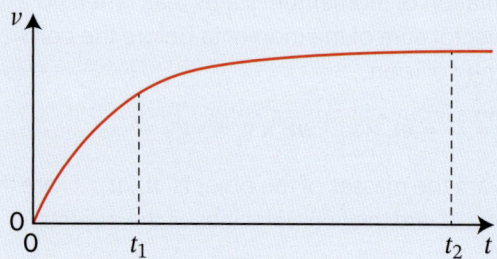

a) Draw a free-body diagram for the ball at:

(i) time t_1

(ii) time t_2.

b) When $t = t_1$, the gradient of the graph is a.

Deduce an expression in terms of the mass of the ball M, acceleration a and acceleration of free fall g for the magnitude of the drag force F_d acting on the ball at $t = t_1$.

c) Deduce, using your answer to part (a)(ii), the magnitude of the drag force acting on the ball when $t = t_2$.

Solution

a) (i) The ball is accelerating, and so there must be a net downward force on it. The weight force acts downwards and there is a smaller drag force upwards.

(ii) The vector lengths should be equal as the velocity is constant. The net force must be zero (N1).

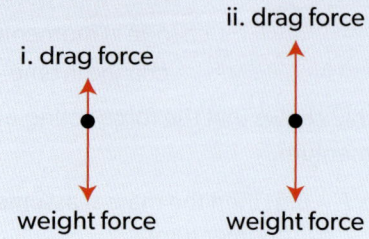

b) The gradient of the graph is equal to the instantaneous acceleration of the ball at $t = t_1$.

Net force $F_{total} = F_{weight} - F_d$

$F_{weight} = Mg$ and $F_{total} = Ma$. So $F_d = M(g - a)$.

c) As the velocity is constant, N1 predicts that the upward and downward forces must be equal. The magnitude of the drag force must equal the magnitude of the weight force, $F_d = Mg$.

Example 7

Two masses, m and M, on a frictionless horizontal table are connected by a compressed spring and released. Mass m moves with velocity v_m.

a) State the change in the momentum of mass m.

b) Determine, in terms of m, M and v_m:

   (i) the velocity of mass M

   (ii) the change in kinetic energy of the masses.

Solution

(a) The initial momentum of the system (both masses) is zero, and so the change in momentum of m is $m \times v_m$.

(b) (i) initial momentum = final momentum. So $0 = mv_m + Mv_M$, where v_M is the velocity of mass M. Therefore, $v_M = -\dfrac{m}{M}v_m$. Mass M moves with a speed $\dfrac{m}{M}v_m$ in the direction opposite to the velocity of mass m.

(ii) The initial kinetic energy is zero.

Total final kinetic energy $= \dfrac{1}{2}mv_m^2 + \dfrac{1}{2}Mv_M^2 = \dfrac{1}{2}\left(mv_m^2 + \dfrac{Mm^2v_m^2}{M^2}\right) = \dfrac{1}{2}m\dfrac{m+M}{M}v_m^2$

This is also the change in kinetic energy.

Example 8

An air-rifle pellet is fired into a wooden block resting on a table.

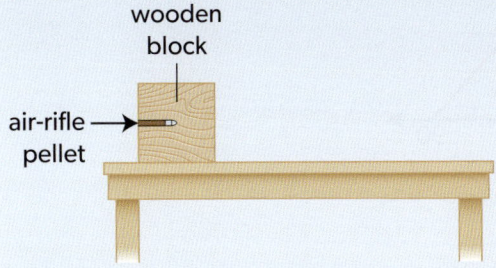

The speed of the block with the embedded pellet immediately after the collision is $4.8\,\text{m s}^{-1}$.

The mass of the pellet is 2.0 g.

The mass of the block is 56 g.

a) Determine the speed of impact of the pellet.

b) Compare the initial kinetic energy of the pellet with the kinetic energy of the pellet and block immediately after the collision.

Solution

a) Initial momentum of the system (pellet and block) $= 0.056 \times 0 + 0.002 \times u_1$

The initial velocity of the pellet is u_1 and movement to the right is taken to be positive.

Final momentum of the system $= (0.056 + 0.002) \times 4.8\,\text{kg m s}^{-1}$

The initial and final momenta are equal. This means that:

$$u_1 = \frac{0.058 \times 4.8}{0.002} = 139.2 \text{ m s}^{-1} \approx 140 \text{ m s}^{-1} \text{ (to 2 s.f.)}$$

b) Initial kinetic energy = $\frac{1}{2} \times 2.0 \times 10^{-3} \times 139.2^2 = 19.4$ J

Final kinetic energy of block and pellet = $\frac{1}{2} \times 58 \times 10^{-3} \times 4.8^2 = 0.67$ J

The collision is inelastic as about 18.7 J of the energy is transferred from kinetic energy during the collision, which is nearly 97% of the initial kinetic energy of the pellet. This energy will appear as sound, deformation of the pellet and of the wood of the block, and some thermal energy. As the block slows down later due to friction from the table, all the remaining kinetic energy will eventually transfer to a thermal form.

Impulse is the change in momentum of an object and is equal to force × time for which the force acts.

The unit of impulse is N s. In fundamental units, this is kg m s^{-1}.

The **area under a graph of force against time** is equal to the impulse and can be used to estimate the momentum change when a force acts on an object.

The change in momentum of an object is also called the impulse that acts on the object.

This is because $F = \frac{\Delta(\text{momentum})}{\Delta t}$ can be written as $\Delta(\text{momentum}) = F \times \Delta t$.

This links to a graph of the variation with time of the force acting on an object. The graph in Example 9 is typical of that often seen when one object collides with another.

Example 9

A ball of mass 0.075 kg moving to the right with a horizontal velocity of 2.2 m s^{-1} strikes a vertical wall and rebounds at right angles to the wall. The duration of the impact is 90 ms.

During the collision, 25% of the ball's initial kinetic energy is transferred to other energy stores.

a) Determine the rebound speed of the ball from the wall.

b) Determine the magnitude and direction of the impulse given to the ball by the wall.

The graph shows how the force F exerted by the wall on the ball varies with time.

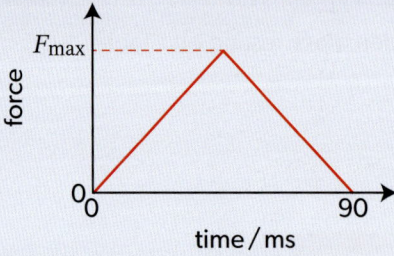

c) Estimate F_{max}.

Nature of science

The kinetic theory of a gas (Topic B.3) is a model that relies heavily on the physics of Theme A. It also relies on a series of simplifying assumptions. This simplification of a system is an important part of the nature of science. You will see it in both the proofs and statements in this text but also in the problems that you are asked to complete.

Solution

a) Initial kinetic energy of the ball = $\frac{1}{2} \times 0.075 \times 2.2^2 = 0.1815$ J

25% of this energy is transferred from kinetic energy. So, $0.75 \times 0.1815 = 0.1361$ J remains.

Rebound speed $v = \sqrt{\frac{2 \times 0.1361}{0.075}} = 1.9$ m s^{-1} and the direction is to the left.

b) The impulse given to the ball by the wall is equal to the change in momentum of the ball and is in the same direction. This is $m(v - u)$. However, care is needed with the sign as u is the initial velocity to the right and v is the final velocity to the left. Therefore, the impulse given to the ball is directed **to the left** and its magnitude is $0.075 \times (1.9 - (-2.2))$ = 0.31 N s.

c) The impulse equals the area under the force–time graph.

$$0.31 = \frac{1}{2} \times 0.090 \times F_{max}, \text{ so } F_{max} = 6.8 \text{ N}$$

The examples given so far feature motion in only one dimension. However, this is an unusual situation. Think of a firework exploding into many bright fragments that move apart in the night sky. In this complex case, the motion of the centre of mass of the fragments does not change (ignoring air resistance) since some fragments move ahead of their previous position and some fall behind (Figure 3).

▲ Figure 3 The motion of the centre of mass of the fragments of an exploding firework continues as though the firework was still in one piece

> **Assessment tip**
>
> At higher level, you may be asked to carry out problems involving momentum changes in two dimensions. In this case, you should resolve the momentum vectors into two convenient directions at 90° to each other.

In the two-dimensional case shown in Figure 4, the best directions to resolve the momentum vectors are in the x- and y-directions. The two components in the y-direction after the collision can be equated (since there was zero initial momentum component in the y-direction). The whole of the initial momentum possessed by the mass m_1 in the x-direction is given to the pair of masses, so that there are three terms to equate.

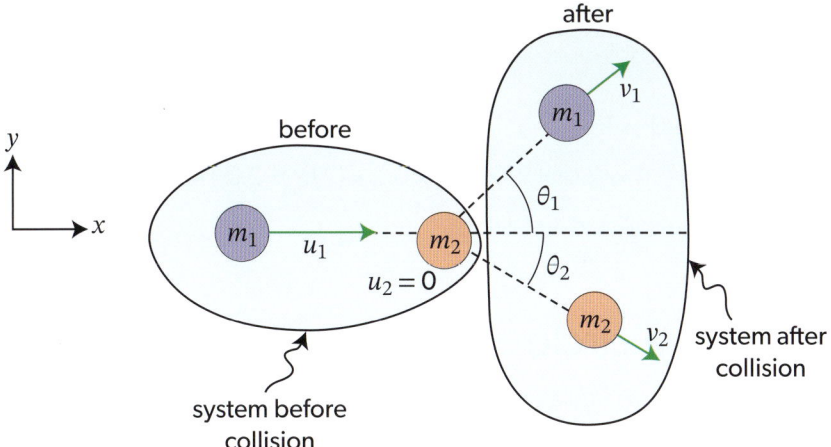

▲ Figure 4 Momentum conservation in two dimensions. The moving mass m_1 collides off-centre with the stationary m_2. After the collision, the masses move apart with an angle of $\theta_1 + \theta_2$. This sum is 90° for an elastic collision and when the masses are the same

Example 10

Ball P, of mass 0.40 kg, travels in the direction of the positive x-axis at a speed of $3.0\,\text{m s}^{-1}$ and collides on a horizontal surface with another ball Q, of mass 0.20 kg, which is initially stationary. The velocity components of P after the collision are $2.0\,\text{m s}^{-1}$ in the x-direction and $1.0\,\text{m s}^{-1}$ in the y-direction.

a) Determine the magnitude and direction of the velocity of ball Q after the collision.

b) Determine whether the collision is elastic.

Solution

a) The velocity components of Q in the x- and y-directions have to be found separately, by considering momentum conservation in each direction.

x-direction: $0.40 \times 3.0 = 0.40 \times 2.0 + 0.20 \times v_x$ therefore $v_x = 2.0\,\text{m s}^{-1}$

y-direction: $0 = 0.40 \times 1.0 + 0.20 \times v_y$ therefore $v_y = -2.0\,\text{m s}^{-1}$

A negative value of v_y means that this velocity component is in the negative y-direction.

Magnitude of the velocity of Q after the collision $= \sqrt{v_x^2 + v_y^2} = \sqrt{2.0^2 + (-2.0)^2}$

$= 2.828\,\text{m s}^{-1} \approx 2.8\,\text{m s}^{-1}$ (to 2 s.f.)

Angle that the velocity makes with the x-axis $= \tan^{-1}\left(\dfrac{|v_y|}{|v_x|}\right) = \tan^{-1}\left(\dfrac{2.0}{2.0}\right) = 45°$

This is shown in the diagram.

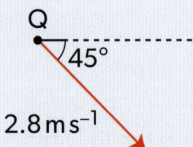

b) Kinetic energy of P before the collision $= \dfrac{1}{2} \times 0.40 \times 3.0^2 = 1.8\,\text{J}$

Combined energy of P and Q after the collision $= \dfrac{1}{2} \times 0.4 \times (2.0^2 + 1.0^2) + \dfrac{1}{2} \times 0.2 \times (2.0^2 + 2.0^2) = 1.8\,\text{J}$

The kinetic energy of the system is unchanged, so the collision is elastic.

Motion in a circle

Quantities involved in circular motion include period, frequency, angular displacement and angular velocity. Newton's first law predicts that a force must act on an object if it is to travel in a circle.

Approaches to learning

Use the links between linear and rotational motion to help you to learn both topics.

Nature of science

The mechanics of motion in a circle is aligned with the mechanics of linear motion. The terms used are similar even though their definitions are modified to reflect the circular nature of the motion. There is a deliberate attempt to link the patterns of the two topics to aid the similar concepts and understanding.

A.2 Forces and momentum

Period T is the time taken for an object to travel once round a circle.

Angular displacement θ is the angle through which an object moves. It is measured in degrees or radians. The angular displacement for a complete circle is 2π rad and the periodic time is T, so $\omega = \dfrac{2\pi}{T}$.

Angular velocity $\omega = \dfrac{\text{angular displacement}}{\text{time}} = \dfrac{\theta}{t}$, where t is the time to travel θ rad. It is a vector quantity but is treated as a scalar in the IB Diploma Programme physics course.

Angular speed has the same definition as angular velocity, but it is a scalar quantity.

Linear speed $v = r\omega$ for a point moving with angular velocity ω at a distance r from an axis of rotation or around the circumference of a circle of radius r.

Circular motion links to the higher level Topic A.4, which considers the kinematics and dynamics of rotating objects.

It also arises in these contexts.

- Gravitational (Topic D.1): The gravitational force between a planet (or the Sun) and a satellite supplies the centripetal force that keeps the satellite in its circular orbit.

- Electrostatic (Topics D.3 and E.1): The mathematics of circular motion was used by Bohr to model the proton–electron system in the hydrogen atom.

- Magnetic (Topic D.3): The force that acts on a charge moving in a magnetic field is at 90° to the plane containing the field direction and the velocity. This is the condition for a centripetal force.

The use of radian measure is described in Topic T.3 as part of the mathematical tools that you need to use in the IB Diploma Programme physics course.

When an object travels in a circle at a constant speed, a force must act on the object. This force is known as the centripetal force. It is directed towards the centre of the circle around which the object moves.

The linear speed is constant but its direction changes, so the linear velocity changes too—that is, the object is accelerated. N1 tells us that velocity change is associated with the force that causes the motion to be in a circle. The force is directed towards the centre of the circle around which the object moves.

Centripetal acceleration $a = \omega^2 r = \dfrac{v^2}{r} = \dfrac{4\pi^2 r}{T^2}$, because $T = \dfrac{2\pi}{\omega}$.

Centripetal force $F = mr\omega^2 = \dfrac{mv^2}{r}$

Assessment tip

Sometimes you will come across the term centrifugal force. This force can only arise when an observer is in a rotating frame of reference. You should **not** use this term in an examination answer—only use centripetal force.

Example 11

An astronaut is rotated horizontally at a constant speed to simulate the forces of take-off.

The centre of mass of the astronaut is 20.0 m from the rotation axis.

a) Explain why a horizontal force acts on the astronaut when the speed is constant.

b) The horizontal force acting on the astronaut is 4.5 times that of normal gravity.

Determine the linear speed of the astronaut.

Solution

a) Velocity is a vector quantity. The velocity of the astronaut is constantly changing because, although the speed is constant, the direction is changing. This means that there is an acceleration and, therefore, a force acting on the astronaut. This is a centripetal force.

b) Centripetal force $= \dfrac{mv^2}{r} = 4.5mg$

$v = \sqrt{4.5gr} = \sqrt{4.5 \times 9.8 \times 20} = 30\,\text{m s}^{-1}$ (to 2 s.f.)

Assessment tip

When answering questions, remember that the centripetal force is provided by the resultant force of **all** the forces that act towards the centre of the circle around which the object is moving.

Example 12

A point mass m at the end of a string moves at a constant speed in a horizontal circle of radius r. The string makes an angle θ with the vertical.

a) Draw a free-body diagram of the forces acting on the mass.

b) Show that the acceleration of the mass is given by $g \tan \theta$.

c) Let $\theta = 30°$, $r = 0.15$ m and $m = 0.050$ kg. Calculate:

   (i) the period of rotation of the mass

   (ii) the tension in the string.

Solution

a) Ignoring air resistance, there are only two forces acting on m: the weight W downwards and the tension F_T along the string. The resultant of these forces is horizontal towards the centre of the circle in which the mass is moving.

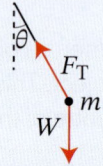

b) The vertical component of tension, $F_T \cos \theta$, is equal to the weight: $F_T \cos \theta = mg$. From this, $F_T = \dfrac{mg}{\cos \theta}$. The horizontal component of tension, $F_T \sin \theta$ is unbalanced and is equal to the net force acting on the mass. From N2:

$$\text{Acceleration} = \frac{F_T \sin \theta}{m} = \frac{1}{m} \times \frac{mg}{\cos \theta} \times \sin \theta = g \frac{\sin \theta}{\cos \theta} = g \tan \theta$$

This is the centripetal acceleration of the mass.

c) (i) $g \tan \theta = \dfrac{4\pi^2 r}{T^2}$, using one of the expressions for the centripetal acceleration. Solving this for the period gives:

$$T = 2\sqrt{\frac{\pi^2 \times 0.15}{9.8 \times \tan 30°}} = 1.0 \text{ s (to 2 s.f.)}$$

(ii) $F_T = \dfrac{0.050 \times 9.8}{\cos 30°} = 0.57$ N

Turning a corner is easier for a vehicle when the surface on which it is travelling is angled up from the horizontal (Figure 5). This is called **banking**.

When the vehicle corners on a horizontal surface, the friction between the tyres and the road surface must be large enough to supply the centripetal force. When the friction is not sufficient, then the vehicle skids and will attempt to go in a straight line (obeying N1). When the surface is banked, the horizontal component of the force normal to the surface provides the centripetal force towards the centre of the circle (Figure 5).

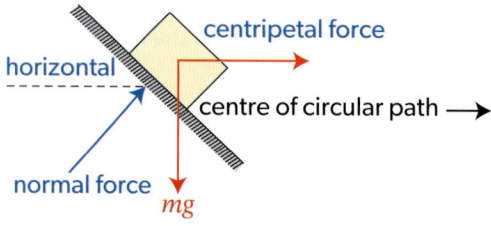

▲ Figure 5 The forces involved in cornering around a banked surface

Aircraft also bank to make a turn. The aircraft tilts out of the horizontal and the lift force now produces a horizontal component to the centre of the circle.

A.2 Forces and momentum

Sample student answer

An unpowered glider moves horizontally at constant speed. The wings of the glider provide a lift force. The diagram shows the lift force acting on the glider and the direction of motion of the glider.

a) Draw the forces acting on the glider to complete the free-body diagram. The dotted lines show the horizontal and vertical directions. [2]

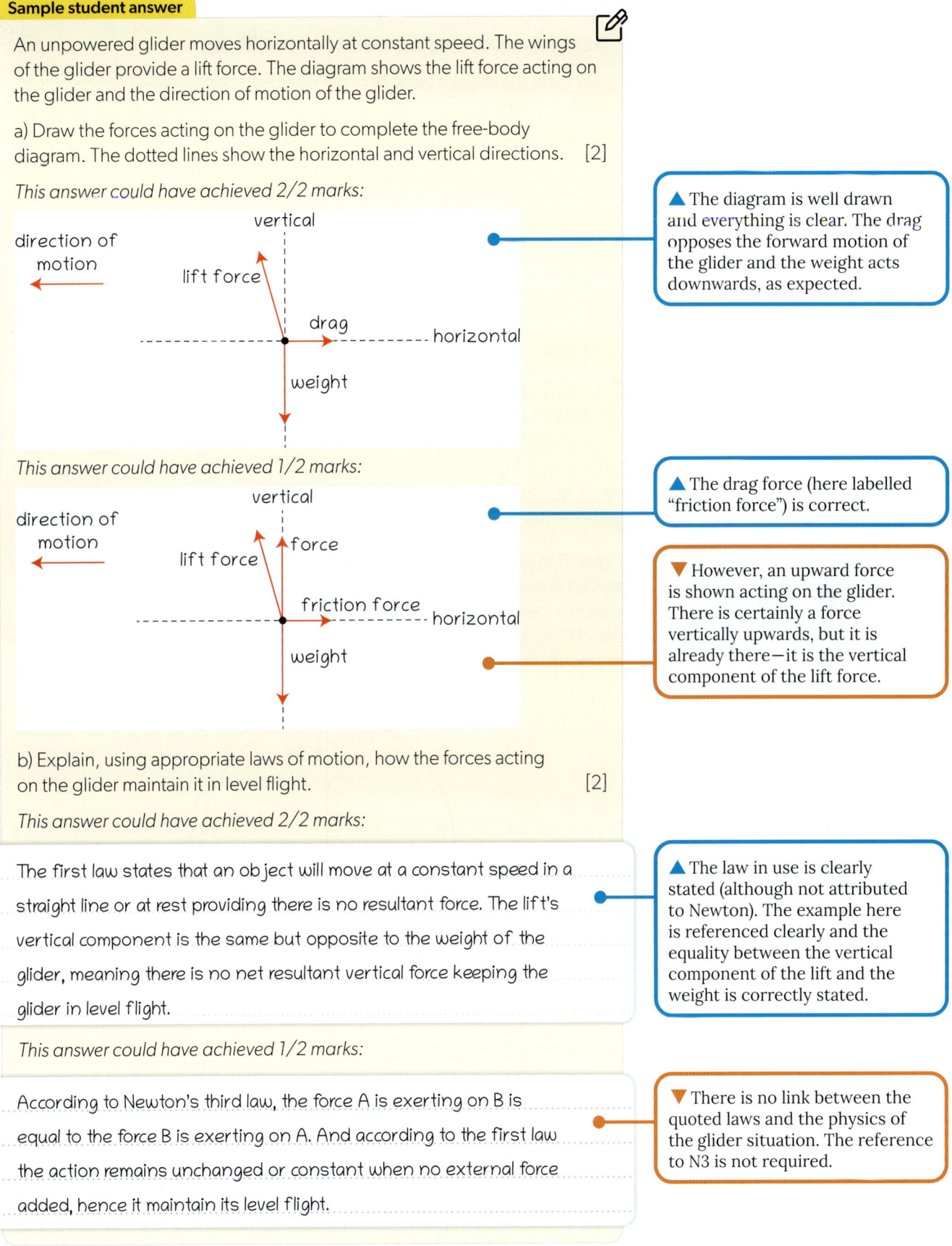

This answer could have achieved 2/2 marks:

▲ The diagram is well drawn and everything is clear. The drag opposes the forward motion of the glider and the weight acts downwards, as expected.

This answer could have achieved 1/2 marks:

▲ The drag force (here labelled "friction force") is correct.

▼ However, an upward force is shown acting on the glider. There is certainly a force vertically upwards, but it is already there—it is the vertical component of the lift force.

b) Explain, using appropriate laws of motion, how the forces acting on the glider maintain it in level flight. [2]

This answer could have achieved 2/2 marks:

The first law states that an object will move at a constant speed in a straight line or at rest providing there is no resultant force. The lift's vertical component is the same but opposite to the weight of the glider, meaning there is no net resultant vertical force keeping the glider in level flight.

▲ The law in use is clearly stated (although not attributed to Newton). The example here is referenced clearly and the equality between the vertical component of the lift and the weight is correctly stated.

This answer could have achieved 1/2 marks:

According to Newton's third law, the force A is exerting on B is equal to the force B is exerting on A. And according to the first law the action remains unchanged or constant when no external force added, hence it maintain its level flight.

▼ There is no link between the quoted laws and the physics of the glider situation. The reference to N3 is not required.

A Space, time and motion

Sample student answer

A cable is wound onto a cylinder of diameter 1.2 m.

Calculate the angular velocity of the cylinder when the linear speed of the cable is 27 m s⁻¹.

State an appropriate unit for your answer. [2]

This answer could have achieved 2/2 marks:

▲ The calculation is correct and well explained. The unit is also correct.

$$\omega = \frac{v}{r} = \frac{27 \times 2\pi}{1.2 \times \pi} = \frac{27}{0.6} = 45 \text{ rad s}^{-1}$$

This answer could have achieved 0/2 marks:

▼ Although there is a reasonable start with a correct identification of the radius and quoting the correct equation, the solution becomes confused when T appears. The unit is incorrect for circular motion.

$$D = 1.2 \text{ m} \quad r = 0.6 \text{ m}$$
$$v = \omega r = \frac{2\pi}{T} \times r = \frac{2\pi}{11} \times 0.6 \text{ m}$$
$$= 0.343 \text{ m s}^{-1}$$

When objects move in a vertical circle, it is necessary to include the effect of gravity in the analysis.

Figure 6 shows what happens when a small object attached to a string is moved in a vertical circle. At points A and D, the weight acts downwards and the centripetal force (horizontal at A and D) is provided only by the string. At B and C, however, the weight and the tension in the string act in the same line (although they act in opposite directions at C).

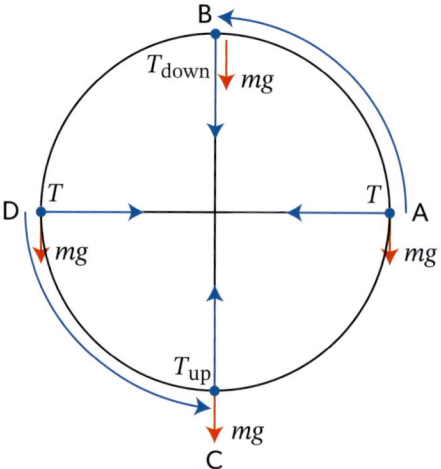

▲ Figure 6 The forces that act when a point mass moves in a vertical circle at constant speed

To maintain the object moving at a constant speed, the tension T in the string must change to take account of the change in direction.

- At A and D, $T = \dfrac{mv^2}{r}$.

- At B, $T_{down} + mg = \dfrac{mv^2}{r}$.

- At C, $T_{up} - mg = \dfrac{mv^2}{r}$.

At point B, when $v < \sqrt{gr}$, the object will not be able to complete its circular motion.

Practice problems

Problem 1
An automobile of mass 950 kg accelerates uniformly from rest to 33 m s^{-1} in 11 s.

a) Calculate the resultant force exerted by the automobile to produce this acceleration.

b) The manufacturer claims a maximum speed of 180 km per hour for the automobile. Explain why an automobile has a maximum speed.

Problem 2
A ball of mass 3.5 g and radius 1.0 cm is submerged in water of density 1.0 × 10^3 kg m^{-3}. The ball is released from rest.

a) Determine the initial acceleration of the ball.

b) Outline the subsequent motion of the ball. Assume that a viscous drag force acts on the ball.

Problem 3
A hammer drives a nail into a block of wood.

The mass of the hammer is 0.75 kg and its velocity just before it hits the nail is 15.0 m s^{-1} vertically downwards. After hitting the nail, the hammer remains in contact with it for 0.10 s. After this time, both the hammer and the nail have stopped moving.

a) Deduce the change in momentum of the hammer during the time it is in contact with the nail.

b) Calculate the average force applied by the hammer to the nail.

Problem 4
A magazine article suggests that wearing seat belts in vehicles can save lives in collisions.

Explain, using the concept of momentum, why this is correct.

Problem 5
A trolley of mass 0.50 kg moves with a horizontal speed of 3.0 m s^{-1}. A bag of sand of mass 1.0 kg is dropped from rest and lands on the trolley.

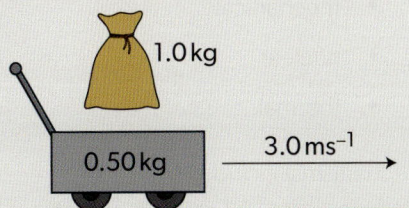

a) Determine the final speed of the system of the trolley and the bag.

b) Calculate the fraction of the initial kinetic energy of the system that is transferred to other forms than kinetic.

Problem 6
Two balls of the same mass collide on a horizontal surface. Ball 1 is moving with an initial velocity of 2.5 m s^{-1} in the positive x-direction. Ball 2 is initially stationary.

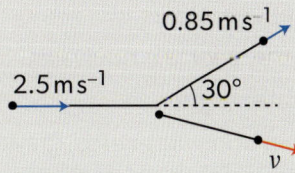

After the collision, ball 1 moves at an angle of 30° to the original direction of motion with a speed of 0.85 m s^{-1}.

a) Determine the magnitude and direction of the velocity v of ball 2 after the collision.

b) Calculate $\dfrac{\text{kinetic energy of the system after the collision}}{\text{kinetic energy of the system before the collision}}$.

Problem 7
A toy train moves with a constant speed on a horizontal circular track of constant radius.

a) State and explain the direction of the horizontal force that acts on the train.

The mass of the train is 0.24 kg. It travels at a speed of 0.19 m s^{-1}. The radius of the track is 1.7 m.

b) Calculate the centripetal force acting on the engine.

Problem 8
A car takes an unbanked turn of radius 50 m on a horizontal road.

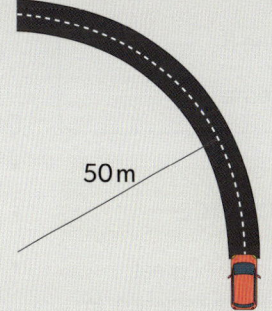

a) State the force that provides the centripetal acceleration for the car.

b) The maximum speed for the car to make the turn without skidding is 70 km h^{-1}. Calculate the coefficient of static friction between the car and the road.

A Space, time and motion

A.3 Work, energy and power

You must know:
- that energy is conserved and energy can transfer between energy stores
- work done by a force is equivalent to an energy transfer
- mechanical energy is:
  - the sum of kinetic energy, gravitational potential energy and elastic potential energy
  - conserved in the absence of resistive forces
- power is the rate of energy transfer
- efficiency is the ratio useful work out : total work in.

You should be able to:
- identify forms of energy stores, including elastic potential, kinetic and gravitational
- calculate the work done on an object as the force acting on the object × its displacement and power as force × speed
- interpret force–distance graphs
- include the effects of resistive and dissipative forces in energy-transfer calculations
- calculate energy transfers that involve:
  - the kinetic energy of translational motion
  - the gravitational potential energy of a mass when close to the Earth's surface
  - the elastic potential energy of a spring
- calculate the efficiency of a device in terms of its overall net energy transfer or its power dissipation
- evaluate and use the energy density of a fuel source
- construct and interpret Sankey diagrams.

Energy transfers and pathways appear throughout the IB Diploma Programme physics course. For example:
- electrical in Topics B.5 and D.4
- heat in Theme B
- waves in Theme C.

Nature of science

Conservation laws are of great significance throughout science. In physics, you will meet conservation of charge, momentum and energy.

In a similar way, there are fundamental constants that are thought not to change with time or place. Conservation rules and known constants enable us to predict the future behaviour of systems—a crucial part of the nature of science.

Energy forms and transfers

Energy is transferred from one energy store to another via an energy pathway.

Assessment tip

Conservation laws are important in science. They recur many times in physics and are a way to learn (and later revise) effectively. Always look for the links between topics in your IB Diploma Programme physics course. Try to link ideas to check your understanding of the whole subject and to ensure that you can apply your knowledge in an unfamiliar context.

Energy stores include:
- chemical
- elastic
- electromagnetic
- gravitational
- kinetic
- mechanical
- nuclear
- thermal.

Energy pathways (or transfer mechanisms) include:
- electric (a charge moving through a potential difference)
- mechanical (a force acting through a distance)
- heating (driven by a temperature difference)
- waves (such as electromagnetic radiation or sound waves).

A.3 Work, energy and power

The rule of conservation of energy is never broken—but you will sometimes have to look carefully to see where some energy goes. In mechanical systems, the energy transferred between stores (such as gravitational to kinetic) is equivalent to the work done.

The total mechanical energy of a system is constant providing that no resistive forces operate in the system.

When an object is moving, energy has been transferred into kinetic energy. Movement of an object within a gravitational field can also lead to transfer of energy. In this case, the term "gravitational potential energy" is used.

Work must be done to extend or compress a spring. This transfers elastic potential energy to the spring.

The equation $W = F \times s$ relates the work done W to the constant force F and the displacement s in the direction of the force. When s and F are not in the same direction, but there is an angle θ between them, then $W = F \times s \cos\theta$.

When the force is not constant, to find the work done you will need to calculate the area under the graph of force against distance.

> The change in **kinetic energy** ΔE_k of an object of (constant) mass m is $\Delta E_k = \frac{1}{2}m(v^2 - u^2)$ when the speed changes from u to v.
>
> The change in **gravitational potential** energy ΔE_p of an object of mass m is $\Delta E_p = mg\Delta h$ when the object is raised through a vertical distance Δh.
>
> The **elastic potential energy** E_H stored in a spring of spring constant k is $E_H = \frac{1}{2}k(\Delta x)^2$ when the spring is extended through a distance Δx from its equilibrium (unstretched) length.
>
> **Work done** is force × distance = $F \times s$. If there is an angle θ between the direction of the force and the displacement, the work done is $Fs \cos\theta$.
>
> The unit of both work and energy is the joule (J), which is equivalent to 1 newton metre (N m). In fundamental units this is $kg\,m^2\,s^{-2}$.

The **mechanical energy** of a system is the sum of all the potential energies and the kinetic energy of the system.

For our purposes, the potential energies are the gravitational potential energy and the elastic potential energy. The gravitational force and the spring force that give rise to these are known as **conservative forces**.

Resistive and dissipative forces such as solid friction, air resistance and damping (Topics A.2 and C.4) are said to be **non-conservative** because the amount of energy transferred to heating by friction depends on the path taken by the system.

Strictly, F and s are vector quantities. They are multiplied here to give a scalar product W. The quantity $\cos\theta$ calculates the component of one quantity in the direction of the other.

This links to Topic T.3.1, which deals with the mathematics you need in the IB Diploma Programme physics course.

Example 1

A boy drags a box to the right across a rough, horizontal surface using a rope that pulls upwards at 25° to the horizontal. The tension in the rope is F.

a) Once the load is moving at a steady speed, the average horizontal frictional force acting on the load is 470 N.

  Calculate the average value of F.

b) The load is moved a horizontal distance of 250 m in 320 s. Calculate the work done on the load by:

  (i) tension F

  (ii) the frictional force.

Solution

a) The frictional force acts to the left. The horizontal component of F to the right is $F \cos 25°$. This component must equal 470 N for the resultant force to be zero with no change in velocity.

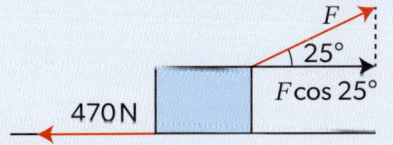

Approaches to learning

Using an energy approach is a convenient way to solve problems when the acceleration of an object is non-uniform—for example, a skier sliding down a slope that has a changing gradient. We know that, when there are no friction losses, $-\Delta E_p = \Delta E_k$ because the mechanical energy is constant. This energy approach can often simplify uniform-acceleration calculations too.

Resolving $F \cos 25° = 470$ gives $F = 519\,\text{N} \approx 520\,\text{N}$ (to 2 s.f., as for the given data).

b) (i) The angle between F and the displacement of the box is 25°.

Work done by tension $= (F \cos 25°) \times \text{distance} = 470 \times 250 = 120\,\text{kJ}$

(ii) The angle between the frictional force and the displacement is 180°.

Work done by friction $= 470 \times 250 \times \cos 180° = -120\,\text{kJ}$

The work done by tension and friction are equal but opposite, which means, of course, that the work done by the net force is zero and the box moves with no change in kinetic energy (constant velocity).

Example 2

A tennis ball is served with an initial speed of $16.0\,\text{m s}^{-1}$ from a height of $2.80\,\text{m}$. Air resistance is negligible. Calculate the speed of the ball as it strikes the ground.

Solution

In the absence of air resistance, mechanical energy is conserved and the ball strikes the ground with kinetic energy $\frac{1}{2}mv^2 = \frac{1}{2}mu^2 + mg\Delta h$, where u is the initial speed and Δh is the change in height.

Final speed $v = \sqrt{16.0^2 + 2 \times 9.8 \times 2.80} = 17.6\,\text{m s}^{-1}$.

Note that the answer does not depend on the mass of the ball or the angle at which is served.

Example 3

A diver climbs onto a diving board and dives from it.

The height of the diving board above the floor is $4.0\,\text{m}$. The mass of the diver is $54\,\text{kg}$.

a) Calculate the gain in gravitational potential energy when the diver climbs to the diving board.

b) The diver enters the water at a speed of $8.0\,\text{m s}^{-1}$. Calculate the kinetic energy of the diver as she enters the water.

c) Suggest why the kinetic energy of the diver in part (b) is different from the gravitational potential energy gained in part (a).

Solution

a) $\Delta E_p = mg\Delta h = 54 \times 9.8 \times 4.0 = 2.1\,\text{kJ}$

b) $\Delta E_k = \frac{1}{2}m(v^2 - u^2)$, where $u = 0$ as the diver starts from rest.

$\Delta E_k = \frac{1}{2}mv^2 = \frac{1}{2} \times 54 \times 8.0^2 = 1.7\,\text{kJ}$

c) A number of factors can be discussed here. Some of these factors explain why the kinetic energy might be lower than the initial gain in the gravitational potential energy, while others explain additional contributions to the kinetic energy.

- Work is done against air resistance.
- The centre of mass of the diver during taking off is not at the same height above the water as the diving board.
- The diver gains gravitational potential energy in taking off.
- Some of the energy usually goes into rotational kinetic energy.

Example 4

A car of mass 1500 kg accelerates from rest on a horizontal road. The graph shows how the driving force provided by the engine of the car varies with the distance moved by the car.

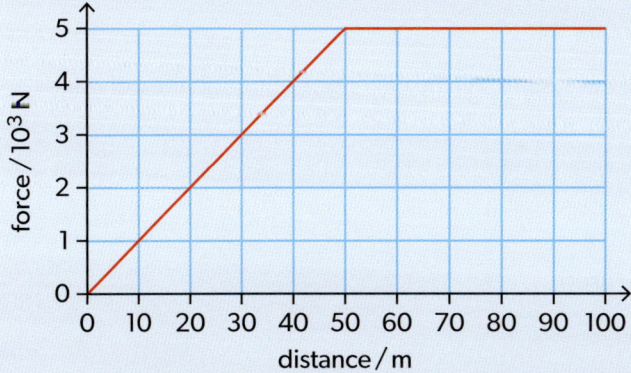

a) Calculate the work done by the driving force in moving the car through a distance of 100 m.

b) The final speed of the car is 20 m s^{-1}. Determine:

   (i) the work done on the car by the resistive force

   (ii) the magnitude of the resistive force, assuming that it is constant.

Solution

a) The work done is equal to the area under the force–distance graph.

   Work done $= \frac{1}{2} \times 5.0 \times 10^3 \times 50 + 5.0 \times 10^3 \times 50 = 3.75 \times 10^5 \approx 380$ kJ

b) (i) The kinetic energy of the car is $\frac{1}{2} \times 1500 \times 20^2 = 300$ kJ and this is equal to the work done on the car by the net force.

   Since (net force) = (driving force) – (resistive force), the work done by the resistive force is equal to the difference between the work of the driving force and the final kinetic energy of the car.

   work of the resistive force = 375 – 300 = 75 kJ

   (ii) Resistive force $= \dfrac{\text{work}}{\text{displacement}} = \dfrac{75 \times 10^3}{100} = 750$ N

Power and efficiency

Energy can be transferred at different rates. Think of two people of equal weight who run up a hill: the quicker person is the more powerful of the two because they transfer the same energy in a shorter time.

work done = force × distance, so power $= \dfrac{\text{force} \times \text{distance}}{\text{time}} = \text{force} \times \dfrac{\text{distance}}{\text{time}}$

Therefore power = force × speed. As you saw in Example 1, not all energy is necessarily transferred from the original source into its final useful form—in most real cases, there is a transfer via friction. A measure of the effectiveness of a transfer is efficiency η. This can be defined in terms of energy or in terms of power.

Power is the rate of transferring energy, the rate of doing work.

Power $= \dfrac{\text{energy transferred}}{\text{time taken for transfer}}$

The unit of power is the watt (W); in fundamental units this is kg m^2 s^{-3}.

Efficiency $= \eta = \dfrac{\text{useful energy transferred}}{\text{total energy input}} = \dfrac{E_{\text{output}}}{E_{\text{input}}} = \dfrac{\text{power output}}{\text{power input}} = \dfrac{P_{\text{output}}}{P_{\text{input}}}$

Approaches to learning

Rate of change—the link between energy and power—is frequently used in science. You meet it many times in this book.

You may choose to approach it as the division of a quantity by time, or you may view it graphically as the gradient of a graph showing the variation with time of a quantity. Both approaches are used in the course.

A Space, time and motion

Example 5

Water in a hydroelectric system falls vertically to a river below at the rate of 12 000 kg every minute.

The water takes 2.0 s to fall this distance. It has zero vertical velocity at the top.

a) Calculate the height through which the water falls.

b) An electrical generator of efficiency 20% is at the foot of the system. All the water goes through the generator.

   Determine the electrical power output of the generator.

c) Outline the energy transfers in this system.

Solution

a) The acceleration g is uniform, so *suvat* equations can be used.

$$s = ut + \frac{1}{2}gt^2 = 0 + \frac{1}{2} \times 9.8 \times 2.0^2 = 19.6 \approx 20\,\text{m} \quad \text{(to 2 s.f.)}.$$

b) From this point, it is best to work in seconds rather than minutes.

Mass of water flowing every second $= \dfrac{12\,000}{60} = 200\,\text{kg}$

Gravitational potential energy transferred every second $= mg\Delta h$

$$= 200 \times 9.8 \times 19.6$$
$$= 38\,416\,\text{J}$$

Thus the power input to the generator is 38 416 W.

As the efficiency is 20%:

power output $= 0.20 \times 38\,416 = 7683\,\text{W} \simeq 7.7\,\text{kW}$ (to 2 s.f.).

c) The water has stored gravitational potential energy at the top of the waterfall. As the water falls, energy is transferred into kinetic energy. At the bottom of the waterfall, the maximum transfer of energy has occurred. The water enters a turbine where the linear kinetic energy of the water is transferred into rotational kinetic energy of the turbine and the dynamo. The dynamo transfers this kinetic energy into electrical energy, wasted thermal energy and frictional losses.

Sample student answer

An electric motor pulls a glider horizontally through a distance of 148 m with a constant force of 1370 N in a time of 11.0 s. The motor has an overall efficiency of 23.0%.

Determine the average power input to the motor. State your answer to an appropriate number of significant figures. [4]

This answer could have achieved 3/4 marks:

$W_{useful} = Fs = 1370\,\text{N} \times 148\,\text{m} = 202\,760\,\text{J}$

$W_{used} = \dfrac{W_{useful}}{0.23} = 881\,565.2\,\text{J}$

$P = \dfrac{W_{used}}{t} = \dfrac{881\,565.2}{11\,\text{s}} = 80\,142.3\,\text{W}$

▼ This is not the easiest way to carry out the problem. First, the energy gained by the glider is calculated and then the efficiency is used to calculate the energy input to the motor. This is then divided by time to give the power input. There are too many significant figures in the answer, which should have been restricted to 2 or 3 given the data in the question. The student has added units in the intermediate steps of the calculation. This is not necessary to gain full credit.

This answer could have achieved 4/4 marks:

Average speed $= \dfrac{148}{11} = 13.45\,\text{m s}^{-1}$

Power $=$ force $\times$ velocity $= 1370 \times 13.45 = 18\,433\,\text{W} = 23\%$

$\dfrac{18\,433}{23} \times 100 = 80\,142.292\,\text{W} = 80.1\,\text{kW}$

▲ The solution uses power = force × speed, which leads to the power of the glider directly. Then it uses a simple efficiency calculation and correct rounding to get the answer with an appropriate number of significant figures.

A.3 Work, energy and power

A thermal-energy power station gives an example of an extended energy transfer.

The initial thermal energy of the power station can be transferred from various energy stores including nuclear, geothermal and fossil fuels. These fuels can be characterized by their energy density.

> **Energy density** is the energy that can be transferred from 1 m³ of the fuel. The unit of energy density is J m⁻³.

Example 6

A coal-fired power station has a power output P. Its efficiency is ε. It burns a mass of coal M of density ρ every second. Derive an expression for the energy density of the coal.

Solution

Energy density is the energy available per unit volume of the fuel.

Use $\rho = \dfrac{M}{V}$.

The volume V of coal consumed every second is $\dfrac{M}{\rho}$.

The power input to the station is $\dfrac{P}{\varepsilon}$, allowing for the inefficiency.

$$\text{Energy density} = \frac{\text{power input}}{\text{coal volume per second}} = \frac{P}{\varepsilon} \times \frac{\rho}{M} = \frac{P\rho}{\varepsilon M}$$

Thermal-energy power stations use the thermal energy to boil water (Topic B.1), creating steam, which rotates a turbine (Topic A.4)—this is a transfer of thermal energy to mechanical (kinetic) energy. The turbine is attached to a dynamo which rotates in a magnetic field (Topic D.4). This causes charge to flow in the dynamo's coil, transferring the kinetic energy to an electrical form. The electrical energy can be transported through a cable network (a grid) to domestic and commercial end-users (Topic B.5).

One way to represent quantitative energy transfers is to use a Sankey diagram (Figure 1).

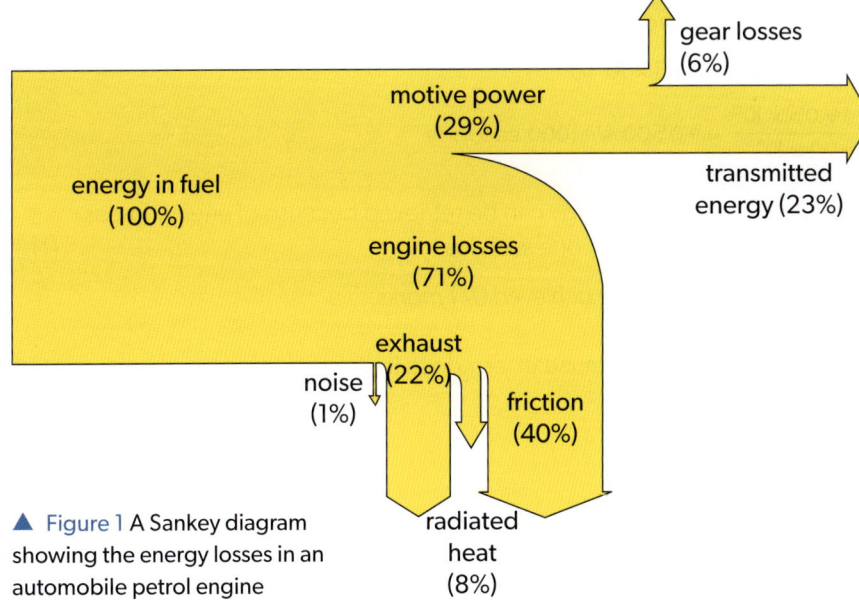

▲ Figure 1 A Sankey diagram showing the energy losses in an automobile petrol engine

Sankey diagrams are flow diagrams that show energy transfers in a system or process. The width of an arrow in the diagram represents the relative size of its contribution to the total energy involved in the transfer. Figure 1 shows a Sankey diagram for an automobile.

Example 7

A wind turbine's blades have a total area 65 m².

The turbine is used in a wind of speed 15 m s⁻¹.

The density of air is 1.3 kg m⁻³.

a) Determine the mass of air incident on the turbine every second.

b) Calculate the total kinetic energy of the air that arrives at the turbine every second.

c) Calculate the power output of the turbine when only 40% of the total kinetic energy of the wind can be converted into electrical energy.

A Space, time and motion

> **Solution**
>
> a) The air volume incident on the turbine in 1 second is equivalent to the volume of a cylinder of cross-sectional area 65 m² and length 15 m:
> $65 \text{ m}^2 \times 15 \text{ m} = 975 \text{ m}^3$.
>
> To calculate the mass of air arriving in one second:
>
> mass = density × volume = $1.3 \text{ kg m}^{-3} \times 975 \text{ m}^3 = 1270 \text{ kg}$
>
> b) Kinetic energy of this air = $\frac{1}{2} \times 1270 \times 15^2 = 140 \text{ kJ}$
>
> c) 0.4 (40%) of this energy can be transferred to electrical energy. Therefore:
>
> Power output = $0.4 \times 140 = 57 \text{ kW}$

Sample student answer

a) A hydroelectric system has four 250 MW generators. The energy density available from the water is $2.7 \times 10^6 \text{ J m}^{-3}$. Determine the maximum time for which the hydroelectric system can maintain full output when a mass of 1.5×10^{10} kg passes through the turbines. [2]

Density of water = 1000 kg m⁻³

This answer could have achieved 2/2 marks:

$2.7 \times 10^6 \text{ Jm}^{-3} \div 1000 = 2.7 \times 10^3 \text{ Jkg}^{-1}$

Total energy available = $2.7 \times 10^3 \text{ Jkg}^{-1} \times 1.5 \times 10^{10} = 4.05 \times 10^{13}$ J

$4 \times 250 \times 10^6 = 1 \times 10^9$ W

$\frac{4.05 \times 10^{13}}{1 \times 10^9} = 40500 \approx 41000$ seconds

▲ A well-presented, correct solution, including a clearly shown answer given to 2 significant figures.

b) Not all the stored energy can be retrieved because of energy losses in the system. Explain one such loss. [1]

This answer could have achieved 0/1 marks:

High temperature will cause energy loss in thermal energy.

▼ There appears to be the suggestion that high temperatures lead to energy loss. There are certainly frictional losses in the turbine bearings and resistive losses in the electrical cables. But this answer is too vague for credit.

Practice problems

Problem 1
A bus travels at constant speed of 6.2 m s⁻¹ along a road inclined upwards at 6.0° to the horizontal. The mass of the bus is 8.5×10^3 kg. The total output power of the engine of the bus is 70 kW and the efficiency of the engine is 35%.

a) Draw a labelled sketch to represent the forces acting on the bus.

b) Calculate the input power to the engine.

c) Determine the rate of increase of gravitational potential energy of the bus.

d) Estimate the magnitude of the resistive forces acting on the bus.

e) The engine of the bus stops working.

  (i) Determine the magnitude of the net force opposing the motion of the bus at the instant at which the engine stops.

  (ii) Discuss, with reference to the air resistance, the change in the net force as the bus slows down.

Problem 2
A body of mass 0.20 kg is dropped from rest from a height of 5.0 m. The body impacts the ground with a speed of 9.0 m s^{-1}. Calculate:

a) the change in the kinetic energy of the body

b) the work done on the body by air resistance.

Problem 3
The motor of an escalator has a maximum power output of 3.5 kW. The escalator carries passengers through a height difference of 7.0 m.

a) Determine the maximum number of passengers of average mass 62 kg that the escalator can carry in one minute.

b) The efficiency of the motor is 75%. Calculate the maximum electrical power input to the motor.

Problem 4
A body of mass 0.45 kg hangs from a spring. The spring is extended through a distance of 5.0 cm from its unstretched length.

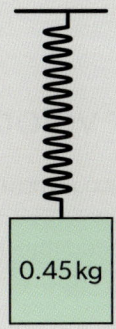

a) Determine the spring constant.

b) Calculate the elastic potential energy stored in the spring.

c) Work of 0.20 J is done to extend the spring further. Calculate the new extension of the spring, relative to its unstretched length.

Problem 5
The Sankey diagram shows energy transfers in a nuclear power station. The electrical power output of the power station is 2.4 GW.

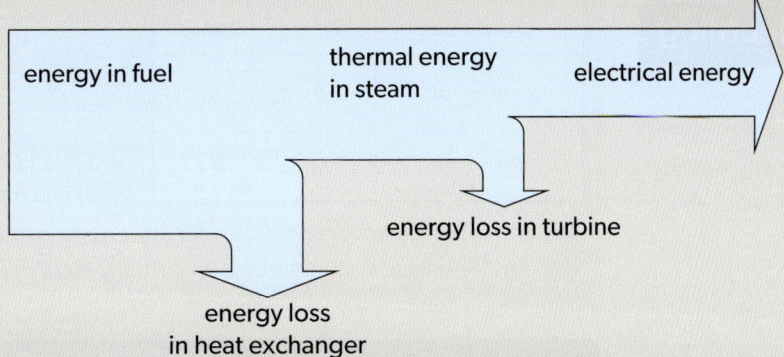

The efficiency of the conversion of nuclear energy to electrical energy is 40%.

a) The power lost in the heat exchanger is 2.2 GW. Calculate the power lost in the turbine.

b) The energy density of the nuclear fuel is 1.5×10^{18} J m^{-3}. Determine the volume of fuel used in the power station in one day.

A.4 Rigid body mechanics

You must know:
- what is meant by angular displacement, angular velocity and angular acceleration
- the equations of rotational motion that apply under conditions of constant angular acceleration
- what is meant by moment of inertia, torque, rotational equilibrium, angular momentum and angular impulse
- that angular momentum is conserved
- that Newton's second law of motion can be applied in a modified form to angular motion
- the distinction between rolling and slipping.

You should be able to:
- calculate torque for a single force
- interpret graphs of the variation with time of angular displacement, angular velocity and torque
- solve problems involving the rotational form of Newton's second law
- solve problems that involve rotational and/or translational equilibrium
- solve problems by treating the quantities in rotational dynamics as analogies for those in linear dynamics.

Topics A.1, A.2 and A.3 cover linear mechanics and the interaction of objects that are treated as points. Topic A.4 deals with objects that have shape and size. Many of the quantities in rotational mechanics have direct analogies in linear mechanics.

You should be confident using the equations from Topic A.1 and be familiar with the circular motion section of Topic A.2.

Approaches to learning

Learn the links between linear quantities and rotational quantities and the way in which they are used. Topic A.4 emphasizes these links.

Angular velocity and acceleration

When an object rotates about an axis with no translational motion and is displaced through an angle $\Delta\theta$ in a time t, it has an **angular velocity** $\omega = \dfrac{\Delta\theta}{t}$.

Remember, from Topic A.2, that $\omega = \dfrac{2\pi}{T}$ and that, from Topic C.1, $\omega = 2\pi f$.

When the initial angular speed ω_i changes to a final angular speed ω_f in a time t, its **angular acceleration** $\alpha = \dfrac{(\omega_f - \omega_i)}{t}$.

The correspondence between linear and rotation equations is shown in Table 1.

Linear equation	Corresponding rotational equation
$v = u + at$	$\omega_f = \omega_i + \alpha t$
$v^2 = u^2 + 2as$	$\omega_f^2 = \omega_i^2 + 2\alpha\Delta\theta$
$s = ut + \dfrac{1}{2}at^2$	$\Delta\theta = \omega_i t + \dfrac{1}{2}\alpha t^2$
$s = \dfrac{v+u}{2}t$	$\Delta\theta = \dfrac{\omega_f + \omega_i}{2}t$

◀ Table 1 Correspondence between linear and rotation equations

The unit for angular velocity (angular speed) is normally written as rad s^{-1}. The unit for angular acceleration is normally rad s^{-2}.

Example 1

A laboratory centrifuge reaches its working angular speed of 1100 rad s^{-1} from rest in 4.2 s.

a) Calculate the angular acceleration of the centrifuge.

b) Determine the number of revolutions the centrifuge makes during the acceleration.

Solution

a) Angular acceleration $= \dfrac{\omega_f - \omega_i}{t} = \dfrac{1100 - 0}{4.2} = 260 \text{ rad s}^{-2}$

b) Angular displacement $= \dfrac{\omega_f + \omega_i}{2} t = \dfrac{1100 + 0}{2} \times 4.2 = 2310 \text{ rad}$

One full revolution corresponds to an angular displacement of 2π.

Number of revolutions $= \dfrac{2310}{2\pi} = 370$ (to 2 s.f.)

Example 2

A merry-go-round rotates with an initial angular velocity of 1.2 rad s^{-1}. It decelerates uniformly and its angular velocity decreases to 0.80 rad s^{-1} during one complete revolution.

Determine:

a) the angular acceleration of the merry-go-round

b) the time to complete the first revolution

c) the additional angle through which the merry-go-round rotates before coming to rest.

Solution

a) The angular displacement $\Delta\theta = 2\pi$ and you can use the equation $\omega_f^2 = \omega_i^2 + 2\alpha\Delta\theta$ to find the unknown angular acceleration α.

$0.80^2 = 1.2^2 + 2 \times 2\pi \times \alpha$

$\alpha = \dfrac{0.80^2 - 1.2^2}{4\pi} = -0.064 \text{ rad s}^{-2}$

b) The time can be calculated in several ways, for example, by solving the equation $\Delta\theta = \dfrac{\omega_f + \omega_i}{2} t$.

Time for the first revolution $= \dfrac{2 \times 2\pi}{1.2 + 0.8} = 6.3 \text{ s}$

c) You can again use the equation $\omega_f^2 = \omega_i^2 + 2\alpha\Delta\theta$, but this time you have to substitute $\omega_f = 0$, $\omega_i = 0.8 \text{ rad s}^{-1}$ and $\alpha = -0.06366... \text{ rad s}^{-2}$ (the unrounded acceleration from part (a)).

Additional angular displacement $\Delta\theta = \dfrac{0.8^2 - 0^2}{2 \times 0.06366...} = 5.0 \text{ rad}$

The merry-go-round makes less than one additional revolution before coming to rest.

Nature of science

The Earth's seasons depend directly on the rotation of the Earth on its axis. This axis is tilted at about 24° relative to the plane of the orbit. When the north pole is towards the Sun, then it is summer in the Northern Hemisphere.

Linking rotational dynamics and the Earth–Sun system shows the global impact of science on the planet.

Moment of inertia

The rotational equivalent of mass is moment of inertia I. Moments of inertia depend on the axes of rotation. For two very small spheres, each of mass m and connected by a light rod of length l about an axis through the centre of the rod, I is given by:

$$I = m\left(\dfrac{l}{2}\right)^2 + m\left(\dfrac{l}{2}\right)^2 = \dfrac{1}{2}ml^2$$

However, when the rotational axis is changed to the centre of one of the spheres, I becomes $I = 0 + ml^2$.

The moment of inertia for a bicycle wheel of mass m and radius r about an axis of rotation through the centre of the wheel is simply mr^2 (assuming that the spokes and the centre hub have negligible mass compared with the rim and tyre).

For a system of objects, the **moment of inertia** $I = \sum mr^2$, where m is the mass of one object and r is the distance from the axis of rotation for that object. The unit of moment of inertia is kg m².

Assessment tip

You will not be required to derive the equation of the moment of inertia of a continuous shape such as a cylinder; the equation will be provided. However, you may be asked to use the equation to calculate the numerical value of the moment of inertia. You could also be asked to calculate I for a collection of individual point masses.

Approaches to learning

Gyroscopes are an example of rotational motion used in day-to-day life. They have many uses, from their important role in navigation of aircraft to the more mundane part that they play in stabilization of moving systems. Gyroscopes spin rapidly and so have a large angular momentum. In this topic, you are using knowledge from linear mechanics to develop skills in a new concept.

When a force F acts at a distance r from a rotation axis, with an angle θ between the force direction and the line connecting the axis with the point where F is applied, the **torque** $\tau = F \times r \sin\theta$.

Angular momentum L is defined as $L = I \times \omega$. It has units of kg m² s⁻¹.

Example 3

Four point masses m are placed in the corners of a square of side length L.

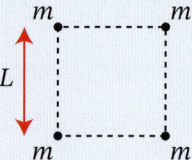

Calculate the moment of inertia of the system about an axis perpendicular to the plane of the square and passing through:

a) the centre of the square
b) one of the masses.

Solution

a) The diagonal of the square has length $\sqrt{2}L$ and so each of the masses is at a distance of $\frac{\sqrt{2}}{2}L$ from the axis of rotation. Each mass contributes a moment of inertia of $m\left(\frac{\sqrt{2}}{2}L\right)^2 = \frac{1}{2}mL^2$.

Moment of inertia $= 4 \times \frac{1}{2}mL^2 = 2mL^2$

b) One of the masses coincides with the axis of rotation, so its distance from the axis is zero, two masses are at a distance L from the axis and the last one is at a distance $\sqrt{2}L$.

Moment of inertia $= m(0)^2 + mL^2 + mL^2 + m(\sqrt{2}L)^2 = 4mL^2$

This is twice the value calculated in part (a).

Torque and angular momentum

The rotational equivalent of linear force is torque. A torque exists when a force F acts at a distance r from an axis of rotation and the torque produces a turning effect. Torque is a vector and, therefore, has a direction—this is defined by a right-hand rule (rather like Fleming's rule in electromagnetism is defined by a left-hand rule). The relationship is shown in Figure 1.

You will not be asked to treat rotational quantities as vectors in the IB Diploma Programme physics course. Always treat them as scalar quantities.

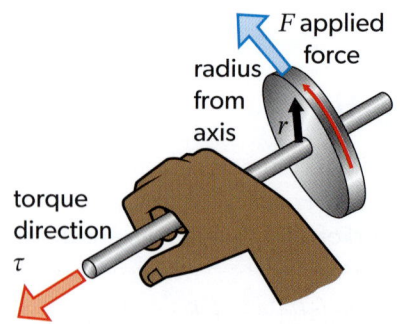

▲ **Figure 1** The right-hand rule that describes torque

The link between Newton's second law, $F = ma$, and $F = \dfrac{\text{change in momentum}}{\text{time taken for change}}$ also applies to rotational motion when angular momentum has been defined.

The torque τ is given by $\tau = \dfrac{\text{change in angular momentum}}{\text{time taken for change}}$.

The angular momentum of a system is conserved unless an external torque acts on the system. This has consequences when the moment of inertia of the system is modified by purely internal means. A common example of this is an ice skater spinning about a vertical axis through the centre of their body. The skater alters the rotational speed by pulling their arms into the body. This reduces the moment of inertia, so the speed of the spin increases to conserve angular momentum. Notice that no external torque acts here because the movement of the arms can only give rise to an internal torque. The energy transfer involved must be produced by the skater.

Rotational kinetic energy

Rotational kinetic energy links to linear kinetic energy. As usual, changes can be inelastic, elastic, or have energy added depending on the changes in momentum of the system or systems involved.

> The **rotational kinetic energy** E_k of a body with moment of inertia I and angular velocity ω is $E_k = \frac{1}{2}I\omega^2$.
>
> The change in rotational kinetic energy E_k of a body with moment of inertia I when the angular velocity changes from ω_i to ω_f is $\Delta E_k = \frac{1}{2}I(\omega_f^2 - \omega_i^2)$.
>
> The unit of E_k is kg m² s⁻², which is the same as a joule.

> **Angular impulse** ΔL is change in angular momentum, which is $\Delta L = \Delta(I \times \omega) = \tau \times \Delta t$. The unit is the same as that for angular momentum.

> Newton's laws of motion can be expressed in rotational contexts.
>
> **Newton's first law for rotation:** Every rotating body continues to rotate at constant angular velocity unless an external torque acts on it.
>
> **Newton's second law for rotation:** Newton's second law in a rotational context is $\tau = I \times \alpha$.
>
> **Newton's third law for rotation:** The action torque and reaction torque acting between two bodies are equal and opposite.

Example 4

A flywheel is accelerated from rest. The flywheel has a moment of inertia of 250 kg m² and takes 8.0 s to accelerate to 90 rev min⁻¹.

a) Calculate the angular acceleration of the flywheel.

b) Calculate the average accelerating torque acting on the flywheel.

c) Calculate the rotational kinetic energy stored in the flywheel at the end of its acceleration.

Solution

a) Final angular velocity of the flywheel = $\frac{90 \times 2\pi}{60}$ = 9.425 rad s⁻¹

Angular acceleration $\alpha = \frac{9.425}{8.0} = 1.178 \approx 1.2$ rad s⁻²

b) Torque $\tau = I\alpha = 250 \times 1.178 = 290$ N m

c) Rotational kinetic energy $= \frac{1}{2}I\omega_f^2 = \frac{1}{2} \times 250 \times 9.425^2 = 1.1 \times 10^4$ J

> **Assessment tip**
>
> There are a number of pitfalls in answering rotational mechanics questions.
>
> - Make sure that your calculator is set to use radians or degrees as appropriate.
>
> - There are a number of possible ways to express rotation and angular speed: in radians or revolutions (1 rev ≡ 2π rad), and in revolutions per second (rad⁻¹) or revolutions per minute.

Example 5

A uniform disc of mass M = 0.80 kg and radius R = 0.20 m rotates at an angular velocity of 18 rad s⁻¹. A resistive torque of 0.040 N m is applied to the disc for a time of 4.0 s.

The moment of inertia of the disc is $\frac{1}{2}MR^2$.

a) Determine the final angular velocity of the disc.

b) Calculate the angular displacement of the disc during the 4.0 s.

c) Calculate the work done by the frictional torque.

Solution

a) Moment of inertia $I = \frac{1}{2} \times 0.80 \times 0.20^2 = 0.016\,\text{kg}\,\text{m}^2$

The final angular velocity can be determined by two different methods: by finding the angular acceleration of the disc, or, alternatively, the angular impulse delivered to the disc. Both methods involve Newton's second law for rotation, in the form $\tau = I\alpha$ or $\tau = \frac{\Delta L}{\Delta t}$.

Method 1: Angular acceleration

Angular acceleration $\alpha = \frac{\tau}{I} = \frac{-0.040}{0.016} = -2.5\,\text{rad}\,\text{s}^{-2}$

The negative sign is used because the frictional torque acts against the direction of rotation of the disc.

Final angular velocity $\omega_f = \omega_i + \alpha t = 18 - 2.5 \times 4.0 = 8.0\,\text{rad}\,\text{s}^{-1}$

Method 2: Angular impulse

Angular impulse $\Delta L = \tau \Delta t = -0.040 \times 4.0 = -0.16\,\text{N}\,\text{m}\,\text{s}$

Again this has a negative sign because the impulse acts against the initial angular velocity of the disc.

Final angular momentum $\Delta L = I\omega_i + \Delta L = 0.016 \times 18 - 0.16 = 0.128\,\text{N}\,\text{m}\,\text{s}$

Final angular velocity $\omega_f = \frac{L}{I} = \frac{0.128}{0.016} = 8.0\,\text{rad}\,\text{s}^{-1}$

which is in agreement with the first method.

b) Angular displacement $= \omega_i t + \frac{1}{2}\alpha t^2 = 18 \times 4.0 - \frac{1}{2} \times 2.5 \times 4.0^2 = 52\,\text{rad}$

which is a little more than eight full revolutions. Note that the answer can also be obtained from $\Delta\theta = \frac{\omega_f + \omega_i}{2}t$.

c) The work done is equal to the change in the rotational kinetic energy of the disc.

Work $= \Delta E_k = \frac{1}{2}I(\omega_f^2 - \omega_i^2) = \frac{1}{2} \times 0.016(8.0^2 - 18^2) = -2.1\,\text{J}$

The work is negative because the energy is transferred from the disc.

In linear mechanics the area under a force–time graph is equivalent to the impulse acting on an object.

In rotational mechanics, by analogy, the area under a **torque–time** graph is the **angular impulse** and also the **change in angular momentum** ΔL.

This is used in Example 6.

Example 6

A bicycle wheel is placed on a repair stand and is accelerated from rest. The graph shows how the torque applied to the wheel varies with time.

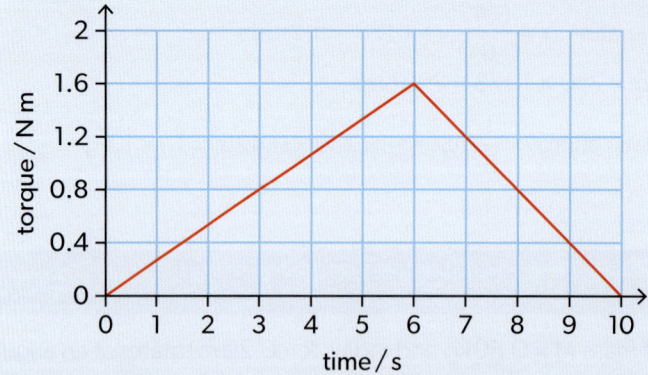

The moment of inertia of the wheel is $0.090\,\text{kg}\,\text{m}^2$.

a) Calculate the final angular momentum of the wheel.

b) Hence, calculate the final angular velocity of the wheel.

The torque is provided by a force F applied tangentially to a sprocket of the wheel. The radius of the sprocket is 4.0 cm.

c) Calculate the maximum force applied to the sprocket.

Solution

a) The area under the graph is equal to the change in the angular momentum of the wheel. The initial angular momentum is zero; therefore the final angular momentum L is equal to the area.

L = area under the graph = $\frac{1}{2} \times 1.6 \times 10 = 8.0\,\text{N m s}$.

b) From $L = I\omega$:

angular velocity $= \frac{L}{I} = \frac{8.0}{0.090} = 89\,\text{rad s}^{-1}$

c) The force is applied at a distance of 0.040 m from the axis of rotation.
From $\tau = Fr$:

maximum force $F_{\text{max}} = \frac{\tau_{\text{max}}}{r} = \frac{1.6}{0.040} = 40\,\text{N}$

Example 7

A flywheel of moment of inertia $I_1 = 0.20\,\text{kg m}^2$ rotates at an initial angular velocity $\omega_1 = 120\,\text{rad s}^{-1}$. Another flywheel, of moment of inertia $I_2 = 0.16\,\text{kg m}^2$, is initially at rest. The flywheels are now coupled together.

a) Calculate the final angular velocity of the flywheels.

b) Calculate the percentage of the initial kinetic energy of the system that is transferred to other forms than kinetic energy.

Solution

a) Assume that no external torques act on the system, and therefore the angular momentum is conserved. Initially, only the first flywheel contributes to the angular momentum: $L_i = I_1 \omega_1$. Once the flywheels are coupled together, they achieve a common angular velocity ω and their final angular momentum is $L_f = (I_1 + I_2)\omega$.

Equating initial and final angular momentum gives $I_1 \omega_1 = (I_1 + I_2)\omega$.

Final angular velocity $\omega = \frac{I_1 \omega_1}{I_1 + I_2} = \frac{0.20 \times 120}{0.20 + 0.16} = 67\,\text{rad s}^{-1}$

b) From $E_k = \frac{L^2}{2I}$:

$$\frac{E_{k,\text{final}}}{E_{k,\text{initial}}} = \frac{\frac{L_f^2}{2(I_1 + I_2)}}{\frac{L_i^2}{2I_1}} = \left(\frac{L_f}{L_i}\right)^2 \times \frac{I_1}{I_1 + I_2}$$

Since $L_f = L_i$:

$E_{k,\text{final}} = \frac{I_1}{I_1 + I_2} E_{k,\text{initial}} = \frac{0.20}{0.20 + 0.16} E_{k,\text{initial}} = 0.56 E_{k,\text{initial}}$

56% of the initial kinetic energy of the first flywheel remains in the system. This means that 44% of the initial kinetic energy is transferred to other energy forms.

When a cylinder **rolls** along horizontal ground, the point of contact between the cylinder and ground is at rest.

When a cylinder **slides** on the ground, the point of contact moves (slips) along the ground.

Because the contact point is at rest in rolling, the coefficient of static friction (Topic A.2) should be used in any calculations.

A cylinder of radius r rolls to the right at linear speed v. The top of the cylinder moves to the right at a speed of $v + r\omega$ and the bottom point must be moving at $v - r\omega$. However, as $v - r\omega = 0$ (the bottom point must be at rest) then $v = r\omega$ and the top of the cylinder moves to the right with linear speed $2v$.

Nature of science

A complex model involving rotational mechanics is needed to study the theoretical behaviour of a rotating neutron star. The model can then be compared with astronomical observations to improve our understanding of these remote stars.

The total kinetic energy of a rolling object is the sum of the linear plus rotational kinetic energies, in other words, $\frac{1}{2}I\omega^2 + \frac{1}{2}m(\omega r)^2$, which simplifies to $E_k = \frac{1}{2}(I + mr^2)\omega^2$ using $v = r\omega$.

When this kinetic energy is gained by the object rolling down a slope with a vertical height change of h, then $mgh = \frac{1}{2}I\omega^2 + \frac{1}{2}mv^2 = \frac{1}{2}(I + mr^2)\omega^2$ to conserve mechanical energy.

Example 8

A sphere of radius R and mass M rolls down a ramp of vertical height h. The initial velocity of the sphere is zero. The moment of inertia of the sphere is $\frac{2}{5}MR^2$.

a) Derive an expression, in terms of h, for the translational velocity v of the sphere at the bottom of the ramp.

b) Determine the ratio $\dfrac{\text{translational kinetic energy of the sphere}}{\text{total kinetic energy of the sphere}}$.

Solution

a) The change in the gravitational potential energy of the sphere is equal to the sum of the rotational kinetic energy and translational kinetic energy of the sphere. In symbols:

$$Mgh = \frac{1}{2}I\omega^2 + \frac{1}{2}Mv^2 = \frac{1}{2} \times \frac{2}{5}MR^2 \times \left(\frac{v}{R}\right)^2 + \frac{1}{2}Mv^2 = \left(\frac{1}{5} + \frac{1}{2}\right)Mv^2$$

$$= \frac{7}{10}Mv^2$$

Therefore, $v = \sqrt{\dfrac{10}{7}gh}$

b) Translational kinetic energy is $\frac{1}{2}M \times \frac{10}{7}gh = \frac{5}{7}Mgh$ and the total kinetic energy is Mgh:

$$\frac{\text{translational kinetic energy}}{\text{total kinetic energy}} = \frac{5}{7}$$

> **Sample student answer**
>
> A satellite approaches a rotating space probe at a negligibly small speed in order to link to it. The satellite does not rotate initially, but after the link they rotate at the same angular speed.
>
> The initial angular speed of the probe is 16 rad s^{-1}.
>
> The moment of inertia of the probe about the common axis is 1.44×10^4 kg m^2.
>
> The moment of inertia of the satellite about the common axis is 4.80×10^3 kg m^2.
>
>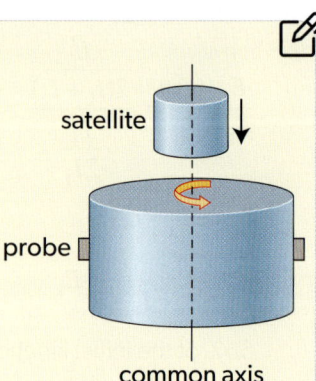
>
> a) Determine the final angular speed of the probe–satellite system. [2]
>
> *This answer could have achieved 2/2 marks:*

▲ A well-presented solution that makes everything obvious to the examiner. The physics principles are stated and the substitutions are clear.

> Angular momentum L is conserved and $L = I\omega$
>
> $I_p \omega_{pi} = I_s \omega_{sf} + I_p \omega_{pf}$
>
> $= \omega_{pf}(I_s + I_p)$ as $\omega_{sf} = \omega_{pf}$
>
> $1.44 \times 10^4 \times 16 = \omega_{pf}(1.44 \times 10^4 + 4.80 \times 10^3)$
>
> $230400 = \omega_{pf} \times 19200$
>
> $\omega_{pf} = 12$ rad/s

b) Calculate the loss of rotational kinetic energy due to the linking of the probe with the satellite. [3]

This answer could have achieved 3/3 marks:

$E_{krot\,i} = \frac{1}{2}I_p\omega^2 = \frac{1}{2} \times 1.44 \times 10^4 \times 16^2 = 1843200$

$E_{krot\,f} = \frac{1}{2}(I_p + I_s) \times \omega_{pf}^2 = \frac{1}{2} \times 19200 \times 12^2 = 1382400$

$E_{krot\,i} - E_{krot\,f} = 460800 \approx 460000\,J\,lost$

460 000 J lost.

▲ Again, the answer is well laid out and accurate. It is easy to award the full credit here. This is a model of how to answer a question.

Practice problems

Problem 1
A potter's wheel is rotating at an angular speed of 5.0 rad s^{-1} with no torque acting.

The potter throws a lump of clay onto the wheel so that the wheel and the clay have a common axis of rotation.

The moment of inertia of the wheel is 1.6 kg m^2 and the moment of inertia of the clay is 0.25 kg m^2, both about the common axis.

The angular speed of the wheel changes suddenly when the clay lands on it. No external angular impulse is added to the system.

a) Calculate the angular speed of the wheel immediately after the clay has been added.

The potter now applies a tangential force to the rim of the wheel over 0.25 of a revolution to return the angular speed to 5.0 rad s^{-1}.

The wheel has a diameter of 0.62 m.

b) Calculate the angular acceleration of the wheel.

c) Deduce the average tangential force applied by the potter.

Problem 2
A flywheel consists of a solid cylinder of mass 1.22 kg and radius 240 mm.

A mass M is connected to the flywheel by a string wrapped around the circumference of the cylinder.

The mass falls from rest and exerts a torque on the flywheel which accelerates uniformly.

After a time of 4.85 s, the velocity of M is 2.36 m s^{-1}.

Moment of inertia for the flywheel = $\frac{1}{2}$(mass) × (radius)^2.

a) Calculate the angular acceleration of the flywheel.

b) Deduce the torque acting on the flywheel.

c) Determine M.

Problem 3
A flywheel of moment of inertia 0.15 kg m^2 rotates counterclockwise at an initial angular velocity 100 rad s^{-1}. A clockwise torque is applied to the flywheel. The graph shows how the torque varies with time.

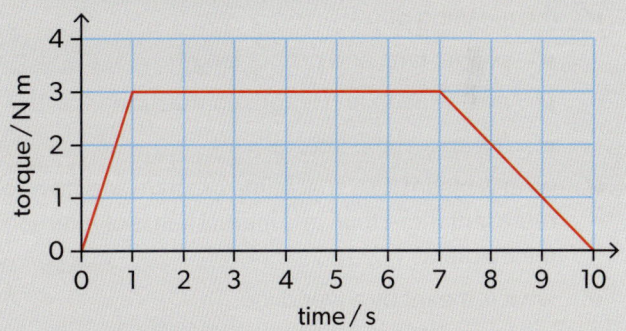

a) Calculate the impulse delivered to the flywheel during the 10 s.

b) Determine the final angular velocity of the flywheel.

c) Calculate the change in the rotational kinetic energy of the flywheel.

d) Determine the time at which the flywheel is instantaneously at rest.

Problem 4
A cylinder of mass M and radius R rolls down a ramp of vertical height h, starting from rest. The moment of inertia of the cylinder is $\frac{1}{2}MR^2$.

a) Derive an expression, in terms of R and h, for the final angular velocity of the cylinder.

b) It is given that M = 50 g and h = 25 cm. Calculate the change in:

(i) the gravitational potential energy of the cylinder

(ii) the rotational kinetic energy of the cylinder.

A Space, time and motion

A.5 Galilean and special relativity

You must know:
- what is meant by a reference frame
- an inertial reference frame is non-accelerating
- the meaning of Galilean relativity
- the Galilean transformation equations lead to a velocity addition equation for speeds much less than the speed of light
- the two postulates of special relativity and their consequences
- the Lorentz transformations:
  - how they arise from the special relativity postulates
  - how they lead to an equation for relativistic velocity addition
- what is meant by an invariant quantity
- the meaning of
  - proper time and proper length
  - time dilation and length contraction
  - the spacetime interval
- what is meant by a spacetime diagram and how to represent more than one inertial frame of reference on the diagram
- what is meant by a worldline.

You should be able to:
- use the Galilean transformation equations for position and time and to solve problems of velocity addition
- use the Lorentz transformations
  - to determine the position and time coordinates of events
  - compare the simultaneity of events
- solve problems involving time dilation and length contraction
- represent on a spacetime diagram
  - an event
  - time dilation
  - length contraction
- determine, for a specific speed, the angle between a world line and the time axis of a spacetime diagram
- discuss simultaneity of events in terms of a spacetime diagram
- explain and use the results of muon decay experiments.

> An **inertial reference frame** is one in which Newton's first law of motion holds. No net force acts on objects that are at rest or moving at a constant velocity in the frame. By implication, the frame cannot be accelerating relative to other inertial reference frames.
>
> The term inertial reference frame is sometimes shortened to "inertial frame".

Galilean relativity

A reference frame defines the motion of an object relative to others. Reference frames consist of an origin together with a set of axes and a set of clocks.

An inertial reference frame is an extension of the reference frame idea.

Albert Einstein was not the first scientist to discuss reference frames. Galileo Galilei (1564–1642) suggested that butterflies in a windowless cabin on a moving ship would always be observed to fly at random. An observer in the cabin could not deduce the motion of the ship by watching the butterflies.

Galileo's ideas suggest that:
- direction is relative
- position is relative
- stationary is not an absolute condition.

These lead to a set of equations (quoted for the x-direction only). When the origin of reference frame A' (Figure 1) is a distance X from the origin of another reference frame A, and the position of an object is x relative to the origin in frame A, the position of the object in frame A' is $x' = x - X$.

A.5 Galilean and special relativity

When the origin of an inertial frame A' is moving relative to the origin of inertial frame A with speed v (and assuming that the origins coincide at time $t = 0$), the speed u of an object in frame A is related to the speed u' of the same object in frame A' by $u' = u - v$. Combining distances and speeds leads to $x' = x - vt$ at time t after the coincidence of the origins. These equations are known as the Galilean transformations. Time is absolute in the Galilean transformation, so $t' = t$.

All observers in inertial frames must make the same deductions about physical law. James Clerk Maxwell connected electrostatics (Topic D.2) and electromagnetism (Topics D.2, D.3 and D.4) by incorporating the speed of light c in a vacuum into his equations. If observers in different inertial frames are to agree about physical laws, then they must all observe identical values for the speed of light in a vacuum. However, this is not what the Galilean transformation predicts. The equation $u' = u - v$ suggests that, if one observer moves at a constant velocity $(u - u')$ relative to another, then their observations of c will differ by v.

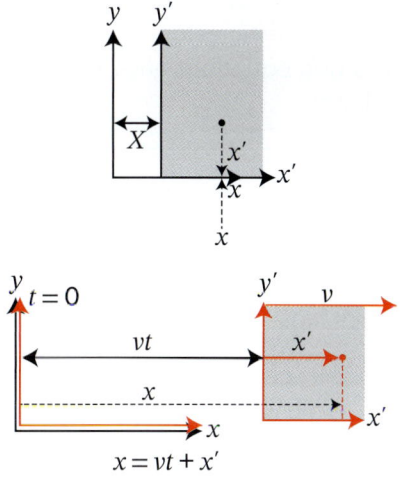

▲ Figure 1 Two frames of reference connected by the Galilean transformation

Special relativity and Lorentz tranformations

In 1887, American scientists Albert Michelson and Edward Morley confirmed that the speed of light was independent of the measurement reference frame within their experimental error. It was this result that led Einstein to suggest the two postulates of special relativity.

Lorentz introduced the gamma factor γ that compared the speed v of an inertial frame with the speed of light c:

$$\gamma = \frac{1}{\sqrt{1 - \frac{v^2}{c^2}}}$$

He then modified the Galilean transformations to $x' = \gamma(x - vt)$ and $t' = \gamma\left(t - \frac{v}{c^2}x\right)$. The difference between time coordinates of two events is the same in all inertial reference frames according to the Galilean transformation, but not according to Lorentz. Suppose that the time difference between the events in one frame is $\Delta t = t_2 - t_1$. In special relativity, the time difference in a second inertial frame moving relative to the first is $\Delta t' = \gamma\left(\Delta t - \frac{v}{c^2}\Delta x\right)$ and depends on the differences Δt and $\Delta x = x_2 - x_1$ in the first frame. This is in contrast with the Galilean transformation, which simply predicts that $\Delta t' = \Delta t$. The way the difference Δx between the x-coordinates is transformed from one inertial frame to another is also modified in the Lorentz transformation compared with the Galilean.

It is useful to write these equations in terms of the factor ct:

$$x' = \gamma\left(x - \frac{v}{c}ct\right) \qquad ct' = \gamma\left(ct - \frac{v}{c}x\right)$$

Writing the transformations in this way emphasizes the symmetry between the x and t equations and makes them easier to learn.

Sometimes, the position and time are known in frame S' and the equivalent values in frame S are needed: in this case, the inverse Lorentz transformations are required.

The **postulates of special relativity** are:

- The laws of physics are the same in all inertial reference frames.
- The speed of light in free space (a vacuum) is the same in all inertial frames of reference.

The construction of the **inverse Lorentz transformations** is straightforward.

- The time equation rearranges to $ct = \gamma\left(ct' + \frac{v}{c}x'\right)$ using $\frac{1}{\gamma^2} = 1 - \frac{v^2}{c^2}$.
- The position equation rearranges to $x = \gamma\left(x' + \frac{v}{c} \times ct'\right)$.

Table 1 on the following page compares Galilean, Lorentz and inverse Lorentz transformations. Notice that all the Lorentz expressions become identical with the Galilean transformations as $\frac{v}{c} \to 0$. The inverse Lorentz transformations are not given in the *Physics data booklet*.

A Space, time and motion

The **relativistic velocity-addition equations** arise as follows:

Frame A is moving relative to frame B with constant velocity v. An object in A is moving at u_A relative to A, and an object in frame B is moving at u_B relative to B.

- The speed of the object in A as measured by an observer in B is $u_A' = \dfrac{u_A + v}{1 + \dfrac{u_A v}{c^2}}$.

- The speed of the object in B as measured in A is $u_B' = \dfrac{u_B - v}{1 - \dfrac{u_B v}{c^2}}$.

Galilean	Lorentz	Inverse Lorentz
$x' = x - vt$	$x' = \gamma(x - vt)$ $\Delta x' = \gamma(\Delta x - v\Delta t)$	$x = \gamma(x' + vt')$
$t' = t$	$t' = \gamma\left(t - \dfrac{vx}{c^2}\right)$ $\Delta t' = \gamma\left(\Delta t - \dfrac{v\Delta x}{c^2}\right)$	$t = \gamma\left(t' + \dfrac{vx'}{c^2}\right)$

▲ **Table 1** Lorentz transformations and inverse Lorentz transformations (for transformation in the x-direction)

The transformations for length and time lead to expressions for the relative velocities when one or both objects are moving at speeds close to that of light. The equations follow from the usual definition of speed from Topic A.1, but use the Lorentz transformations.

Example 1

Two rockets, A and B, move along the same straight line when viewed from Earth.

A is travelling away from Earth at speed $0.80c$ relative to Earth.

B is travelling away from A at speed $0.60c$ relative to A.

a) Calculate the velocity of B relative to the Earth according to Galilean relativity.

b) Calculate the velocity of B relative to the Earth according to the theory of special relativity.

c) Comment on your answers.

Solution

a) Using $u' = u - v$ with appropriate signs gives $0.60c + 0.80c = 1.40c$

b) Using the relativistic velocity addition equation:

$$u' = \dfrac{u-v}{1 - \dfrac{uv}{c^2}} = \dfrac{0.60 + 0.80}{1 + 0.60 \times 0.80}c = 0.95c$$

c) In part (a), the speed exceeds c which is not possible under the Einstein postulates. Galilean relativity is not valid (except as an approximation at low speeds when $v \ll c$).

Assessment tip

The relativistic velocity-addition equations are given in the *Physics data booklet* in the form $u' = \dfrac{u - v}{1 - \dfrac{uv}{c^2}}$. When you use this equation, you must adjust the signs. One way to do this is to imagine what the signs would be in the equivalent Galilean transformation and to remember that the signs in the numerator and denominator are always the same.

Nature of science

Discoveries are made from time to time that cause major shifts in our scientific models.

Examples include:

- Galileo Galilei's observation of Jupiter's moons, which supported the Copernican view of the Solar System
- Isaac Newton's work on gravity
- Albert Einstein's recognition that time is not an absolute quantity but depends on the motion and position of the observer.

Example 2

A reference frame S' is moving at a relative speed $0.75c$ in the direction of the positive x-axis of another reference frame S. The clocks of S and S' show $t' = t = 0$ when $x' = x = 0$.

According to the S reference frame, event A occurs at $x_A = 0$, $t_A = 20$ ns, and event B occurs at $x_B = 12$ m, $t_B = 25$ ns.

Calculate, according to the S' reference frame:

a) the spacetime coordinates of A

b) the time interval between A and B.

Solution

a) The Lorentz factor for this situation is $\gamma = \dfrac{1}{\sqrt{1 - 0.75^2}} = 1.51$.

Use Lorentz transformation equations to calculate x'_A and t'_A.

$x'_A = \gamma(x_A - vt_A) = 1.51 \times (0 - 0.75 \times 3 \times 10^8 \times 20 \times 10^{-9}) = -6.8\,\text{m}$

$t'_A = \gamma\left(t_A - \dfrac{vx_A}{c^2}\right) = 1.51 \times \left(20 \times 10^{-9} - \dfrac{0.75 \times 0}{3 \times 10^8}\right) = 30\,\text{ns}$

b) $t'_B - t'_A = \gamma\left((t_B - t_A) - \dfrac{v(x_B - x_A)}{c^2}\right) = 1.51 \times \left(5 \times 10^{-9} - \dfrac{0.75 \times 12}{3 \times 10^8}\right)$

$= -38\,\text{ns}$

The calculated value is negative, which implies that, in the S' frame, event B occurs before event A. According to observers in S', the events happen in the opposite order than in the S reference frame!

The need to use appropriate transformations does not, however, mean that every physical quantity is different in different inertial reference frames. Some quantities are invariant (unchanged) between frames.

Example 3

Event A has spacetime coordinates $x_A = -8.0\,\text{m}$, $t_A = 16\,\text{ns}$ and event B has spacetime coordinates $x_B = 2.0\,\text{m}$, $t_B = 36\,\text{ns}$, all according to an inertial reference frame S.

a) Calculate the spacetime interval Δs^2 between A and B.

b) Events A and B are simultaneous in an inertial reference frame S'. Determine the distance between A and B according to measurements in S'.

Solution

a) $\Delta s^2 = (c\Delta t)^2 - (\Delta x)^2 = (3 \times 10^8 \times (36 - 16) \times 10^{-9})^2 - (2.0 - (-8.0))^2 = -64\,\text{m}^2$

b) In the S' frame, $\Delta t' = 0$ and the spacetime interval between A and B is $\Delta s'^2 = 0^2 - (\Delta x')^2 = -(\Delta x')^2$.

The spacetime interval is invariant; hence $\Delta s'^2 = \Delta s^2 = -64\,\text{m}^2$.

$-(\Delta x')^2 = -64\,\text{m}^2$; therefore $\Delta x' = 8.0\,\text{m}$.

In the S' frame, A and B occur simultaneously 8.0 m from each other.

The spacetime interval Δs^2 can be either positive or negative.

When $\Delta s^2 < 0$, the events can be simultaneous in some inertial reference frame (Δt can be zero), but they cannot happen at the same point in space. Such intervals are called **space-like**.

When $\Delta s^2 > 0$, the events can happen at the same point in some inertial reference frame (Δx can be zero), but they cannot happen simultaneously. Such intervals are called **time-like**.

Δs^2 always represents a single quantity, which is usually expressed in m² and should **not** be treated as the square of a quantity Δs.

The **invariant quantities** are the following.

- **Proper time interval** Δt_0 (often shortened to "proper time") is the time interval between two events occurring at the same place in an inertial reference frame. It is the shortest possible time that can be observed between the two events. In any other reference frame, the time between two events is **dilated** (takes longer) and the time interval Δt in this new frame is given by $\Delta t = \gamma \Delta t_0$.

- **Proper length** L_0 is the length of an object as measured by an observer at rest relative to the object. Any other observation of the object will result in a length measurement L that is shorter (**contracted**) than the proper length: $L = \dfrac{L_0}{\gamma}$.

- **Spacetime interval**
$\Delta s^2 = (c\Delta t)^2 - (\Delta x)^2 - (\Delta y)^2 - (\Delta z)^2$.
The IB Diploma Programme physics course limits problems to one dimension, so this becomes
$\Delta s^2 = (c\Delta t)^2 - (\Delta x)^2$. For any pair of inertial frames,
$\Delta s^2 = (c\Delta t)^2 - (\Delta x)^2 = (c\Delta t')^2 - (\Delta x')^2$.

- For completeness, **electric charge** (see Topic B.5 and D.2) and **rest mass** are also invariant. Rest mass is not discussed in the IB Diploma Programme physics course.

A Space, time and motion

> A **spacetime diagram** consists of two axes: one for distance (x-axis) and one for time (drawn on a conventional y-axis). The x-axis can be in units such as metres, kilometres or light years.
>
> The y-axis is ct also measured in distance units (metres, kilometres, light seconds or light years) representing the distance travelled by light in a particular time. On this basis, the y-axis grid on Figure 2 could have been labelled $t/\mu s$ and marked as 6.67, 13.3 and 20.0 instead of 2 km, 4 km and 6 km.
>
> A **worldline** is a sequence of spacetime events—the line in spacetime that joins all the positions of an object throughout its existence.

> ### Approaches to learning
>
> The idea of an observer pervades this topic. But this is not an observer in the sense of someone who makes measurements and observations. Indeed, the theoreticians who worked on relativity theory in its early days probably had a different view from scientists today. We tend to think now of a team of observers, each of whom can make observations in their local area. Everyone on the team has their own synchronized clock in order to report data to each other. This is possibly the ultimate gedanken (thought) experiment!

Spacetime diagrams

A spacetime diagram (also known as a Minkowski diagram) gives a visual comparison of events that occur in one or more reference frames (Figure 2).

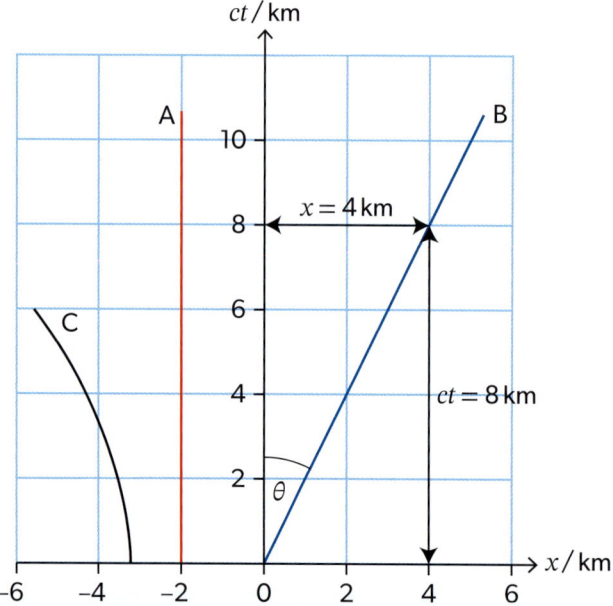

▲ Figure 2 A spacetime diagram drawn in the reference frame of A

Consider Figure 2.

- Particle A is stationary in this reference frame and is situated −2.0 km from the origin of the diagram. The single line parallel to the time axis is called the **worldline** of this particle. The worldline is parallel to the time axis because the particle is stationary in this frame.

- Particle B is moving at a constant velocity of $0.5c$. At $ct = 0$, it was at the origin ($x = 0$, $ct = 0$) of the reference frame of the spacetime diagram. The worldline of this particle is a straight line on the spacetime diagram. The speed of B relative to A is $+0.5c$ (in A's frame) because, when the distance from the origin is 4.0 km, the time is $ct = 8.0$ km and therefore $t = \dfrac{8.0}{c}$ and $v = \dfrac{4.0}{\left(\dfrac{8.0}{c}\right)} = \dfrac{1}{2}c$.

- Particle C is accelerating from rest relative to A. It began at the point (−3.2 km, 0). The worldline for acceleration is a curve.

- The angle θ between the B worldline and the ct-axis gives the speed of B relative to A: $\tan \theta = \dfrac{\text{opposite}}{\text{adjacent}} = \dfrac{X}{cT} = \dfrac{v}{c}$, so angle $\theta = \tan^{-1}\left(\dfrac{v}{c}\right)$.

For Figure 2, θ is measured to be 27° and $\tan 27° = 0.5$, so $v = 0.5c$.

Figure 3 shows that the spacetime diagram for a particle G can be displayed on the spacetime diagram drawn for the reference frame of another particle F.

- G is moving at $0.67c$ relative to F and is at the F origin (coincident with the G origin) at $ct = 0$ in both reference frames.

- As far as F is concerned, the worldline for G is the same as the time axis for G because G is stationary in its own frame.

- The G time and distance axes both rotate from the F axes, as shown in Figure 3. The G axes are labelled as ct' and x'. This is a good way to compare two inertial reference frames as it makes paradoxes in simultaneity easier to understand.

A.5 Galilean and special relativity

A final element of Figure 3 is a light cone which has been drawn from the event E that occurs at a position $x = 1.0$ km and a time $ct = 5.0$ km. This shows the path of light in the reference frame of F. As nothing can exceed the speed of light, all ct'-axes must lie between the F ct-axis and the light cone. The intersection of the two sets of axes gives us important information about the way F and G perceive the timing of E in their reference frames:

- W is the time ($ct = 5$ km) at which E happens in the reference frame of F.
- X (the intersection of the light cone with ct) is the time when an observer in F on the worldline through the origin sees E happen.
- Y (the intersection of the light cone with ct') is the time when an observer in G on the world line of ct' sees E happen.
- Z (the intersection of the ct'-axis with a line constructed through E and parallel to x') is when E happens in frame G.

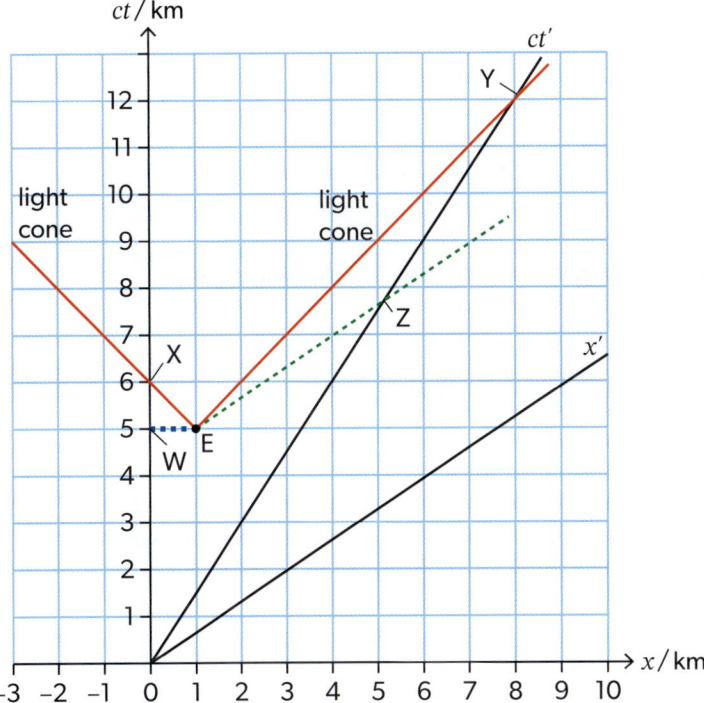

▲ Figure 3 Events and the light cone on a spacetime diagram

Observers in the two frames do not agree about the timing of events. This is a consequence of the constancy of the speed of light to all observers. Relativistic simultaneity is often explored through the use of paradox, as in the example shown in Figure 4 where there is only partial agreement about the result of a race.

The race between two runners is adjudicated by a judge. Athlete A is slower than athlete B. To allow for this, A runs to a closer finishing line than B. The spacetime diagram for the race, drawn in the judge's reference frame, is shown in Figure 4. The axes for A (x_A, ct_A) and B (x_B, ct_B) are shown together with the worldlines for the finishing posts. Parts of the light cones for events P and Q are also shown.

In Figure 4:
- P and Q are the events where an athlete crosses a finishing line (P is A finishing the event and Q is B finishing the event).
- In B's frame, when B crosses the B finishing line, athlete A is at R.
- In A's frame, when A crosses the A finishing line, athlete B is at S.
- Q and R are the same time according to B, and P and S are the same time according to A so both athletes believe that B has won the race.

A Space, time and motion

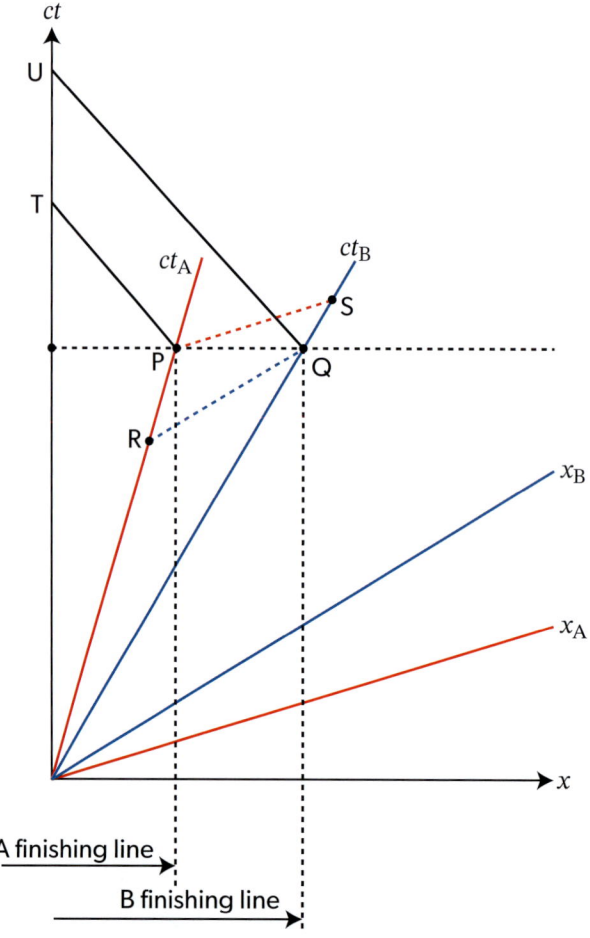

▲ Figure 4 Spacetime diagram for the race between A and B

- In the judge's frame the athletes cross their respective finishing lines simultaneously. However, the judge will not observe these events as simultaneous because the light cones from the finishing events do not arrive at the ct origin at the same instant. The observations of the finish occur at T and U so that the judge will **see** A cross the A finishing line first.
- The view of both A and B on this simultaneity requires them to keep running after they have crossed the finishing line. When an athlete stops, the simultaneity of this athlete immediately changes.

Example 4

Four light beacons are used to guide a rocket A to a docking station. The events P, R, Q and S are flashes from the four beacons and are shown in the spacetime diagram.

The diagram shows the reference frames of Earth (ct, x) and A (ct', x').

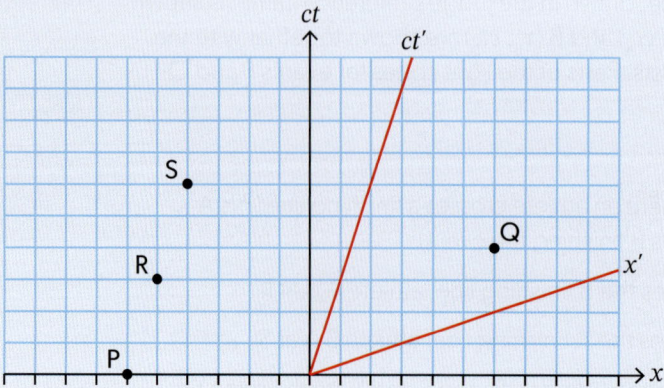

Deduce the order in which:

a) the beacons flash in the reference frame of A

b) an Earth observer sees the beacon flashing.

Solution

a) Construct lines through the events parallel to x' and consider the order in which the lines cross the ct'-axis. These show that P and Q occur at the same time and that R then flashes before S.

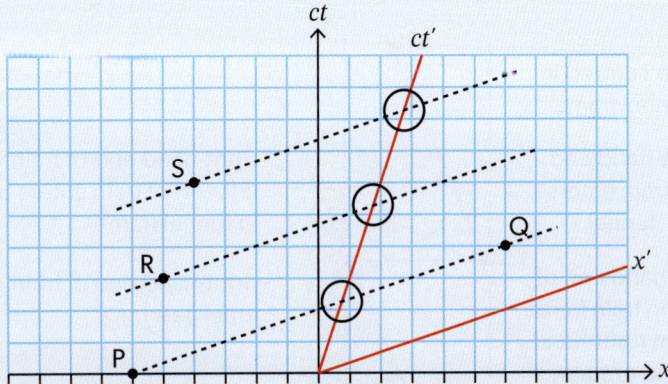

b) Construct light cones from the events. These cross the ct-axis (and are seen on Earth) in the order P, R, Q and S simultaneously.

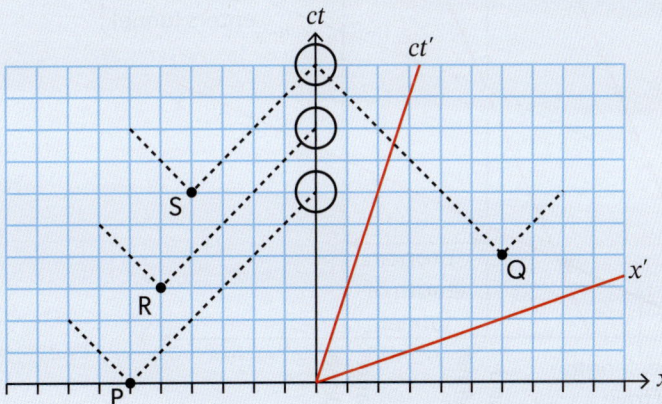

Example 5

A train of proper length 100 m is moving to the right towards a tunnel of proper length 80 m that has doors at each end. The speed of the train is $0.80c$.

Explain why, in the tunnel reference frame, the doors can be shut briefly and simultaneously on the train, but the tunnel is too short for this to happen in the reference frame of the train.

Solution

γ for the train is $\frac{5}{3} \approx 1.67$, so, in the rest frame of the tunnel, the length of the train is 60 m.

It will obviously fit into the tunnel in the tunnel reference frame. The tunnel doors are shut when the train is in the centre of the tunnel (with 20 m space in front of it).

From the train frame of reference, the tunnel is contracted and approaches the train (moving to the left). The contracted length of the tunnel is 48 m and so the train is too long for the tunnel.

However, both tunnel doors can still shut for an instant. This is because the clocks (observers) in the tunnel and on the train that measure the timing of the doors shutting cannot ever agree.

A Space, time and motion

Suppose there is a pair of clocks at each end of the tunnel. These clocks are synchronized in the tunnel frame. In the train frame, the tunnel clocks disagree. The one at the right-hand end of the tunnel is ahead of the left-hand clock by:

$$\frac{vL}{c^2} = \frac{0.8 \times 80}{3 \times 10^8} = 210 \, \text{ns}$$

In the time interval between the right-hand clock and the left-hand clock reading zero according to the train observers, the left-hand door of the tunnel will have moved past the left-hand end of the train.

This problem is made clear in a spacetime diagram. The tunnel frame is $ct - x$, the train frame is $ct' - x'$.

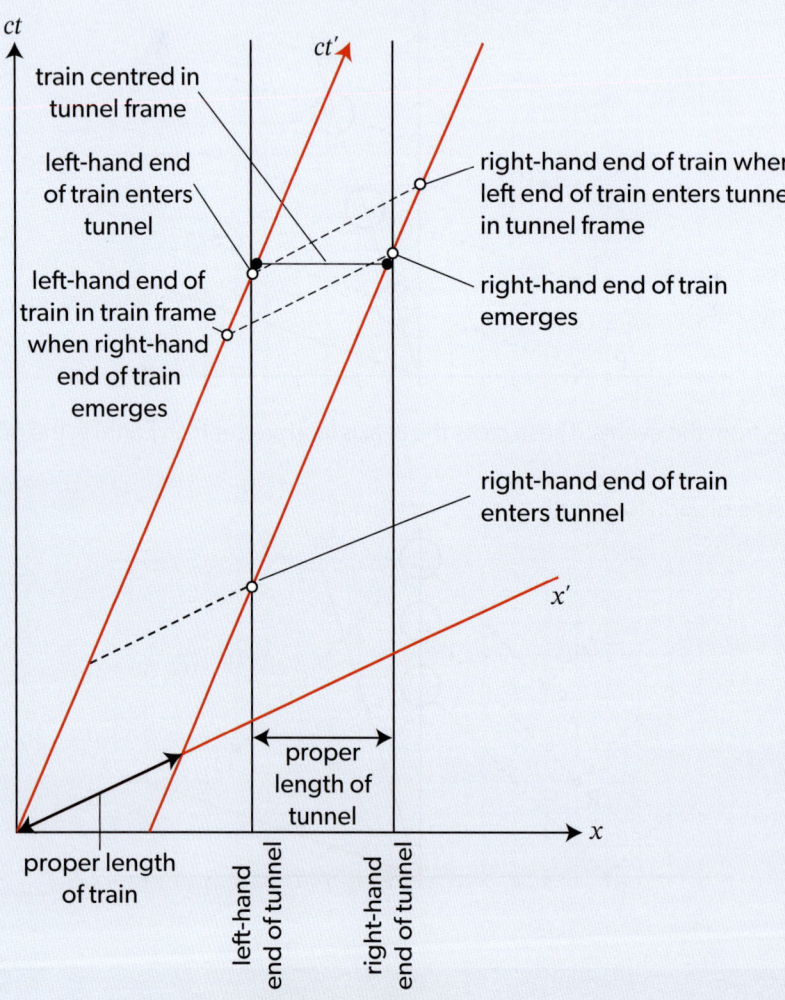

Sample student answer

An observer on Earth watches a rocket A. The spacetime diagram shows part of the motion of A in the reference frame of the Earth observer. Three flashing light beacons, X, Y and Z, are used to guide rocket A. The flash events are shown on the spacetime diagram. The diagram shows the axes for the reference frames of Earth and of rocket A. The Earth observer is at the origin.

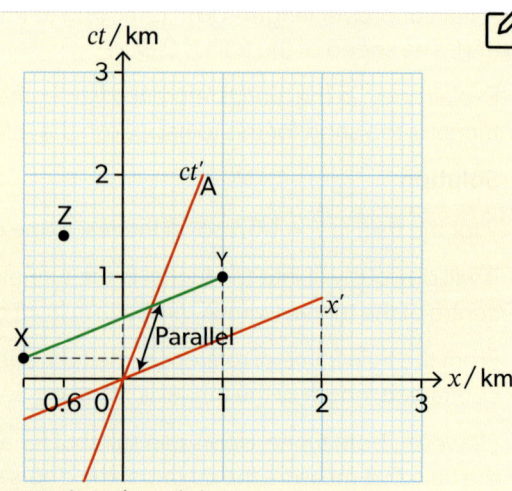

a) Using the graph above, deduce the order in which the beacons flash in the reference frame of the rocket. [2]

This answer could have achieved 2/2 marks:

> Beacons X and Y would happen at the same time before Z, because my green worldline is parallel to the x'-axis, meaning that in the reference frame of A, they would happen at the same time. Z would flash later, since it is further up the ct'-axis than X and Y.

▲ This is a complete answer that shows a good understanding of the implications of spacetime diagrams. The annotations they have drawn on the diagram are well done and the student has taken time to indicate (with the word 'parallel') the essentials of the construction.

b) Using the graph above, deduce the order in which the Earth observer sees the beacons flash. [2]

This answer could have achieved 2/2 marks:

> Assuming that all beacons flash light going at the same velocity, X will be seen first by the observer on the Earth. Y and Z would arrive simultaneously, because Y flashes 0.4 km before, but has to travel 0.4 km, hence they, Y and Z, arrive at the same time after X.

▲ This is correct and gains full marks...

▼ ...however, the student could have made this clearer on the diagram. Light cones drawn from Z and Y intersect the worldline at $ct = 2$ km, which means that the events Y and Z are seen simultaneously.

The muon decay experiment

This observation involves time dilation (or length contraction) for its explanation. Muons are produced at the top of the atmosphere when high-energy cosmic rays interact with air molecules. The muons have a known probability of decay after formation and a known speed that is close to the speed of light. It is possible to predict the fraction of the muons that ought to reach the ground. In practice, many more muons than this reach the ground according to Earth observers. This is because the muons' probability of decay applies to a reference frame in which they are at rest. However, to us (the Earth observers) in a frame moving relative to the muon frame at about $0.98c$, this muon decay lifetime is dilated and appears much longer than the proper lifetime.

The muon observation can also be explained in terms of length contraction. The distance from the top of the atmosphere is measured by Earth observers to be a few tens of kilometres. However, to the muons, the distance is contracted (to about 1 km) as they are moving fast relative to our reference frame. We know how many muons are expected to decay for every metre they travel in our reference frame, but the distance as measured by an observer in the muon frame is much shorter. Therefore fewer muons are removed from the beam as it travels from the atmosphere to the surface.

Measurements made in one reference frame are invalid in another frame unless we apply the Lorentz transformations. There is also the issue of measuring time in two inertial frames that are not at rest relative to each other.

When two clocks are in the same inertial frame but are separated by 10 m, an observer near to one clock will notice that the other clock differs by 3×10^{-7} s. This is not a simple problem because moving the clock closer (assuming that they are not in the same place) means that the clock must be accelerated. During this motion, the clock is no longer in an inertial reference frame. The way to move a clock from one frame to another is to do so infinitely slowly or to use a third clock (also moving very slowly) and transfer one clock reading to the second clock.

A Space, time and motion

> ### Example 6
>
> Muons are particles with a proper lifetime of 2.2 μs. They are produced at a height of 3.0 km above the surface of Earth and move downwards to the surface at a speed of 0.98c. Explain, with calculations, why the detection of the muons at the surface of Earth provides evidence for relativity.
>
> **Solution**
>
> One of the conclusions of special relativity is that, to a moving observer, time is dilated.
>
> Assuming Galilean relativity, muons must reach the ground in $\frac{3 \times 10^3}{0.98 \times 3 \times 10^8} = 10\,\mu s$. This is several times longer than the lifetime of the muons, so very few of them if any should arrive undecayed.
>
> Special relativity predicts that γ for this relative speed is $\frac{1}{\sqrt{1-0.98^2}} = 5.0$.
>
> To an observer in the frame of reference of Earth, the mean lifetime of the muons should be $5.0 \times 2.2 = 11\,\mu s$, which is comparable with their expected time of travel. Therefore, a significant fraction should have reached the ground before they decayed. This is the observed outcome, and it confirms the special relativity theory.

Sample student answer

An electron is emitted from a nucleus with a speed of 0.975c as observed in a laboratory. The electron is detected at a distance of 0.800 m from the emitting nucleus as measured in the laboratory.

a) For the reference frame of the electron, calculate the distance travelled by the detector. [2]

This answer could have achieved 2/2 marks:

$$\gamma = \frac{1}{\sqrt{1 - \frac{(0.975c)^2}{c^2}}} = 4.5$$

$$L = \frac{L_0}{\gamma} = \frac{0.800}{4.5} = 0.178\,m$$

b) For the reference frame of the laboratory, calculate the time taken for the electron to reach the detector after its emission from the nucleus. [2]

This answer could have achieved 2/2 marks:

$$v = \frac{d}{t} \text{ so } t = \frac{d}{v} = \frac{0.800}{0.975c} = 2.74 \times 10^{-9}\,s$$

c) For the reference frame of the electron, calculate the time between its emission at the nucleus and its detection. [2]

This answer could have achieved 2/2 marks:

$$\Delta t = \gamma \Delta t_0 \text{ so } \Delta t_0 = \frac{\Delta t}{\gamma} = \frac{2.74 \times 10^{-9}}{4.50} = 6.08 \times 10^{-10}\,s$$

d) Outline why the answer to part (c) represents a proper time interval. [1]

This answer could have achieved 1/1 marks:

Because it is the time interval between two observers in the same point in space.

▲ This sequence of answers, to parts (a)–(d), scores full marks. The solutions in parts (a)–(c) are clear and easy to follow. The written response to part (d) is an acceptable alternative definition for proper time interval to the one given on page 43. It would also be acceptable to say that it is the shortest time interval that it is possible to observe.

Practice problems

Problem 1
A space station is at rest relative to Earth. A spaceship passes Earth travelling towards the space station at a constant velocity with $\gamma = 1.25$.

a) Calculate, in terms of c, the speed of the spaceship relative to Earth.

b) In the reference frame of Earth, the space station is at a distance of 5.00×10^{10} m. Calculate the time of travel of the spaceship to the space station according to clocks on:

   (i) Earth

   (ii) the spaceship.

c) As the spaceship passes Earth, a radio signal is emitted from Earth that is reflected by the space station and later observed on the spaceship.

   Sketch a spacetime diagram to show these events.

Problem 2
a) Outline what is meant by proper length.

b) A pion decays in a proper time of 46 ns. It is moving with a velocity of $0.95c$ relative to an observer. Calculate the decay time of the pion as measured by the observer.

Problem 3
According to measurements in an inertial reference frame S, events P and Q have spacetime coordinates $x_P = 2800$ m, $t_P = -2.0$ μs and $x_Q = 300$ m, $t_Q = 10$ μs.

a) Calculate the spacetime interval between P and Q.

b) P and Q occur at the same x-coordinate in an inertial reference frame S′. Calculate the time difference between P and Q in S′.

Problem 4
An electron moves at a speed of $0.980c$ relative to a laboratory. The electron is detected at a distance of 0.800 m from its source in the laboratory frame.

a) Calculate the distance travelled by the detector in the electron frame.

b) Calculate the time taken for the electron to reach the detector from the source in the laboratory frame.

c) Calculate the time taken by the electron to move between its source and the detector in the electron frame.

d) Suggest which of your answers to parts (b) and (c) is a proper time interval.

Problem 5
A spaceship leaves Earth with a speed $0.70c$.

a) Draw a spacetime diagram for the Earth's frame including the motion of the spaceship.

b) Label your diagram with the angle between the worldline of the spaceship and that of the Earth.

Problem 6

A spaceship moves away from Earth at a constant velocity. The spacetime diagram shows the worldline of Earth in the reference frame (x, ct) of the spaceship.

A supply capsule is sent from Earth to the spaceship. Event S is the capsule leaving Earth.

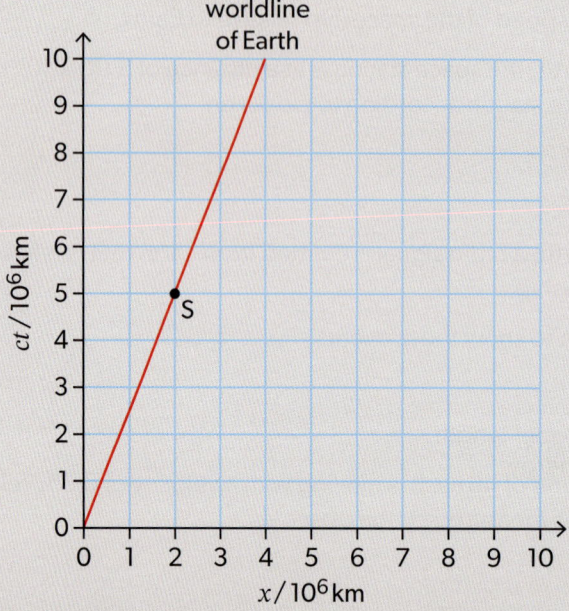

a) State the relative speed of Earth and the spaceship.

b) Calculate the spacetime coordinates (x', ct') of S in the reference frame of Earth.

c) The speed of the supply capsule relative to Earth is $0.75c$. Calculate the speed of the capsule relative to the spaceship.

d) Draw, on the diagram, the worldline of the supply capsule.

e) Hence, calculate the time at which the supply capsule arrives at the spaceship, as measured by the clocks on the spaceship. State the answer in seconds.

B The particulate nature of matter

B.1 Thermal energy transfers

You must know:
- ✔ the molecular theory of solids, liquids and gases
- ✔ what is meant by a phase change
- ✔ density = $\frac{\text{mass}}{\text{volume}}$, $\rho = \frac{m}{V}$
- ✔ what is meant by a temperature scale
- ✔ the meaning of internal energy
- ✔ the concepts of temperature and absolute temperature
- ✔ what is meant by specific heat capacity and specific latent heat
- ✔ the three methods of thermal energy transfer:
  - ✔ convection
  - ✔ conduction
  - ✔ radiation
- ✔ the nature of black-body radiation
- ✔ what is meant by apparent brightness and luminosity.

You should be able to:
- ✔ describe the molecular differences between solids, liquids and gases
- ✔ describe a phase change in terms of molecular behaviour
- ✔ use Kelvin and Celsius temperature scales and convert between them
- ✔ sketch and interpret graphs showing how the temperature of a substance varies with time and with energy transferred to or from the substance
- ✔ describe the mechanisms that govern conduction, convection and radiation
- ✔ use thermal conductivity to solve problems of thermal conduction
- ✔ solve problems involving the use of Wien's displacement law and the Stefan–Boltzmann law.

There are four known states of matter. Three of these states are solids, liquids and gases. The fourth state, plasma, only exists at very high temperatures (and is not part of the IB Diploma Programme physics course). All substances are made up of atoms and molecules and the differences between states is due to the nature of the bonding between them. Normally, when thermal energy is transferred to a solid, it first changes to a liquid and then a gas. These are called phase changes.

Phase changes are linked to the latent heat energy transfers discussed later in this topic. Phase changes are:
- melting and freezing (changes between solid and liquid)
- boiling and condensing (changes between liquid and gas).

Phase changes can be demonstrated when thermal energy is transferred to a mass of a substance that is initially in the solid state. The graph in Figure 1 shows the variation with transferred energy of the temperature of the substance.

The graph can also be plotted as temperature against time (when the energy is input to the substance at a constant rate).

The term **particle** is used to model **atoms** and **molecules** as small point-like objects without size or shape.

Gases have individual particles that are normally independent of each other and move freely within a container to fill it completely. Pressure arises as the particles interact with the container walls.

The particles of a **liquid** only exchange positions with nearest neighbours. This enables a liquid to have a definite volume but to be able to flow within its container.

Solids allow little, if any, movement between particles. These rarely exchange positions with each other. Solids have a fixed shape.

B The particulate nature of matter

> **Temperature** is a measure of the average kinetic energy of a collection of moving atoms and molecules.
>
> The two fixed points for the **Kelvin** scale are 0 K and 273.16 K. These are absolute zero (the temperature at which atoms and molecules have no kinetic energy, equal to −273 °C) and the triple point of water (equal to 273.16 K or 0.01 °C). The **triple point** is where all three phases of water co-exist in a sealed container—it occurs at a unique temperature and pressure.
>
> In 1848, Lord Kelvin defined one degree to be identical in both the Kelvin and Celsius scales. To convert from a Celsius temperature to a Kelvin temperature, add 273. To convert from a Kelvin temperature to a Celsius temperature, subtract 273.

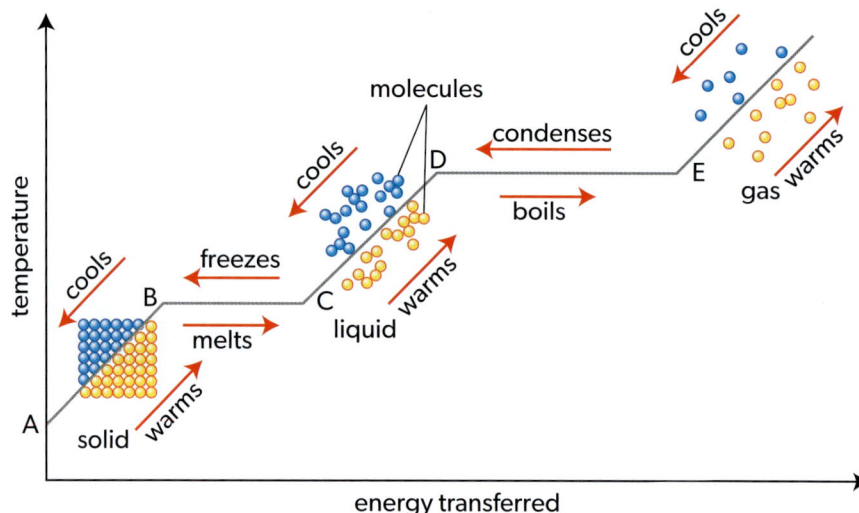

▲ Figure 1 The variation of temperature with energy transferred to a substance

When energy is transferred to a substance without any change of phase, the temperature of the substance rises. When the substance cools, energy is transferred away from it.

Temperature is sometimes described as the "degree of hotness" of a body. At the microscopic level, temperature is related to the motion of the atoms and molecules.

Temperature is defined using a temperature scale. The present-day scientific scale is the Kelvin scale. It is also known as the absolute temperature scale. All temperature scales require two fixed points (temperatures that are defined in terms of the properties of a substance).

The connection between temperature and the motion of the particles in a substance gives a direct link between the macroscopic and microscopic energy descriptions. When the average kinetic energy of one group of particles is higher than that of a second group, the first group has a higher temperature.

> ## Approaches to learning
>
> Always try to link quantities and concepts across themes and topics.
>
> The value of the Boltzmann constant is 1.38×10^{-23} J K^{-1}. The unit gives a clue to the meaning of k_B—it is the energy required to increase the "temperature" of the average particle by 1 K.
>
> The Boltzmann constant is strongly linked to the kinetic theory introduced in Topic B.3. Another interpretation of k_B is provided in that topic. There is further use of k_B in Topic B.4 (AHL only).
>
> The quantity $\overline{E}_k = \frac{3}{2} k_B T$ is used again in Topic B.3, where this expression is an outcome of the kinetic model of a gas.

Nature of science

Modelling is important in physics. Physicists use macroscopic and microscopic models. The behaviour of bulk materials (macroscopic) can be modelled using latent heat, heat capacity (Topic B.1) and the gas laws (Topic B.3). Alternatively, kinetic theory (also in Topic B.3) models the motion of atoms and molecules—imagined as infinitesimally small particles (microscopic).

Use the differences and similarities of models to assist your learning.

In mathematical terms, the kinetic energy of the average particle is given by $\overline{E}_k = \frac{3}{2} k_B T$, where k_B is the Boltzmann constant and T is the absolute (kelvin) temperature.

B.1 Thermal energy transfers

The internal energy of a substance is equal to the sum of the random kinetic energy of a collection of particles plus the potential energy that arises from intermolecular forces. This potential energy is negative because energy must be added to the particles to separate them from each other. The stronger the intermolecular forces, the more negative the potential energy must be.

Assessment tip

Celsius temperature is always expressed in °C. The kelvin unit never has a ° sign and is written as K. Temperature differences are either written as K or deg (meaning "change of degrees"). When you are giving an answer that is a **change** in temperature, never use °C.

Similarly, the unit of specific heat capacity is $J\,kg^{-1}\,K^{-1}$ not $J\,kg^{-1}\,°C^{-1}$.

Example 1

Outline the difference in internal energy of a piece of metal and the internal energy of an ideal gas.

Solution

The internal energy of the metal equals the total kinetic energy of the atoms plus the potential energy of the system. This potential energy arises from the bonds between the metal atoms.

The molecules of an ideal gas have only kinetic energy. One of the assumptions of an ideal gas is that the particles in it do not interact through molecular bonds, so there is no potential energy to consider.

Example 2

Sketch a graph to show the relationship between the internal energy of an ideal gas and its temperature measured in degrees Celsius. Explain the key features of this graph.

Solution

The graph shows that, at 0 °C, there is some internal energy in the gas. This is because the internal energy of the gas is directly proportional to the absolute temperature. Absolute zero is at 0 K and this is the intercept on the temperature axis. In Celsius, this value is −273 °C.

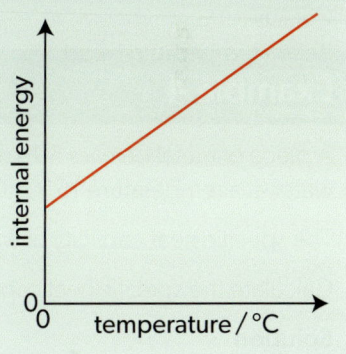

When energy is added to a substance and its state does not change, its temperature rises—this is a heat capacity change. The internal energy of the substance is increasing as a result of the increase in the random kinetic energy of the particles. Meanwhile, the potential energy is largely unchanged (this is only approximately true, especially when expansion or contraction occurs).

When the state of a substance is changing, the temperature is constant—this is a latent heat change. Energy is being transferred into the potential form and the kinetic energy is constant.

The **specific heat capacity** c of a substance is the energy required to change the temperature of 1 kg of the substance by 1 deg (or 1 K). The unit of specific heat capacity is $J\,kg^{-1}\,K^{-1}$. In fundamental units, this is $m^2\,s^{-2}\,K^{-1}$.

$c = \dfrac{Q}{m \times \Delta T}$, where Q is the energy transferred (in J), m is the mass (in kg) and ΔT is the change in temperature (in K).

The **specific latent heat** L of a substance is the energy required to change the phase of 1 kg of the substance (without change in temperature). The unit of specific latent heat is $J\,kg^{-1}$. In fundamental SI units, this is $m^2\,s^{-2}$.

$L = \dfrac{Q}{m}$, where Q is the energy transferred (in J) and m is the mass (in kg). You must specify the type of phase change that is occurring by writing, for example, for a substance going from water to ice, "The specific latent heat of freezing of ice is 0.34 MJ kg^{-1}".

B The particulate nature of matter

Example 3

The internal energy of a piece of zinc is increased by 1.5 kJ by heating. Its subsequent increase in temperature is 11 deg. The piece of zinc has mass 0.35 kg.

a) Explain the meaning of internal energy and heating.

b) Calculate the specific heat capacity of zinc.

Solution

a) The internal energy is the sum of the potential energy and the kinetic energy of the zinc atoms. It can also be described as the amount of energy stored in the zinc.

Heating is the process of transferring energy using a non-mechanical or thermal pathway from an energy source to the zinc. The zinc is acting as an energy sink.

b) $c = \dfrac{Q}{m\Delta T} = \dfrac{1500}{0.35 \times 11} = 390 \text{ J kg}^{-1}\text{K}^{-1}$

The specific heat capacity of a material can be determined using the method of mixtures. A hot solid of known temperature is added to a cold liquid, also at known temperature. The masses of the solid and liquid are known. The resulting mixture of solid and liquid reaches a final, measured temperature. When one of the two specific heat capacities is known, the other can be determined.

Example 4

A piece of metal of mass 50 g and temperature 95 °C is dropped into a thermally insulated container with 120 g of water at a temperature 15 °C. The final temperature of the system is 18 °C.

The specific heat capacity of water is 4200 J kg^{-1} K^{-1}.

Calculate the specific heat capacity of the metal.

Solution

The container is insulated so there are no thermal energy transfers other than between the metal and the water. The energy lost by the metal is equal to the energy gained by the water:

$-m_{metal} c_{metal} \Delta T_{metal} = m_{water} c_{water} \Delta T_{water}$

$\Delta T_{metal} = 18 - 95 = -77 \text{ K} \qquad \Delta T_{water} = 18 - 15 = 3.0 \text{ K}$

$c_{metal} = \dfrac{120 \times 10^{-3}}{-50 \times 10^{-3}} \times \dfrac{3.0}{-77} \times 4200 = 390 \text{ J kg}^{-1}\text{K}^{-1}$

Assessment tip

Heat capacity and latent heat calculations often involve multiple steps.

Present any multi-step solution clearly with an explicit description of each step. An examiner can then give you partial credit if you have made an error elsewhere.

Example 5

A 300 g sample of aluminium is initially at a temperature of 20 °C. The sample is melted in an electric furnace that transfers a power of 1.2 kW to the sample. The following data are given about aluminium:

Melting temperature = 660 °C

Specific heat capacity = 900 J kg^{-1} K^{-1}

Specific latent heat of fusion = 390 kJ kg^{-1}

Calculate the time required to melt the sample.

B.1 Thermal energy transfers

Solution

E_1 is the energy needed to increase the temperature of the sample to the melting point:

$E_1 = 0.300 \times 900 \times (660 - 20) = 173 \text{ kJ}$

Once the sample has reached the melting point, the additional energy needed to melt it is E_2: $E_2 = 0.300 \times 390 \times 10^3 = 117 \text{ kJ}$

Total energy transferred to the sample $= E_1 + E_2 = 290 \text{ kJ}$

Time required for the furnace to transfer this energy $= \dfrac{290 \text{ kJ}}{1.2 \text{ kW}}$

$= 242 \text{ s} = 4 \text{ min } 2 \text{ s}$

Sample student answer

In an experiment to determine the specific latent heat of fusion of ice, an ice cube is dropped into water contained in a well-insulated calorimeter of negligible specific heat capacity. The following data are available.

- Mass of ice cube = 25 g
- Mass of water = 350 g
- Initial temperature of ice cube = 0 °C
- Initial temperature of water = 18 °C
- Final temperature of water = 12 °C
- Specific heat capacity of water = 4200 J kg⁻¹ K⁻¹

a) Using the data, estimate the specific latent heat of fusion ice. [4]

This answer could have achieved 2/4 marks:

> $Q = mc\Delta T$ $m = 0.350 \text{ kg}$ $c = 4200 \text{ J/kg.K}$ $\Delta T = 6K$
> $Q = 0.350 \text{ kg} \times 4200 \times 6K = 8.82 \text{ kJ}$
> $m = 0.025 \text{ kg}$ $\Delta T = 12K$
> $0.025 \text{ kg} \times 4200 \times 12K = 1.26 \text{ kJ}$ $c = 4200 \text{ J/kg.K}$
> $7.56 \text{ kJ} = 0.375 \times 4200 \times ?$
> $\dfrac{7.56 \text{ kJ}}{1.575 \text{ kg.K}} = 4.8 \text{ °K} + 273 = 277.8 \text{ °K}$

▲ There are correct calculations of the energy lost by the cooling water and the (heat capacity) energy gained by the ice once it has melted. Then there is a recognition that 7.56 kJ is the energy available to melt the ice.

▼ The equation is incorrect (it is for heat capacity not latent heat).

▼ The student uses degree signs in this solution. You should try to avoid these as the ° sign should only be used with Celsius temperatures.

b) The experiment is repeated using the same mass of ice. This time, the ice is crushed.

Suggest the effect of this, if any, on the time it takes the water to reach its final temperature. [1]

This answer could have achieved 0/1 marks:

> The time it takes to reach the final temperature will decrease.

▼ The answer is correct, but the answer fails to appreciate the significance of the command term "suggest". This means to propose a hypothesis and it requires some explanation of the proposal. The answer needs to go on to say that the surface area of the ice increases when crushed, so the water and the ice can interact more quickly.

Thermal energy is transferred by three processes:
- conduction—the principal mechanism in solids
- convection—the principal mechanism in fluids
- radiation—the only mechanism for transfer in a vacuum.

B The particulate nature of matter

Approaches to learning

As you learn about conduction, remember that its ideas apply to both electrical and thermal phenomena. Linking these ideas in your mind will help you to understand both more thoroughly.

Conduction

Conduction occurs through collisions between electrons and atoms, and through intermolecular interaction.

All atoms possess kinetic energy at temperatures above absolute zero. An atom moving about its fixed position can transfer energy to other atoms close by when it has a greater energy (temperature) than its neighbours. In this way, energy can be transferred from high-temperature regions of a solid to low-temperature regions. This is the mechanism that drives the energy transfer by conduction.

Energy transfer and temperature difference are linked using the thermal conductivity of a material.

Thermal conductivity $k = -\dfrac{\text{rate of energy transfer through material}}{\text{cross-sectional area of material} \times \text{temperature gradient across material}}$

The unit of thermal conductivity is $W\,m^{-1}\,K^{-1}$.

The negative sign in the equation indicates that the positive direction of energy transfer is from the higher temperature to the lower temperature (the opposite to the direction of the temperature gradient).

In symbols this is written as:

$$\dfrac{Q}{\Delta t} = -kA\dfrac{\Delta T}{\Delta x}$$

where $\dfrac{Q}{\Delta t}$ is the rate of energy transfer, A is the area of the material and $\dfrac{\Delta T}{\Delta x}$ is the temperature gradient across the material.

In the diagram, this is $\dfrac{Q}{\Delta t} = kA\dfrac{T_1 - T_2}{\Delta x}$, where $T_1 > T_2$.

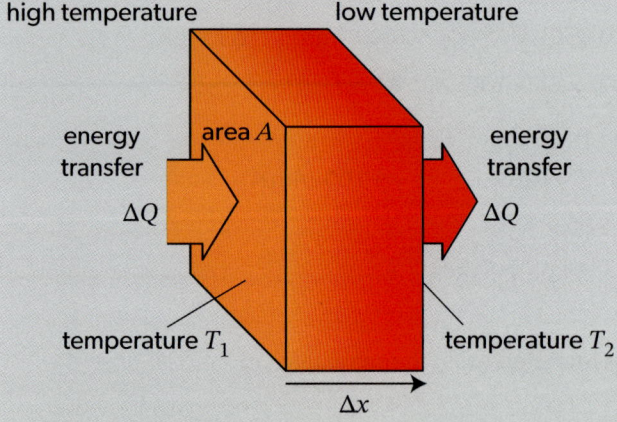

Example 6

A temperature difference of 150 K is maintained between the ends of a cylindrical metal rod of length 50 cm and diameter 1.5 cm. The rate of thermal energy transfer along the rod is 21 W. The rod is insulated so that no thermal energy is transferred through its curved surface.

a) Outline the mechanism by which thermal energy is transferred along the rod.

b) Calculate the thermal conductivity of the material of the rod.

Solution

a) The atoms at the hot end of the rod vibrate with a relatively large kinetic energy. As they interact with neighbouring atoms and with free electrons in the rod, the energy associated with atomic vibrations is exchanged between the particles and passed towards the cold end of the rod. In metals, free electrons contribute significantly to thermal conduction because of their high mobility.

b) Temperature gradient in rod $= \dfrac{\Delta T}{\Delta x} = -\dfrac{150}{0.50} = -300 \text{ K m}^{-1}$

The minus sign signifies that the temperature decreases in the direction of thermal energy transfer, towards the cold end of the rod.

Cross-sectional area of the rod $A = \pi \left(\dfrac{0.015}{2}\right)^2 = 1.767 \times 10^{-4} \text{ m}^2$

Thermal conductivity $= -\dfrac{21}{1.767 \times 10^{-4} \times (-300)} = 400 \text{ W m}^{-1} \text{ K}^{-1}$

Example 7

The walls of a cabin are made of pinewood logs of thickness 7.0 cm. Thermal conductivity of pinewood is 0.11 W m⁻¹ K⁻¹.

On a winter day, the temperature in the cabin is 15 °C and the air temperature outside is −10 °C.

Calculate the rate of thermal energy transfer through a wall of surface area 12 m².

Solution

Temperature gradient $= \dfrac{-10-15}{0.070} = -357 \text{ K m}^{-1}$

Rate of thermal energy transfer $= -kA\dfrac{\Delta T}{\Delta x} = -0.11 \times 12 \times (-357) = 470 \text{ W}$

Convection

Convection occurs in fluids. The hot areas of a fluid are less dense than the cold areas, so the particles rise from the hot areas to the cold areas. The denser, cold areas then fall into the hot areas, creating a convection current.

Figure 2 shows a convection cycle in operation. Water is heated in the bottom right corner of a square glass tube of constant cross-section. The small volume of hot, less dense, water rises and the convection current begins. A coloured dye shows the movement of the water in the tube.

Density $= \dfrac{\text{mass of substance}}{\text{volume of substance}}$
The units of density are kg m⁻³.

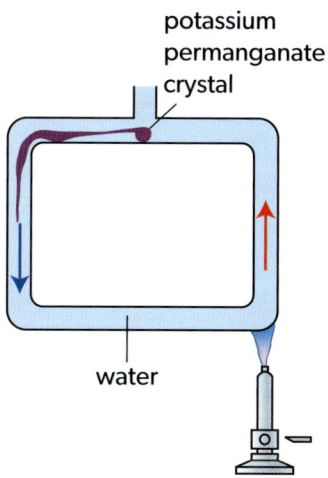

▲ Figure 2 Convection in a liquid

Example 8

Energy loss from a house can be reduced by placing foam between the inner and outer house walls. This is called cavity-wall insulation.

Explain how heat transfer from inside a warm house to the cold exterior can be reduced by:

a) a cavity wall with no foam

b) a cavity wall with foam.

Solution

a) Air is a good insulator, so energy transferred through the inner wall is not conducted through the cavity. Convection currents will, however, cause air movement within the cavity. The resulting convective heat transfer can diminish the insulating effect of the cavity.

b) A foam is a mixture of gas and solid. The gas is fixed in position within the solid and cannot convect. The solid and gas parts of the foam are both poor conductors and do not allow effective energy transfer.

Thermal radiation

Thermal radiation is emitted as electromagnetic waves by all objects at temperatures greater than 0 K (absolute zero). Electromagnetic waves can travel through a vacuum.

All objects radiate electromagnetic radiation. They also absorb electromagnetic radiation incident on them. A black body absorbs all the radiation incident on it. A black body also emits radiation with a pattern characterized by its temperature—this is called black-body radiation.

Figure 3 shows the variation with wavelength of the relative intensity of the radiation emitted by a black body. The wavelength peak shifts to shorter wavelengths as the temperature of the black body increases.

> **Intensity** I of radiation is the power P emitted per unit area: $I = \frac{P}{A}$, where A is the surface area of the emitting object.
>
> The area under the curve in Figure 3 is given by the **Stefan–Boltzmann law**. This predicts that the intensity output by a black-body radiator is $I = \sigma T^4$, where σ is the Stefan–Boltzmann constant (5.67×10^{-8} W m^{-2} K^{-4}).
>
> In some areas of physics (for example, astrophysics), P is known as the **luminosity** L of a body: $L = A\sigma T^4$. Luminosity has the unit of W (or J s^{-1}).
>
> **Wien's displacement law** predicts the characteristic peak in the graph of relative intensity against wavelength:
> $\lambda_{max}/\text{m} = \frac{2.90 \times 10^{-3}}{T/\text{K}}$
> The wavelength must be in metres and the temperature must be in kelvin.
>
> **Apparent brightness** b is an astronomical term used for the amount of light from a star or other body that reaches the Earth. It is the intensity of the radiation at the orbit of the Earth. As it is an intensity, it has the units W m^{-2}. For example, the apparent brightness of the Sun is known as the solar constant (Topic B.2) and is equal to about 1400 W m^{-2}. The next brightest star is Sirius, which has an apparent brightness of $b \approx 10^{-7}$ W m^{-2}.

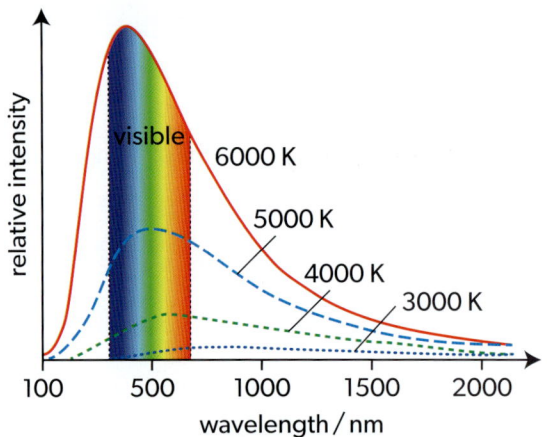

▲ **Figure 3** The relative intensity of radiation from a black body varies with wavelength

Example 9

The filament of a halogen light bulb has a surface area of 1.5×10^{-5} m^2. The light bulb radiates a power of 80 W.

a) Calculate the absolute temperature of the filament, assuming that it is a black-body radiator.

b) Calculate the peak wavelength of the radiation emitted by the filament. State the region of the electromagnetic spectrum in which the peak wavelength is located.

Solution

a) The temperature of the filament can be found from the Stefan–Boltzmann law.

$80 = 5.67 \times 10^{-8} \times 1.5 \times 10^{-5} \times T^4$

$T = \sqrt[4]{\frac{80}{5.67 \times 10^{-8} \times 1.5 \times 10^{-5}}} = 3100\,\text{K (to 2 s.f.)}$

b) $\lambda_{max} = \frac{2.90 \times 10^{-3}}{3100} = 930\,\text{nm}$

The peak wavelength is in the infrared region of the electromagnetic spectrum (see Topic C.2).

Example 10

The peak wavelength of the electromagnetic radiation emitted by the Sun is 500 nm.

a) Calculate:

(i) the surface temperature of the Sun

(ii) the power radiated by 1 m² of the surface of the Sun.

The radius of the Sun is 7.0×10^8 m and the average distance between the Sun and Earth is 1.5×10^{11} m.

b) Calculate:

(i) the luminosity of the Sun

(ii) the apparent brightness of the Sun at the location of Earth.

Solution

a) (i) $T = \dfrac{2.9 \times 10^{-3}}{500 \times 10^{-9}} = 5800$ K

(ii) $I = \sigma T^4 = 5.67 \times 10^{-8} \times 5800^4 = 6.4 \times 10^7$ W

As usual, we assume that the Sun is a black-body radiator.

b) (i) The luminosity is the total power radiated by the Sun, which can be calculated by multiplying the answer to part (a)(ii) by the surface area of the Sun.

$L = 4\pi (7.0 \times 10^8)^2 \times 6.4 \times 10^7 = 3.9 \times 10^{26}$ W

(ii) The intensity of solar radiation follows the inverse-square law with the distance from the Sun. At the location of Earth:

apparent brightness of the Sun $b = \dfrac{3.9 \times 10^{26}}{4\pi (1.5 \times 10^{11})^2} = 1400$ W

This quantity can also be measured directly and is usually called the solar constant (see Topic B.2).

Practice problems

Problem 1

A container holds a mixture of monatomic helium and neon gases at an initial temperature of 50 °C.

The average random speed of helium atoms in the mixture is 1400 m s⁻¹. The atomic mass of neon is five times greater than that of helium.

a) Calculate the average speed of neon atoms in the mixture.

The mixture is now heated and its temperature increases to 450 °C. The average kinetic energy of the atoms increases from E_i to E_f during the heating.

b) Calculate $\dfrac{E_f}{E_i}$.

Problem 2

X and Y are two solids with the same mass at the same initial temperature. Their temperatures are raised by the same amount. They both remain solid.

The specific heat capacity of X is greater than that of Y.

a) Explain which substance has the greater increase in internal energy.

Cold water, initially at a temperature of 14 °C, flows over an insulated heating element in a domestic water heater. The heating element transfers energy at a rate of 7.2 kW. The water leaves the heater at a temperature of 40 °C.

The specific heat capacity of water is 4.2 kJ kg^{-1} K^{-1}.

b) (i) Estimate the rate of flow of the water.

   (ii) Suggest one reason why your answer to part (b)(i) is an estimate.

Problem 3

An ice cube of mass 30 g and temperature −18 °C is dropped into a thermally insulated container that holds 250 g of water at 20 °C.

Specific heat capacity of ice = 2.1 kJ kg^{-1} K^{-1}

Specific latent heat of fusion of ice = 334 kJ kg^{-1}

Specific heat capacity of water = 4.2 kJ kg^{-1} K^{-1}

a) Calculate the energy required to warm the ice cube to the melting point and melt it completely.

b) Determine the final equilibrium temperature of the system.

Problem 4

One end of a metal bar is kept in a mixture of ice and water at 0 °C and the other end is at a constant temperature of 180 °C. The length of the bar is 0.45 m and its cross-sectional area is 4.0 × 10^{-4} m^2.

a) Calculate the temperature gradient in the rod.

The ice melts at a rate of 1.1 × 10^{-4} kg s^{-1}. Specific latent heat of fusion of ice is 334 kJ kg^{-1}.

b) Calculate:

   (i) the rate of energy transfer through the bar

   (ii) the thermal conductivity of the material of the bar.

Problem 5

Star X has surface temperature 1.6 × 10^4 K and radius 2.7 × 10^9 m.
The distance from star X to Earth is 6.1 × 10^{18} m.

a) Calculate, for star X:

   (i) the peak wavelength of electromagnetic radiation

   (ii) the luminosity

   (iii) the apparent brightness.

Star Y has the same physical properties as star X, but its apparent brightness is 10% of that of star X.

b) Calculate the distance from star Y to Earth.

Problem 6

The solar intensity arriving from the Sun at the radius of the Earth's orbit is 1400 W m^{-2}.

Mean radius of the Earth's orbit around the Sun = 1.5 × 10^{11} m

Radius of the Sun = 7.0 × 10^8 m

a) Estimate the total output power of the Sun.

b) Use your estimate in part (a) to deduce the temperature of the Sun.

B.2 Greenhouse effect

You should know:
✔ that energy is conserved in the context of the Earth's climate
✔ what is meant by emissivity and albedo
✔ what is meant by the greenhouse effect and the enhanced greenhouse effect
✔ the main greenhouse gases and their origins.

You should be able to:
✔ solve problems involving albedo, emissivity, the solar constant and the Earth's average temperature
✔ explain the greenhouse effect in terms of resonance and molecular energy levels.

The Earth's climate is affected by many factors, including incident and emitted radiation at the surface and elsewhere, and also by the composition of the atmosphere. The modelling of global warming and climate change is an important area of research.

Radiation arrives at the Earth's surface from the Sun. The spectrum of this black-body radiation is determined by the temperature of the outer layer of the Sun's photosphere (about 5700–5800 K). This temperature is taken to be the surface temperature of the Sun.

The energy per second transferred to Earth from the Sun is the overall difference between the incoming solar radiation and the radiation that Earth re-emits back into space.

However, the average power that arrives at the surface is smaller than 1360 W m^{-2}. The energy from the Sun falls on only **half** of the area of the Earth's sphere at any one time. The energy is then transferred to the **whole** of Earth's surface (Figure 1). The energy arriving comes through a disc of radius R, where R is the radius of Earth.

disc of radius R area collecting radiation is πR^2

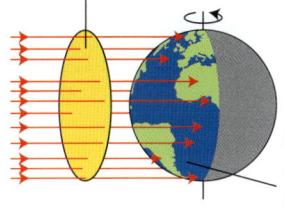

radiation falls on half of the surface of the Earth, which has a radius R

▲ Figure 1 The geometry of radiation incident on a sphere

This disc area is πR^2 and the energy coming through it is spread over the whole of the Earth's surface. This sphere surface has an area of $4\pi R^2$.

The average incident intensity $I_{surface}$ during 24 hours at any point on the surface must be:

$$I_{surface} = \frac{\text{power passing through the disc}}{\text{total area of Earth's surface}} = \frac{S \times \pi R^2}{4\pi R^2} = \frac{S}{4}$$

$I_{surface}$ therefore has a maximum value of $0.25 \times 1360 = 340$ W.

The actual mean power that is incident on each square metre of the Earth's surface is lower than this calculated value of 340 W. This is because the Sun's energy arriving at the surface is reduced for the following reasons.

> **Assessment tip**
>
> Estimates vary for the temperature of the photosphere depending, for example, on the wavelength that is used. Always use the value quoted in a question rather than one you have memorized.

> The **solar constant** S is the power that is incident on one square metre at the top of the Earth's atmosphere—this value is 1360 W m^{-2}.

> **Approaches to learning**
>
> The ideas in this topic link to a number of areas in the course, principally the astrophysics ideas from Topic E.5. Use links such as this creatively to energize your learning and revision as you prepare for assessments.

B The particulate nature of matter

Black-body radiation is a continuous spectrum that depends only on the temperature of the radiator. Its main properties are described in Topic B.1.

The variation of intensity with wavelength for a black body was shown in Topic B.1 Figure 3 (page 60).

The Stefan–Boltzmann law (Topic B.1, page 60) predicts the fourth-power dependence of the total emitted power on temperature. Doubling the temperature of a black body from 3000 K to 6000 K means that the emitted power is increased by a factor of 16.

- Radiation is absorbed and scattered by the atmosphere.
- The radiation passes through a greater thickness of atmosphere at dawn and dusk than in the middle of the day—more scattering and absorption take place when the thickness of the atmosphere is greater.

The theoretical black body from Topic B.1 cannot be realized in practice (even though some objects, such as stars, come very close to the ideal). Real objects are known as grey bodies—their radiated emission is less than that of an equivalent black body at the same temperature. The extent to which an emitter is imperfect compared with a black body is described by its emissivity, e.

$$\text{Emissivity} = \frac{\text{energy radiated from the surface of an object}}{\text{energy radiated from a black body at the same temperature as the object}}$$

$$e = \frac{\text{power radiated per unit area from the surface of an object}}{\sigma T^4}$$

- For a black body, $e = 1$.
- For a perfectly reflecting and non-radiating object, $e = 0$.

Emissivity has no units because it is a ratio of energies.

The Earth's surface is a grey body and scatters some of the incident energy back into the atmosphere. The extent to which the surface does this is known as its albedo.

$$\text{Albedo} = \frac{\text{energy scattered by a given surface in a given time}}{\text{total energy incident on the surface in the same time}}$$

$$= \frac{\text{total scattered power}}{\text{total incident power}}$$

Example 1

The diagram shows, for a particular region on the surface of Earth, the intensities of the incident and reflected solar radiation and the intensity of the radiation emitted by Earth.

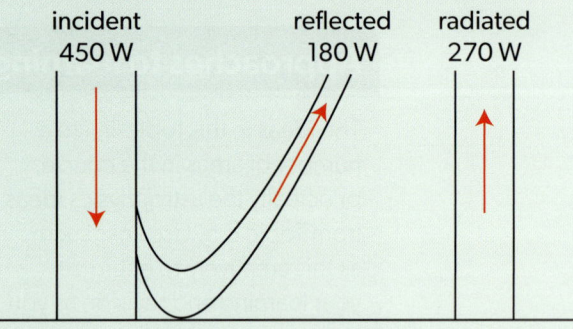

incident 450 W   reflected 180 W   radiated 270 W

a) Outline why the surface temperature of the region remains constant.
b) Calculate the albedo of the region.
c) The surface temperature of the region is 20 °C. Determine the emissivity of the region.

Solution

a) The sum of the radiated and reflected intensities is equal to the incident intensity, so the surface of the region is in equilibrium with the incoming solar radiation and there is no net energy transfer to or from the surface.

b) $\text{Albedo} = \dfrac{\text{reflected intensity}}{\text{incident intensity}} = \dfrac{180}{450} = 0.40$

c) At surface temperature 20 °C, intensity radiated by a black body $= \sigma T^4 = 5.67 \times 10^{-8}(273+20)^4 = 418 \text{ W m}^{-2}$

Emissivity of the region $= \dfrac{270}{418} = 0.65$

B.2 Greenhouse effect

The global annual mean albedo for Earth is 0.3. This means that 70% of the incident power from the Sun is absorbed by the planet. The value for the albedo shows both daily and seasonal variations for any one location on Earth. The albedo value is affected by:

- the latitude of the location—this is important because this determines the angle of the Sun in the sky
- cloud coverage.

The atmosphere is effectively transparent to most of the peak wavelength from the Sun (about 500 nm, which is in the green part of the visible spectrum) although there is some scattering of the shorter (blue) wavelengths.

The mean albedo value (0.3) tells us that about 70% of this energy is absorbed by the Earth itself—the rest (30%) is scattered from the surface and the atmosphere. The Earth itself also emits radiation because it is at a temperature greater than 0 K.

For the average temperature of the planet to be constant, there must be a dynamic energy equilibrium between the incident and emitted radiation.

However, the Earth's temperature is much lower than that of the Sun. The peak wavelength emitted by Earth is in the infrared region—radiation with a longer wavelength. Certain atmospheric gases (called greenhouse gases) are opaque to infrared (IR) and ultraviolet (UV) wavelengths. The energy from these photons is trapped inside the Earth–atmosphere system. This contributes to an increased average temperature for the Earth. This overall warming of the system through atmospheric absorption is known as the greenhouse effect.

> The **greenhouse gases** are water vapour (H_2O), carbon dioxide (CO_2), methane (CH_4) and dinitrogen monoxide (sometimes called nitrous oxide) (N_2O).

When the concentrations of greenhouse gases in the atmosphere increase, more of the outgoing infrared radiation is absorbed and more energy remains in the system. The overall emissivity has become smaller. This increases the temperature of the planet, and this temperature increase affects both climate and sea level.

The effect of an increased concentration of greenhouse gases is known as the enhanced greenhouse effect.

> Photons and their energies are discussed in Topic E.1.
>
> The energy E of a photon is $E = hf$, where f is the frequency of the photon and h is the Planck constant (6.63×10^{-34} J s). The smaller the wavelength, the greater the energy because $c = f\lambda$, where λ is the wavelength and c is the speed of light (Topic C.2).

The IR and UV wavelengths are absorbed by the atmosphere.

UV photons are highly energetic and can ionize (break apart) the gas molecules in the atmosphere to their separate atoms or ionized atoms. This leads, for example, to the creation of ozone (O_3—a three-atom form of oxygen).

IR wavelengths are much less energetic and do not break molecules apart. But when a particular photon frequency matches the vibrational frequency of a gas molecule in the atmosphere, then resonance occurs.

A carbon dioxide (CO_2) molecule has the single carbon atom connected between two oxygen atoms so that, normally, both carbon–oxygen bonds are in a straight line. Figure 2 shows the possible vibrational modes for CO_2.

> **Resonance** is the stimulation of a vibrating system into oscillation by the arrival of energy. This phenomenon is described in Topic C.4.

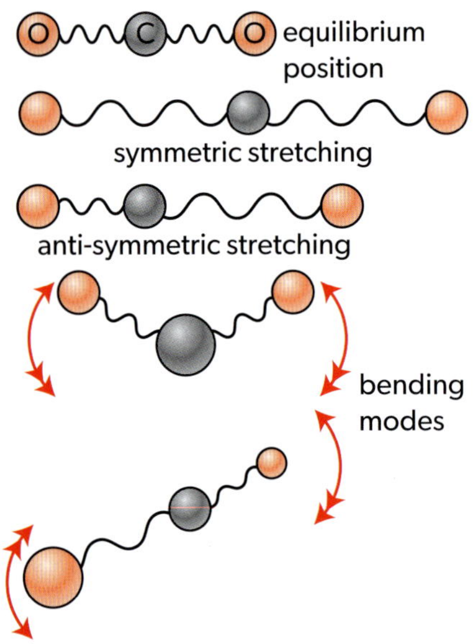

Figure 2 Some of the possible vibrational modes for CO_2

Each of these modes (and there are others) has a characteristic frequency. A photon that arrives at the molecule with this characteristic frequency will stimulate the bonds into vibration. As a result, the charge distributions within the whole molecule change during the vibration cycle. This is an acceleration of the charges. Accelerated charges emit electromagnetic waves (Topic E.1). This electromagnetic radiation is emitted in all directions, not necessarily in the direction of the incident photon. The molecule has become (effectively) a point source of radiation.

Another alternative explanation is that the incident photons stimulate the molecules into higher energy states. (The two explanations are linked because this energy state is the energy of the vibrational state.) As the molecules fall back to their original, lower-energy state, radiation must be re-emitted. Once again, this happens in all directions, so that the molecule is radiating much of its energy in directions other than in the direction from Earth. Such radiation is said to be isotropic.

Oxygen and nitrogen do not contribute to the greenhouse effect because their molecules are symmetrical and the charge symmetry is maintained when they interact with incident radiation. This does not allow the re-radiation in all directions as is the case for the greenhouse gases.

Nature of science

The greenhouse effect itself is beneficial because it has raised the temperature of the Earth and we have evolved for these conditions. The **enhanced** greenhouse effect leads to an increase in temperature above this value because of the release of extra greenhouse gases. This branch of science has a global impact for everyone.

Example 2

a) Determine the average global temperature of Earth if it were:

   (i) a black body

   (ii) a body with an albedo of 0.30 and no atmosphere

   (iii) a body with an albedo of 0.30 whose atmosphere re-emits 39% of the outgoing thermal radiation back to the surface.

b) Outline the mechanism by which some of the thermal radiation emitted by Earth is returned to the surface.

Solution

a) In all calculations, we assume that there is an equilibrium between incoming and outgoing radiation intensity.

   (i) The average incoming intensity is $\frac{S}{4} = 340$ W. Assuming that Earth is a black body of temperature T in equilibrium with incoming radiation, the average radiated intensity must be equal to the incoming intensity:

   $$\sigma T^4 = 340 \text{ W}$$

   $$T = \sqrt[4]{\frac{340}{5.67 \times 10^{-8}}} = 278 \text{ K}$$

   This is equivalent to about 5 °C, which is much less than the actual temperature of Earth.

   (ii) In this model, only 70% of the intensity of incoming solar radiation is absorbed by Earth and re-emitted as thermal radiation:

   $$T = \sqrt[4]{\frac{(1 - 0.30) \times 340}{5.67 \times 10^{-8}}} = 255 \text{ K}$$

   This is equivalent to −18 °C, which suggests that without the greenhouse effect, much of Earth's surface would be frozen.

   (iii) The remaining 61% of the thermal radiation leaves Earth and is radiated into space. The emissivity of Earth in this model is effectively 0.61.

> The energy balance equation must include both the albedo and the emissivity of Earth:
>
> $$(1-0.30) \times \frac{S}{4} = 0.61 \sigma T^4$$
>
> $$T = \sqrt[4]{\frac{(1-0.30) \times 340}{0.61 \times 5.67 \times 10^{-8}}} = 288 \text{ K}$$
>
> This is very close to the actual average temperature of Earth of about 15 °C.
>
> b) Molecules of greenhouse gases (GHG) present in the atmosphere have resonant frequencies in the infrared region, which matches the frequency range of the thermal radiation of Earth and leads to a resonant absorption of this radiation. The absorbed energy is then re-emitted by GHG molecules in random directions, including back towards the surface.

Climate models attempt to predict the effect of increasing concentrations of greenhouse gases in the atmosphere. These gases have been released through fossil-fuel burning and other human activities. Modelling the surface–atmosphere energy balance is complex, but current models suggest that:

- global warming is occurring
- ice and snow cover at the poles will decrease, which will decrease the average albedo of the Earth and therefore increase the overall rate at which energy is absorbed at the surface
- ocean water temperatures will increase so that the CO_2 dissolved in seawater will be released—this will lead to a positive feedback for the CO_2 levels in the atmosphere.

International discussions on this issue take place annually at UN Climate Change Conferences.

Example 3

Vesta is a minor planet that orbits the Sun at a mean distance of 3.5×10^{11} m. The albedo of Vesta is 0.42.

Estimate the mean temperature of Vesta.

Solution

Intensity of solar radiation at the position of Vesta $= \left(\dfrac{1.5 \times 10^{11}}{3.5 \times 10^{11}}\right)^2 \times 1360$

$= 250 \text{ W}$

Average intensity incident at the surface of Vesta $= \dfrac{1}{4} \times \left(\dfrac{1.5 \times 10^{11}}{3.5 \times 10^{11}}\right)^2 \times 1360$

$= 62.4 \text{ W}$

where the usual factor $\dfrac{1}{4}$ is due to the projected area of Vesta being equal to the quarter of its total area.

The energy balance equation for the surface of Vesta, including its albedo, is $(1 - 0.42) \times 62.4 = \sigma T^4$.

From here, $T = \sqrt[4]{\dfrac{0.58 \times 62.4}{5.67 \times 10^{-8}}} = 160 \text{ K}$.

Assessment tip

You may have to solve problems that involve equilibrium temperature of bodies in the solar system other than Earth. The intensity I of solar radiation at the position of the body is usually calculated from $I = \dfrac{L_\odot}{4\pi d^2}$, where d is the distance from the body to the Sun and $L_\odot = 3.8 \times 10^{26}$ W is the power output (luminosity) of the Sun (the subscript ⊙ stands for the Sun here and in topics such as E.5). Alternatively, $I = \left(\dfrac{d_{\text{Earth}}}{d}\right)^2 \times S$, where $S = 1360 \text{ W m}^{-2}$ is the solar constant and $d_{\text{Earth}} = 1.5 \times 10^{11}$ m is the mean Earth–Sun distance. This distance is known as the astronomical unit (see Topic E.5). The values of S and the astronomical unit are given in the *Physics data booklet*.

B The particulate nature of matter

Sample student answer

The Sun has a radius of 7.0×10^8 m and is at a distance of 1.5×10^{11} m from the Earth. The surface temperature of the Sun is 5700 K.

a) Show that the intensity of the solar radiation incident on the upper atmosphere of the Earth is approximately $1400 \, W \, m^{-2}$. [2]

This answer could have achieved 1/2 marks:

> $e = 1$
>
> $\sigma = 5.67 \times 10^{-8} \, W \, m^{-2} \, K^{-4} \quad T = 5700 \, K$
>
> $A = (7.0 \times 10^8)^2 \, \pi \, m^2$
>
> $P = e\sigma A T^4 = 9.216 \times 10^{25} \, W$
>
> $I = \dfrac{P}{A} = \dfrac{9.877 \times 10^{25}}{(1.5 \times 10^{11})^2 \times \pi} = 1397 \, W \, m^{-2} \simeq 1400 \, W \, m^{-2}$

▼ The student uses the Stefan–Boltzmann equation correctly to show the total output power of the Sun. Then there is an error because there is a confusion in the calculation of the area of the sphere at the Earth's orbit. This is needed to calculate the power per unit area. The area of this sphere is $4\pi r^2$ not πr^2. The same mistake occurs in the calculation of the area of the Earth.

b) The albedo of the atmosphere is 0.30. Deduce that the average intensity over the entire surface of the Earth is $245 \, W \, m^{-2}$. [2]

This answer could have achieved 0/2 marks:

> $0.3 = \dfrac{x}{1400}$
>
> $x = 0.3 \times 1400$

▼ The student should first try to show that the average intensity with no atmospheric absorption is one-quarter of $1400 \, W \, m^{-2}$ and then use the albedo expression to calculate the final answer.

c) Estimate the average surface temperature of the Earth. [2]

This answer could have achieved 0/2 marks:

> $T^4 = \dfrac{P}{e\sigma A}$
>
> $\dfrac{T^4}{A} = \dfrac{P}{e\sigma} = \dfrac{1400}{0.3 \times 5.67 \times 10^{-8}} = 8.23 \times 10^{10}$
>
> $T = \sqrt[4]{8.23 \times 10^{10}} = 535.6 \, K \simeq 536 \, K$

▼ The student should use the Stefan–Boltzmann equation again as $\sigma T^4 = 245$, which would lead to an answer of 256 K. The 1400 value from (a) is the wrong emitted intensity for the surface. This was the answer to (b). There is also a mistake in re-arranging the expression, but the student does not get a mark for using this as the 1440 is the intensity of the radiation not its power.

Practice problems

Problem 1

The diagram shows a simplified model of the energy balance in the Earth surface–atmosphere system. The arrows represent the intensities of radiation.

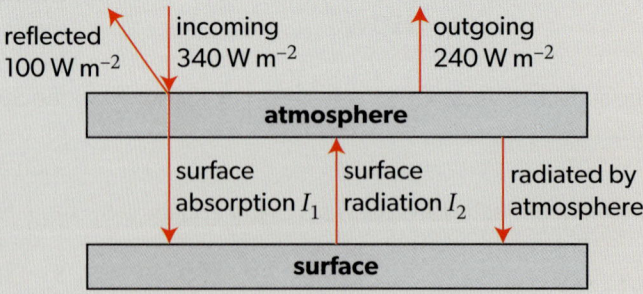

a) Calculate:

  (i) the albedo of Earth

  (ii) the intensity I_1 of radiation absorbed by the surface.

The average global temperature of the surface of Earth is 288 K.

b) (i) Calculate the intensity I_2 of the radiation emitted by the surface.

   (ii) Outline one other way by which thermal energy is transferred from the surface to the atmosphere.

   (iii) Estimate the emissivity of Earth.

c) Explain the effect of the enhanced greenhouse effect on the energy balance of the Earth surface–atmosphere system.

Problem 2

When the concentration of carbon dioxide in the atmosphere doubles, the albedo of the Earth increases by 0.01.

Average intensity received at Earth from the Sun = 340 W m^{-2}

Average albedo = 0.30

a) Determine the change in the intensity of the radiation being reflected into space by the Earth.

b) State one reason why the answer to part (a) is an estimate.

Problem 3

Venus is a planet in the solar system that orbits the Sun at an average distance of 1.1×10^{11} m.

a) Calculate the intensity of solar radiation at the location of Venus.
   The power output of the Sun is 3.8×10^{26} W.

The albedo of Venus is 0.76.

b) (i) State what is meant by albedo.

   (ii) Estimate the average temperature of Venus, ignoring the greenhouse effect of its atmosphere.

The actual average temperature of the surface of Venus is 740 K.

c) Calculate the ratio $= \dfrac{\text{average intensity radiated by the surface of Venus}}{\text{average intensity of solar radiation absorbed by Venus}}$.

B.3 Gas laws

You should know:
- what is meant by pressure, mole, molar mass and the Avogadro constant
- what is meant by an ideal gas
- the gas laws for constant volume, constant pressure and constant temperature
- the ideal gas law equation and the equations that govern the behaviour of an ideal gas
- the differences between an ideal and a real gas
- the conditions for which an ideal gas can be used as a good approximation for a real gas.

You should be able to:
- sketch and interpret pressure–volume, pressure–temperature and volume–temperature graphs
- solve problems using $\dfrac{PV}{T}$ = constant and $PV = Nk_BT = nRT$
- recall the assumptions of the kinetic model
- relate gas pressure to change of momentum by particles as they collide at a surface
- show how the kinetic model assumptions lead to a theoretical model.

Pressure arises with all phases of matter. The pressure on the walls of a gas container depends on the rate at which the gas particles transfer momentum when they collide with the wall.

Pressure $P = \dfrac{\text{force}}{\text{area}}$

Its unit is $N\,m^{-2}$, which is the same as a pascal (Pa). In fundamental SI units, this is $kg\,m^{-1}\,s^{-2}$.

In **solids**, a normal force F applies through a contact area A between the solid and the surface on which it rests.

In a **liquid** of density ρ, the pressure at depth h is $\rho g h$.

The mole is the fundamental SI unit for the quantity of matter of a substance. It corresponds to the mass of a substance that contains 6.022×10^{23} particles of the substance.

Nature of science

Quantity of matter is the way scientists compare numbers of objects. A mole of atoms, a mole of electrons, a mole of ions, always gets you about 6.02×10^{23} objects (atoms, electrons and ions, respectively). You could meet the concept of the mole anywhere in the course but especially throughout Themes B and E.

The **mole** (abbreviated **mol**) is the SI unit of substance. One mole is defined to contain $6.022\,140\,76 \times 10^{23}$ elementary entities of the substance exactly. "Exactly" means that there are no further numbers after the final "6".

This number is known as the **Avogadro constant** N_A.

The number n of moles of a substance is $n = \dfrac{\text{number of molecules}}{N_A} = \dfrac{N}{N_A}$

The definition of the mole in terms of elementary entities has only been used since 2019. Before then, the Avogadro number was defined in terms of the kilogram as the number of atoms in a mass of 12 g of the isotope carbon-12. The $12\,g\,mol^{-1}$ for this isotope of carbon is known as its **molar mass**. For other chemical elements, the mass of one mole of atoms, in grams, is approximately equal to the atomic mass number of the element. (For a discussion of the meaning of isotope see Topic E.3.)

Real gases have behaviours close to those of an ideal gas only for **low pressures, low densities** and **moderate temperatures**. Treat gases as ideal unless told otherwise in an examination.

An equation of state describes a gas using three variables: pressure P, volume V and temperature T. Equations of state are possible for real gases, but they need extra terms to account for high densities and particle interactions.

The equation of state for an ideal gas is $PV = nRT$ where n is the number of moles. This can be summarized in the **general gas equation** for two states, 1 and 2, of a gas:

$$\dfrac{P_1 V_1}{T_1} = \dfrac{P_2 V_2}{T_2}$$

The equation can also be written as $PV = \dfrac{N}{N_A} RT$ and $PV = N k_B T$, where R is the (ideal) gas constant ($8.31\,J\,mol^{-1}\,K^{-1}$), k_B is the Boltzmann constant ($1.38 \times 10^{-23}\,J\,K^{-1}$) and N is the number of molecules in the gas.

Assessment tip

Notice that both sides of the equation of state have the units of energy. You can think of R as being analogous to the specific heat capacity of one mole of a gas, with k_B being analogous to the specific heat capacity of one molecule in the gas.

Example 1

An ideal gas in a container of volume 1.2×10^{-5} m^3 has a pressure of 1.5×10^5 Pa at a temperature of 50 °C.

Calculate:

a) the number of molecules of gas in the container

b) the number of moles of gas in the container.

Solution

The temperature must be in kelvin, so $50 + 273 = 323$ K.

a) $N = \dfrac{PV}{k_B T} = \dfrac{1.5 \times 10^5 \times 1.2 \times 10^{-5}}{1.38 \times 10^{-23} \times 323} = 4.0 \times 10^{20}$ molecules

b) $n = \dfrac{PV}{RT} = \dfrac{1.5 \times 10^5 \times 1.2 \times 10^{-5}}{8.31 \times 323} = 6.7 \times 10^{-4}$ mol

Assessment tip

When using the general gas equation, the units of pressure and volume **must** match on both sides of the equation.

The **only** unit allowed for temperature in the equation is kelvin.

Historically, gas behaviour was identified through experiments that involved three separate **gas laws** (see Figure 1). These, taken together, combine to give the equation of state.

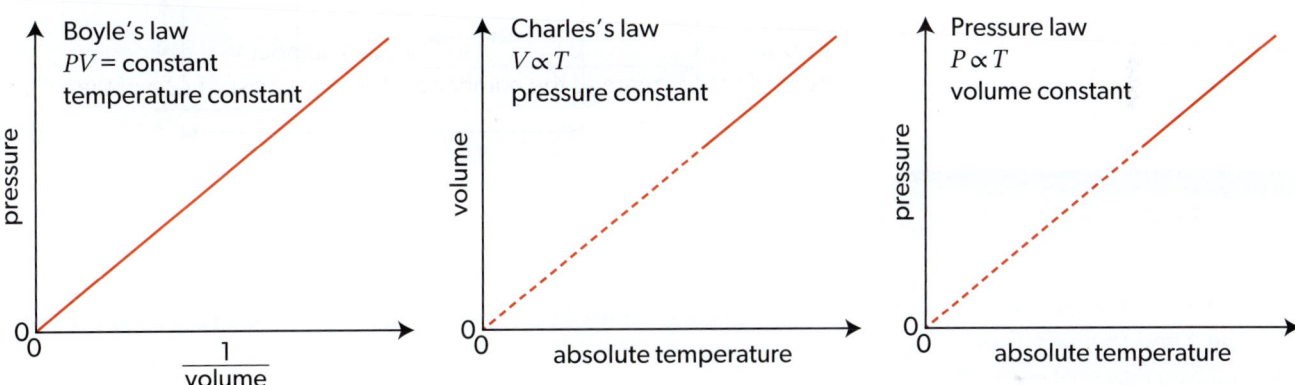

▲ Figure 1 The results from three experiments in which two quantities of a fixed mass of gas are varied with the other quantity being unchanged

The kinetic model of a gas

The kinetic model of a gas is based on the following assumptions about gas particles and their behaviour.

1. A gas consists of particles. The total volume of the particles is negligible compared with the total volume of the gas (or the average distance between particles is much greater than their individual size).
2. Particles have the same mass.
3. Particles are in constant, random motion.
4. Particles collide elastically with each other and the walls of the container.
5. Interactions between particles can be ignored (that is, they do not exert any force on each other).
6. The time for a particle collision is negligible compared with the time between collisions.
7. Gravity can be ignored.

The behaviour of real gases at extremes of temperature and pressure is complicated. A simplified model of an ideal gas is used. In this model, the particles are assumed to collide elastically, with no intermolecular forces.

Assessment tip

Know the meaning of these assumptions and recognize how they affect the kinetic model. The model leads to the relationship between the pressure of the gas and the mean square speed of the particles. The steps in the derivation of the model are given in Example 2.

B The particulate nature of matter

Nature of science

The gas laws were suggested by scientists in the 18th century following experimental work. Such laws are said to be **empirical**. On the other hand, the kinetic model of a gas stems from a **theoretical** standpoint involving assumptions about the gas. These macroscopic and microscopic approaches fit together to confirm our view of gas properties.

Other areas in the course that use empirical rules include Ohm's Law in Topic B.5 and the Rydberg formula in Topic E.1.

Approaches to learning

The kinetic theory is a good way to help you to link ideas from Theme A and Theme B. Even though you may not be asked questions about the whole of the proof in Example 2, you should be able to explain each step and how it arises.

Assessment tip

Part (c)(iii) in Example 2 is a "show that" question. You must make every step clear, **including the final result**.

For a real gas:
- there are effects between the molecules and the walls
- there are effects between molecules when they are very close to each other immediately before and after an intermolecular collision
- within a few degrees of absolute zero, other effects become important to make gases non-ideal.

These effects become noticeable when gas molecules are close together for much of the time. In other words, when the gas has a high density or a high pressure. At high speeds (in other words, at high temperatures) the molecules spend significant fractions of their time colliding with other molecules.

These conditions contravene assumptions 1, 5 and 6 behind the kinetic model of a gas.

Example 2

A particle of mass m moves with velocity u in a box with side lengths x, y and z.

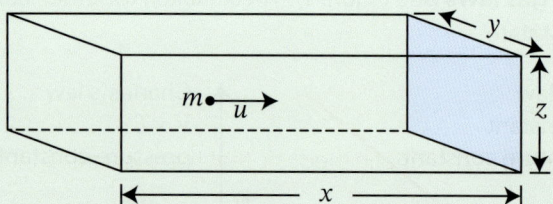

The particle strikes the end faces at right angles and makes repeated elastic collisions with opposite faces.

a) (i) Calculate the time t between collisions with the shaded face.

   (ii) State the expression for the change in momentum per collision with the shaded face.

b) The box contains N identical particles, all moving parallel to the x-direction with speeds $u_1, u_2, \ldots, u_N$. Each particle makes elastic collisions at the end faces. Show that the average force F on the shaded face is given by $F = \dfrac{Nm}{x}\overline{u^2}$, where $\overline{u^2} = \dfrac{u_1^2 + u_2^2 + \cdots + u_N^2}{N}$ is the average squared speed of the particles in the x-direction.

c) The model is refined so that N particles move randomly in the box. The average squared speed in the y-direction is $\overline{v^2}$. The average squared speed in the z-direction is $\overline{w^2}$.

   (i) Deduce an expression for the average squared speed $\overline{c^2}$ of the particles in terms of $\overline{u^2}$, $\overline{v^2}$ and $\overline{w^2}$.

   (ii) Deduce an expression for F in terms of $\overline{c^2}$.

   (iii) Show that $PV = \dfrac{Nm\overline{c^2}}{3}$, where P is the pressure of the gas and V is its volume.

Solution

a) (i) The particle travels a distance $2x$ at a speed u, so $t = \dfrac{2x}{u}$.

   (ii) Change in momentum for each collision at the shaded face = $\Delta p = 2mu$

b) For a single particle moving with speed u_i:

$$\text{force} = \text{rate of change of momentum} = \left(\dfrac{\Delta p}{\Delta t}\right) = \dfrac{2mu_i}{\left(\dfrac{2x}{u_i}\right)} = \dfrac{m}{x}u_i^2$$

So, for N particles:

force $F = \dfrac{m}{x}(u_1^2 + u_2^2 + \cdots + u_N^2) = \dfrac{Nm}{x} \times \left(\dfrac{u_1^2 + u_2^2 + \cdots + u_N^2}{N}\right) = \dfrac{Nm}{x}\overline{u^2}$

c) (i) For one particle, the magnitude of its velocity c_i is given by
$c_i^2 = u_i^2 + v_i^2 + w_i^2$.

For all the particles, $\overline{c^2} = \overline{u^2} + \overline{v^2} + \overline{w^2}$.

(ii) The average squared speeds must be the same in each direction as gases appear the same in whichever direction we look (they are said to be isotropic). As a result:

$\overline{u^2} = \overline{v^2} = \overline{w^2} = \dfrac{\overline{c^2}}{3}$

Therefore, $F = \dfrac{1}{3}\dfrac{Nm\overline{c^2}}{x}$

(iii) $P = \dfrac{F}{A} = \dfrac{1}{3}\dfrac{Nm\overline{c^2}}{x} \times \dfrac{1}{yz} = \dfrac{1}{3}\dfrac{Nm\overline{c^2}}{xyz}$

As xyz is the volume V of the box:

$P = \dfrac{1}{3}\dfrac{Nm\overline{c^2}}{V}$ or $PV = \dfrac{Nm\overline{c^2}}{3}$

An important result of the kinetic theory is that there is a relationship between the pressure P and the average translational speed $\overline{v^2}$ of the gas molecules.

Theory shows that $P = \dfrac{1}{3}\dfrac{Nm\overline{v^2}}{V}$. The total mass of the gas is Nm so that $\dfrac{Nm}{V}$ is the gas density ρ. Thus, $P = \dfrac{1}{3}\rho\overline{v^2}$.

The equation of state links the pressure and volume of a gas to the particle speeds. The temperature can also link directly to the average squared speeds:

$Nk_BT = \dfrac{Nm\overline{v^2}}{3}$

so $m\overline{v^2} = 3k_BT$ and $\dfrac{1}{2}m\overline{v^2} = \dfrac{3}{2}k_BT$.

The left-hand side of the final equation is the average kinetic energy of a gas particle. The right-hand side is a measure of the kinetic energy $\overline{E}_k$ of a gas molecule. The units of k_BT are joules.

Finally, because the internal energy U of an ideal gas is the total kinetic energy of the molecules with no contribution from potential energy, $U = \dfrac{3}{2}nRT$ and $U = \dfrac{3}{2}Nk_BT$.

Example 3

A cylinder of fixed volume contains 15 mol of an ideal gas at a pressure of 490 kPa and a temperature of 27 °C.

a) Determine the volume of the cylinder.

b) Calculate the average kinetic energy of a gas molecule in the cylinder.

The symbol $\overline{v^2}$ occurs throughout this derivation of the kinetic theory. The meaning of the symbol is "the average of the (particle speeds)^2".

This is not the same as the "the squared average of the particle velocities" (which would be written $(\overline{v})^2$). For a gas the average of the particle velocities is zero. On squaring the individual values of velocity for each particle, the direction information is lost so that (velocity)^2 is really a scalar; hence "(particle speeds)^2".

Some areas of physics take this one step further by taking the square root of the average of the quantity^2, that is, $\sqrt{\overline{x^2}}$. This is known as the "root mean square of x".

Solution

a) Use $PV = nRT$ to give:
$$V = \frac{15 \times 8.31 \times (27 + 273)}{4.9 \times 10^5} = 0.076 \text{ m}^3$$

b) Using $\overline{E_K} = \frac{3}{2}k_B T$:
$$\overline{E_K} = \frac{3}{2} \times 1.38 \times 10^{-23} \times 300 = 6.2 \times 10^{-21} \text{ J}$$

Example 4

A sample of 0.040 mol of a monatomic ideal gas has a temperature of 300 K and a volume of 9.0×10^{-4} m^3.

a) Calculate the pressure of the gas.

b) The molar mass of the gas is 20 g mol^{-1}. Calculate:

   (i) the density of the gas

   (ii) the average speed of the particles of the gas.

c) The gas is heated at constant pressure. The temperature of the gas increases to 400 K.

   Determine the final volume of the gas.

Solution

a) From the ideal gas equation:
$$P = \frac{nRT}{V} = \frac{0.040 \times 8.31 \times 300}{9.0 \times 10^{-4}} = 1.1 \times 10^5 \text{ Pa}$$

b) (i) The mass of the gas is $0.040 \times 20 \times 10^{-3} = 8.0 \times 10^{-4}$ kg, so:
$$\text{density } \rho = \frac{8.0 \times 10^{-4}}{9.0 \times 10^{-4}} = 0.89 \text{ kg m}^{-3}$$

(ii) $P = \frac{1}{3}\rho v^2$, so:
$$v = \sqrt{\frac{3P}{\rho}} = \sqrt{\frac{3 \times 1.1 \times 10^5}{0.89}} = 610 \text{ m s}^{-1}$$

c) At constant pressure, $V \propto T$ and so $\frac{V_f}{V_i} = \frac{T_f}{T_i}$, where indices **i** and **f** denote the initial and final values of volume and temperature.
$$\text{Final volume} = V_i \frac{T_f}{T_i} = 9.0 \times 10^{-4} \times \frac{400}{300} = 1.2 \times 10^{-3} \text{ m}^3$$

The kinetic theory applies only to an ideal monatomic gas.

Real gases are non-ideal. For example, they can be compressed into a liquid at high pressures and low temperatures and volumes.

For an ideal gas, the graph showing how $\frac{PV}{RT}$ varies with P is a straight line parallel to the P-axis. Real gases deviate from this behaviour as shown in Figure 2. The real gas (broken lines) deviates from the ideal case (unbroken line) and begins to approach the behaviour of an incompressible liquid at high pressures (which would be a line parallel to the $\frac{PV}{RT}$-axis).

B.3 Gas laws

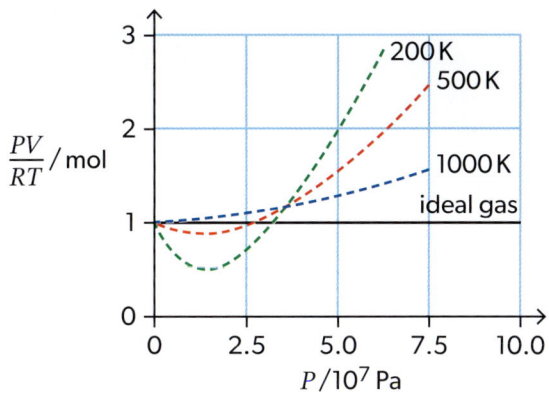

▲ Figure 2 The behaviour of one mole of a real gas at various temperatures compared with the behaviour of an ideal gas

Sample student answer

0.46 mol of an ideal monatomic gas is trapped in a cylinder. The gas has a volume of 21 m³ and a pressure of 1.4 Pa.

a) State how the internal energy of an ideal gas differs from that of a real gas. [1]

This answer could have achieved 0/1 marks:

> Internal energy is constant as molecules move at constant velocity.

▼ The kinetic model does not assume constant velocity and this answer does not make it clear whether it is referring to the ideal or the real case.

This answer could have achieved 1/1 marks:

> Ideal gas ignores intermolecular force between molecules in between collisions. So there is no potential energy and contains kinetic energy only.

▲ The distinction between ideal and real gases is clear even though the word "real" does not appear in the answer. The deduction that no intermolecular force implies no potential energy is correct and is reinforced by the statement about kinetic energy.

b) Determine, in kelvin, the temperature of the gas in the cylinder. [2]

This answer could have achieved 2/2 marks:

> $pV = nRT$
> $1.4 \times 21 = 0.46 \times 8.31 \times T$
> $T = \dfrac{29.4}{3.8226} = 7.7$ kelvin
> The temperature of the gas is 7.7 kelvin.

▲ The answer begins with a clear statement of the gas equation to be used. The substitution is clear (because the numbers are substituted in the same order as in the equation) and the answer is quoted to an appropriate number of significant figures.

Practice problems

Problem 1
a) Distinguish between thermal energy and internal energy.

b) Outline, with reference to the particles, the difference in internal energies of a metal and an ideal gas.

Problem 2
Use the kinetic model to explain why:

a) the pressure of an ideal gas increases when heated at constant volume

b) the volume of an ideal gas increases when heated at constant pressure.

Problem 3
A quantity of 0.25 mol of an ideal gas has a pressure of 1.05×10^5 Pa at a temperature of 27 °C.

a) Calculate the volume occupied by the gas.

When the gas is compressed to $\frac{1}{20}$ of its original volume, the pressure rises to 7.00×10^6 Pa.

b) Calculate the temperature of the gas after the compression.

Problem 4
The pressure in a container is increased using a bicycle pump. The volume of the container is 1.30×10^{-3} m^3.

The pump contains 1.80×10^{-4} m^3 of air at a pressure of 100 kPa and a temperature of 300 K.

Assume that the air acts as an ideal gas.

Assume that all the air molecules from the pump are transferred into the container when the pump is pushed in.

The air in the container is at an initial pressure of 150 kPa and a temperature of 300 K.

a) (i) Calculate, in mol, the initial quantity of gas in the container.

   (ii) Calculate, in mol, the quantity of gas transferred to the container every time air is pumped into it.

   The temperature of the gas in the container returns to 300 K after the pump has been used.

   (iii) Calculate the pressure in the container after the pump has transferred one pump-full of air into the container and its temperature has returned to 300 K.

b) Explain, with reference to the kinetic model of an ideal gas, why the gas in the container has pressure and why this pressure will increase when gas molecules are transferred to the container.

Problem 5
Air in a container has a density of 1.24 kg m^{-3} at a pressure of 1.01×10^5 Pa and a temperature of 300 K.

a) Calculate the mean translational kinetic energy of an air molecule in the container.

b) Calculate the mean translational speed for the air molecules.

The temperature of the air in the container is increased to 320 K.

c) Explain why some of the molecules will have speeds much less than that calculated in part (b).

B.4 Thermodynamics

You must know:

- what is meant by isobaric, isothermal, isovolumetric and adiabatic processes
- the first and second laws of thermodynamics
- alternative definitions of entropy, and the link between the second law of thermodynamics and entropy
- the definition of cyclic processes and that cyclic gas processes can be used to run a heat engine
- what is meant by a Carnot cycle
- that the Carnot cycle sets a limit on the maximum efficiency of a heat engine
- the definition of thermal efficiency.

You should be able to:

- describe the first law of thermodynamics as a statement of conservation of energy that relates the internal energy of a system to the transfer of energy
- describe the second law of thermodynamics using the Clausius interpretation and the Kelvin (Joule–Kelvin) interpretation, and describe entropy change in terms of reversible or irreversible processes in isolated systems
- solve problems using $W = P\Delta V$ for a closed system
- solve problems using $\Delta U = \frac{3}{2} N k_B \Delta T$
- solve problems using the first law of thermodynamics expressed as $Q = W + \Delta U$
- solve problems involving entropy changes and describe processes in terms of entropy change
- solve problems involving adiabatic changes for a monatomic ideal gas for which $PV^{\frac{5}{3}}$ = constant
- solve problems involving thermal efficiency.

As with Topic B.3, this topic covers the behaviour of gases, but it also takes a broader view of all thermodynamic systems. The general properties of systems are considered in terms of the changes they undergo and how these have an impact on the rest of the universe.

A pressure–volume (PV) diagram (e.g. Figure 1) shows the changes in the pressure and volume of a gas as it moves between two or more states or around a closed cycle.

The work done by, or on, the system is the equivalent of the area under a PV graph. This can be evaluated numerically by counting squares under the graph.

The first law of thermodynamics is an expression of the conservation of energy as it applies to a system—examples of systems include a gas that is acted on by its surroundings, or a heat engine such as a refrigerator. The first law applies to a closed system.

The law is written as $Q = W + \Delta U$. The terms in the equation when positive are:

- Q—the energy transferred into the system from the surroundings
- W—the work done by the system on the surroundings
- ΔU—the change in the internal energy of the system.

The work done on or by the system is always related to the change in volume:

- when a gas expands, work is always done **by** the (gas) system ($W > 0$)
- when a gas is compressed, work is always done **on** the (gas) system ($W < 0$)
- when there is no change in volume, no work is done ($W = 0$).

> A **thermodynamic system** defines the items of interest in a particular context. The **system** is separated from its **surroundings** by a boundary. The system together with the surroundings constitutes the **universe**.
>
> An **isolated system** is one that cannot interact with its surroundings. For example, an insulated box from which, or to, no energy can be transferred.
>
> A **closed system** is one where matter (e.g. a gas) cannot escape from or enter the system, but where an energy transfer into or out of the system is possible.

> **Assessment tip**
>
> In the IB Diploma Programme physics course:
>
> Work done **on** a system is **negative** ($W < 0$). Work done **by** a system is **positive** ($W > 0$).
>
> Energy transferred **into** a system from the surroundings is **positive** ($Q > 0$). Energy transferred **from** a system to the surroundings is **negative** ($Q < 0$).
>
> There are other ways to write the first law of thermodynamics with different definitions of the signs. You may see these in other books.

This equation can be applied to the four gas changes shown in Figure 1. For all four changes, imagine that an ideal gas (system) is trapped inside a cylinder with a piston at one end (the cylinder walls and piston are the boundary) with the surroundings being everything else in the universe.

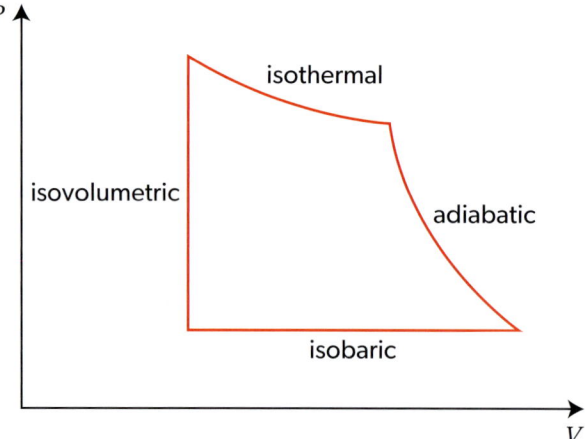

▲ Figure 1 Four possible changes that can be made to the state of a gas

- An **isobaric change** is one carried out at **constant pressure**.

  The work done by the system on the surroundings is at constant pressure. The energy transfer W when a piston of area A is moved through distance x is $P \times A \times x$. But $A \times x$ is the change in volume ΔV and therefore $W = P\Delta V$.

  The first law becomes $Q = \Delta U + P\Delta V$.

- **Isothermal changes** are carried out at **constant temperature**.

  There is no change to U and hence $\Delta U = 0$, so the internal energy of the gas does not change (as explained in Topic B.3).

  For an isothermal change, $Q = W$.

  Any thermal energy transferred **into** the system must appear as external work done **by** the system on the surroundings.

- An **isovolumetric change** is one carried out at **constant volume**.

  The term W is zero because no work is done by or on the system (gas): $Q = \Delta U$.

- Adiabatic means that no energy is transferred into or out of the system.

  Therefore, for an **adiabatic change**, $Q = 0$ so that $\Delta U = -W$.

  Any external work done by the system must come from internal energy. Put simply, when work is done **on** the surroundings by the system, the internal energy and hence the temperature of the system must **decrease**.

Example 1

0.0640 mol of an ideal gas is enclosed in a cylinder by a frictionless piston.

Two isotherms are shown on the PV diagram for 300 K and 500 K.

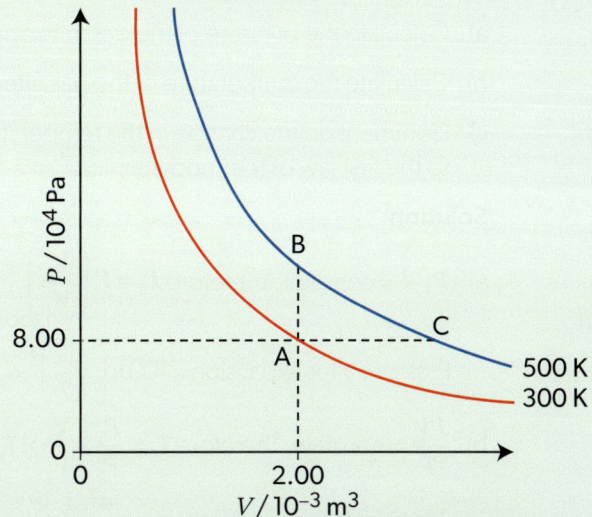

a) Explain how the first law of thermodynamics applies when the state of the gas is changed from:

   (i) A to B at constant volume

   (ii) A to C at constant pressure.

b) Calculate the heat energy absorbed by the gas in the change from:

   (i) A to B

   (ii) A to C.

Solution

a) The first law of thermodynamics is written as $Q = W + \Delta U$, where Q is the energy entering the gas from the surroundings, W is the work done by the gas and ΔU is the change in the internal energy of the gas.

   (i) The change from A to B is at constant volume, so $Q = \Delta U$ as no work is done by the system on the gas ($W = 0$). The temperature increases to reflect the change ΔU.

   (ii) The change from A to C is at constant pressure and so the piston must move to allow this.

   $Q = W + \Delta U$

   In this case, the temperature increases and work is done in expansion; therefore both ΔU and W are positive.

b) (i) $\Delta U = \frac{3}{2}nR\Delta T = 1.5 \times 0.0640 \times 8.31 \times (500 - 300) = 160\,\text{J}$

   No work is done; therefore $Q = 160\,\text{J}$.

   (ii) State C has the same temperature as B; therefore the change in internal energy from A to C is the same as that from A to B, $\Delta U = 160\,\text{J}$.

   To calculate the work done by the gas, we need to know its final volume in state C.

   $V_C = V_A \dfrac{T_C}{T_A} = 2.00 \times 10^{-3} \times \dfrac{500}{300} = 3.33 \times 10^{-3}\,\text{m}^3$

   $W = P\Delta V = 8.00 \times 10^4 \times (3.33 - 2.00) \times 10^{-3} = 107\,\text{J}$

   So $Q = W + \Delta U = 266\,\text{J}$ (this answer to 3 s.f. results from adding unrounded values for W and ΔQ).

B The particulate nature of matter

When the state of an ideal monatomic gas changes from (P_1, V_1) to (P_2, V_2) in an adiabatic change, $PV^{\frac{5}{3}}$ = constant. Therefore $P_1 V_1^{\frac{5}{3}} = P_2 V_2^{\frac{5}{3}}$.

This is why the gradient of an adiabatic change on a PV diagram is steeper for the same gas than when it undergoes an isothermal change (for which PV = constant).

(The exponent is different when the gas has more than one atom in the molecule.)

Example 2

An ideal monatomic gas is in an expansion pump at an initial pressure of 100 kPa and a temperature of 313 K. When the pump is operated, the gas expands adiabatically to 1.7 times its original volume.

a) Calculate the pressure of the gas in the pump after the expansion.
b) Calculate the temperature of the gas after the expansion.
c) Comment on the change in the temperature of the gas with reference to the first law of thermodynamics.

Solution

a) $PV^{\frac{5}{3}}$ = constant; therefore $P_2 = P_1 \left(\dfrac{V_1}{V_2}\right)^{\frac{5}{3}}$, where $\dfrac{V_1}{V_2} = \dfrac{1}{1.7}$.

Pressure after expansion $= 100 \times \left(\dfrac{1}{1.7}\right)^{\frac{5}{3}} = 41.3$ kPa

b) $\dfrac{PV}{T}$ = constant, therefore $T_2 = \dfrac{P_2}{P_1} \times \dfrac{V_2}{V_1} \times T_1$.

Temperature after expansion $= \dfrac{41.3}{100} \times 1.7 \times 313 = 220$ K

c) The decrease in the temperature is consistent with the first law, because the gas has done positive work when expanding ($W > 0$) and no thermal energy has been transferred to the gas ($Q = 0$), so its internal energy has decreased ($\Delta U = -W < 0$).

In Topic A.3, you saw that any engine has an efficiency η given by $\eta = \dfrac{\text{useful work output}}{\text{input energy}}$.

The Carnot cycle

A heat engine is a device that operates around a cycle of temperature changes. After one whole cycle, the engine must return to its original state.

Carnot gave the first description of a **theoretical** heat engine—this is known as the Carnot cycle. Energy is transferred into the working fluid (a gas in this case) at a high temperature and energy is transferred out to the surroundings at a lower temperature. The remainder of the energy does work on the system.

The Carnot cycle is ideal, reversible and closed (Figure 2).

Energy Q_1 is supplied to the gas trapped in a cylinder by a piston from a hot reservoir that is at a high temperature T_{hot}. As gas expands, the piston will move until the pressure of the gas is the same as atmospheric pressure. The gas has done work under these conditions. However, the gas is to work in a cycle, so it must now go back to its original state. This can only happen when an amount of energy Q_2 is rejected to a cold reservoir at a low temperature T_{cold}.

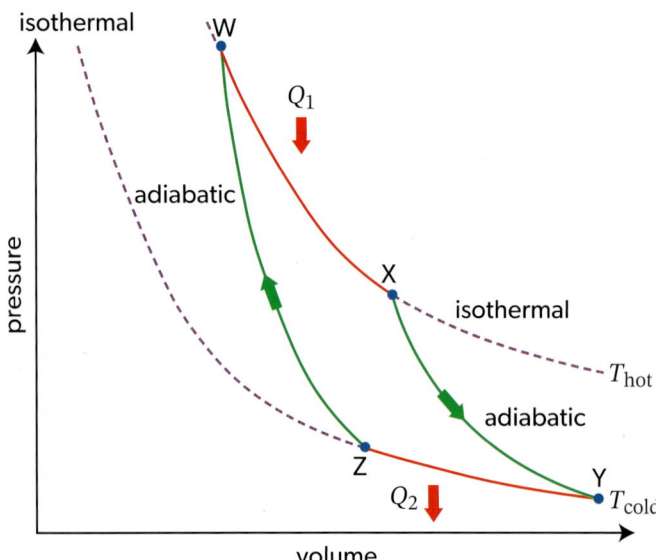

▲ Figure 2 The Carnot cycle

The cycle consists of two isothermal and two adiabatic changes.

- W to X: The gas is at temperature T_{hot} and expands isothermally absorbing energy Q_1. There is no change to the internal energy of the gas because the change is isothermal.
- X to Y: The gas expands adiabatically and the temperature decreases to T_{cold}. The gas loses internal energy and continues to do work on the atmosphere—the piston moves as the gas expands. $Q = 0$.
- Y to Z: The gas is now compressed isothermally to Z with no change in its internal energy. The work done on the gas is ejected as energy.
- Y to W: The gas is compressed, again adiabatically and all the work done on the gas increases its internal energy to return it to T_{hot}.

The net work done by the gas on the surroundings in one cycle is the area enclosed by the curve. This cycle is reversible and it can return to a previous energy state—this means that the cycle must be operated infinitely slowly, which is one reason why the Carnot cycle is theoretical and cannot be achieved in practice.

The **thermal efficiency of the Carnot heat engine** is given by:

$$\eta_{Carnot} = \frac{\text{useful work output}}{\text{input energy}} = \frac{Q_1 - Q_2}{Q_1}$$

This assumes that all the energy $Q_1 - Q_2$ is transferred into useful work and there are no losses to friction and so on (which is another reason why Carnot's engine is only theoretical).

The thermal efficiency for the Carnot cycle can also be written as:

$$\eta_{Carnot} = \frac{T_{hot} - T_{cold}}{T_{hot}} = 1 - \frac{T_{cold}}{T_{hot}}$$

Example 3

The table shows some measurements made on an experimental heat engine.

Temperature of heat source	830 °C
Temperature of cooling system	17 °C
Heat energy supplied per second	78 J
Power output of heat engine	15 W

a) Calculate the maximum possible efficiency of an engine operating between these temperatures.

b) Suggest whether the actual efficiency of the heat engine approaches your answer to part (a).

Solution

a) $\eta = 1 - \dfrac{T_{cold}}{T_{hot}} = 1 - \dfrac{290}{1103} = 74\%$

b) $\eta = \dfrac{P_{out}}{P_{in}} = \dfrac{15}{78} = 19\%$

The engine is significantly less efficient than the maximum theoretical value.

Transferring energy into work

The first law of thermodynamics equates work and energy transfer but says nothing about whether the transfer can occur.

The second law of thermodynamics defines the situations in which energy can be transferred into work. The second law applies to isolated systems (defined at the beginning of this topic).

There are a number of ways to state the second law. Three of these are required in the IB Diploma Programme physics course: the Clausius statement, the Kelvin statement and the entropy formulation.

- The Clausius statement of the second law: Energy cannot flow spontaneously from an object at a low temperature to an object at a higher temperature without external work being done on the system.

Assessment tip

When sketching PV graphs, make sure that the relative gradients of the isothermal and adiabatic changes are correct (see Figure 2).

Remember that the area underneath a PV graph is the energy transferred as work. This can be work done on the gas or work done by the gas, depending on the direction of the state change.

When the gas is taken around a complete cycle (that is, it ends in the same state as it began) the work transferred is the area **inside** the cycle on the graph. The direction of energy transfer depends on the direction of the cycle.

When the gas expands, it is doing work. When it is compressed, work is done on it. Thus, moving to the right on the PV graph means that the gas is doing work. $P \times V$ has the units of energy.

Take care with PV graphs. In particular, when you are working out energy transfers, allow for the presence of a false origin on the V-axis.

- The Kelvin (Joule–Kelvin) statement of the second law: Energy cannot be extracted from a reservoir and transferred entirely into work.
- The entropy formulation statement (due to Boltzmann): For any real process the entropy of the universe must not decrease.

Entropy is defined in terms of the energy ΔQ absorbed by a system and the temperature T in kelvin at which the energy transfer occurs. Entropy is sometimes regarded as a measure of the disorder in a system. Any real process tends to increase disorder.

> In the energy formulation statement, the change in entropy ΔS of the system is $\Delta S = \dfrac{\Delta Q}{T}$. The unit of entropy is J K^{-1}.
>
> For this equation to hold, the energy must be transferred isothermally. Or ΔQ must be small enough for T to be approximately constant.

In a crystal of common salt (sodium chloride), the salt atoms are highly ordered. Dissolve the crystal in water and there are now many possible arrangements for the ions in the solution. The entropy of the system has increased. To restore the order to the crystal, the water must be evaporated, either naturally or by heating. This process of decreasing the entropy of the dissolved salt back to that of the solid will cause other entropy increases in the universe—the second law tells us that the total of these changes always increases.

This is a macroscopic interpretation of entropy. However, it can be described using a microscopic viewpoint too.

One way to understand entropy on a microscopic level is to play a game. The game is played with six counters labelled 1 to 6, a six-sided die and two boxes. When the die is thrown, the number that is uppermost dictates which counter is moved into the other box. The first few moves are shown in Figure 3.

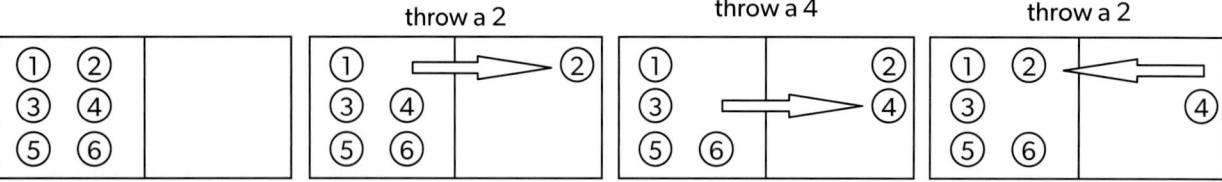

▲ Figure 3 This game explores the microstates of a system

> **Entropy** is defined using **microstates** by $S = k_B \ln \Omega$, where Ω is the number of different microstates of the system in a particular configuration (the **macrostate**).
>
> Note that this definition is for S not ΔS.

Each individual possible arrangement is a microstate. For this game, one possible microstate is where counters 1, 3 and 5 are in the left-hand box and 2, 4 and 6 are in the right-hand box. Another microstate is where counters 1, 2 and 3 are in the left-hand box and 4, 5 and 6 are in the right-hand box. These are different microstates because the actual counters are different. However, looked at purely in numerical terms, there are three counters in each box in both microstates. The case where there are three counters in each box is a macrostate. The macrostate with three counters in each box is the most common macrostate for this system since there are 20 ways to achieve it. However, there is only one microstate where there are no counters in the left-hand box.

This simple game models the expansion of a gas when it expands in a vacuum to double its volume. When N molecules expand from a volume of V_1 to V_2, then the ratio of arrangements $\dfrac{\Omega_2}{\Omega_1} = \left(\dfrac{V_2}{V_1}\right)^N$.

Because $\Omega_1 = 1$ (only one arrangement possible), then $\Omega_2 = \left(\dfrac{V_2}{V_1}\right)^N$ which leads to $\Delta S = k_B \times N \times \dfrac{\Delta V}{V}$. By comparing this with $PV = Nk_BT$, you can see that $P\Delta V = T\Delta S$, so that $\Delta S = \dfrac{\Delta Q}{T}$.

The two definitions of entropy are equivalent.

Example 4

0.40 kg of ice at a temperature of 0 °C is placed in a room with air temperature 20 °C. The system of the ice and the room is isolated.

The ice melts completely. The specific latent heat of fusion of ice is 330 kJ kg⁻¹.

a) Calculate the change in the entropy of:

  (i) the ice

  (ii) the system of the ice and the air in the room.

b) Comment on the results of part (a) with reference to the second law of thermodynamics.

Solution

a) (i) The energy transferred to the ice is $0.40 \times 330 = 132$ kJ.

$$\Delta S_{ice} = \dfrac{132 \times 10^3}{273} = 484 \text{ J K}^{-1}$$

  (ii) The energy gained by the ice during melting is equal to the energy lost by the air. We assume that the room is large enough for its air temperature to remain approximately constant, so that the entropy change of the air can be modelled by:

$$\Delta S_{air} = \dfrac{-132 \times 10^3}{273 + 20} = -451 \text{ J K}^{-1}$$

  Energy is removed from the air, so its entropy change is negative.

  Change in entropy of the system $\Delta S = \Delta S_{ice} + \Delta S_{air} = 484 - 451 = 33 \text{ J K}^{-1}$

b) The second law of thermodynamics applies to isolated systems, but not to their individual parts such as the ice or the air alone. The entropy of the air has decreased, but the entropy of the ice has increased by a greater amount, so that the net entropy of the isolated system of the ice and the room has also increased. This is consistent with the second law of thermodynamics and suggests that melting is an irreversible process.

B The particulate nature of matter

Sample student answer

A heat engine operates on the cycle shown in the pressure–volume diagram. The cycle consists of an isothermal expansion AB, an isovolumetric change BC and an adiabatic compression CA. The volume at B is double the volume at A. The gas is an ideal monatomic gas.

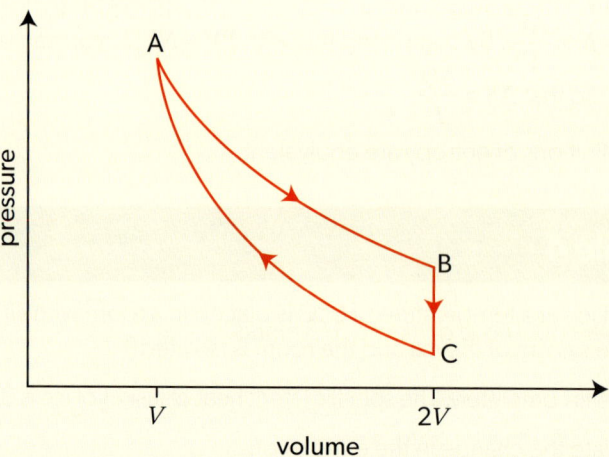

At A the pressure of the gas is 4.00×10^6 Pa, the temperature is 612 K and the volume is 1.50×10^{-4} m³. The work done by the gas during the isothermal expansion is 416 J.

a) (i) Justify why the thermal energy supplied during the expansion AB is 416 J. [1]

This answer could have achieved 1/1 marks:

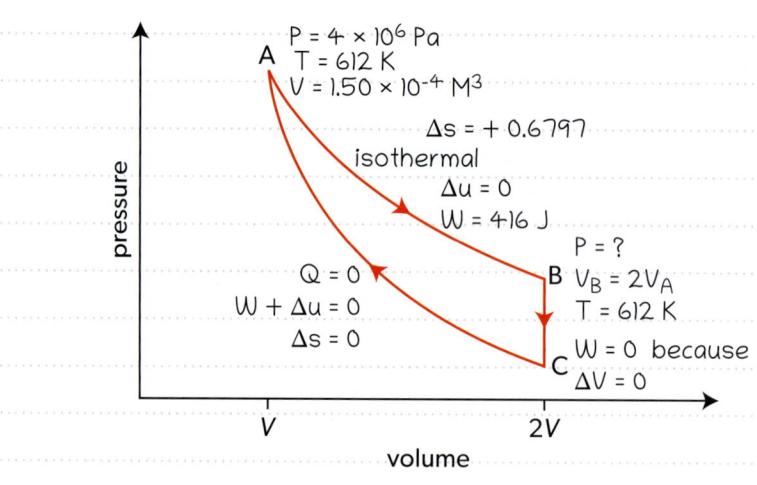

▲ The answer correctly identifies that there is no change in U and therefore $Q = W$ so that Q is also 416 J.

Because it is isothermal, $\Delta T = 0$ and $\Delta u = \frac{3}{2}nR\Delta T$
∴ $\Delta u = 0$ $Q = \Delta u + W$ and if $\Delta u = 0$, $Q = W$.
In this case $W = 416$ J, so $Q = 416$ J.

The temperature of the gas at C is 386 K.

(ii) Show that the thermal energy removed from the gas for the change BC is approximately 330 J. [2]

This answer could have achieved 2/2 marks:

▲ BC is at constant volume so no change in W for this part of the cycle. $Q = \Delta U$ and a calculation using $\frac{3}{2}nR\Delta T$ confirms the result. This is a clear answer.

$\Delta v = 0$, $W = 0$ so $Q = \Delta u$ $nR = \frac{PV}{T} = 0.980$ $\Delta u = \frac{3}{2}nR\Delta T$
$\Delta T = 386 - 612 = -226$ K
$\Delta u = \frac{3}{2} \times 0.98 \times (-226) = -332.227 = -330$ J
$Q = -330$ J, so 330 J is taken out of the gas.

(iii) Determine the efficiency of the heat engine. [2]

This answer could have achieved 2/2 marks:

$e = \dfrac{W}{Q}$ $Q_{in} = 416\,J$

Useful work done = $416 - \Delta u_{AC}$ $\Delta u_{AC} = -\Delta u_{BC} = +330$

$\therefore W = 416 - 330 = 86\,J$

$e = \dfrac{W}{Q} = \dfrac{86}{416} = 0.207$

0.207 or 20.7%

▲ Once again, clear and well presented work. It should be your aim to achieve this sort of quality in your examination answers.

b) State and explain at which point in the cycle ABCA the entropy of the gas is the largest. [3]

This answer could have achieved 3/3 marks:

B would have the highest entropy. Entropy difference is calculated using $\Delta S = \dfrac{\Delta Q}{\Delta T}$ and for the change AB, $\Delta S_{AB} = +0.680$. This means that the system gains entropy from A → B. A and C have the same entropy because $\Delta Q = 0$, as the change is adiabatic. Therefore, if the entropy at B is higher than at A, it will also be higher than that at C. Therefore B has the largest entropy.

▲ The answer is stated clearly and there is a chain of argument that supports the answer. Again, a model answer.

Practice problems

Problem 1

A monatomic ideal gas is enclosed in a cylinder of initial volume $6.0 \times 10^{-4}\,m^3$. The gas is initially at a pressure of 110 kPa and a temperature of 290 K.

a) Calculate the quantity of gas in the cylinder. State an appropriate unit for your answer.

b) The gas is compressed quickly by a piston to a pressure of 210 kPa and a volume of $4.0 \times 10^{-4}\,m^3$.

   (i) Suggest, with a calculation, whether the gas is compressed isothermally.

   (ii) Explain why the compression may be adiabatic.

   (iii) Estimate the work done on the gas, assuming that the compression is adiabatic.

c) The compression is repeated very slowly.

   Discuss the entropy change that takes place in the cylinder and its surroundings as the air is compressed.

Problem 2

A quantity of 0.16 mol of a monatomic ideal gas expands at a constant pressure of 1.2×10^5 Pa from an initial volume of $4.0 \times 10^{-3}\,m^3$ to a volume of $6.0 \times 10^{-3}\,m^3$.

a) Calculate:

   (i) the initial temperature of the gas

   (ii) the work done by the gas during the expansion.

b) Determine the energy transferred to the gas during the expansion.

The gas is now compressed isothermally to the original volume.

c) (i) Explain why energy is removed from the gas during the compression.

   (ii) Calculate the final pressure of the gas.

Problem 3

The diagram shows a Carnot cycle ABCDA for an ideal gas.

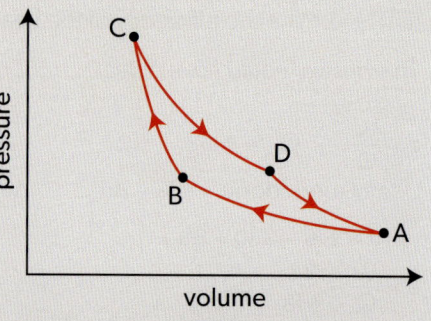

a) State during which part of the cycle the energy is transferred to the gas.

b) Explain the change in entropy of the gas during the compression:

   (i) AB

   (ii) BC.

The cycle models a heat engine. The efficiency of the cycle is 0.60. The temperature of the gas at A is 360 K.

c) Calculate the temperature of the gas at C.

The engine rejects thermal energy into a cold reservoir at a rate of 100 J per cycle.

d) Determine:

   (i) the work done by the engine during one cycle

   (ii) the energy transferred to the gas from a hot reservoir during one cycle.

Problem 4

A heat engine whose working substance is a monatomic ideal gas operates on the cycle ABCA shown in the diagram.

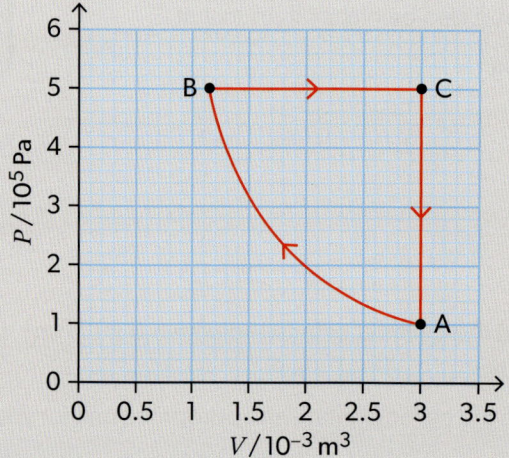

The change AB is an adiabatic compression from an initial pressure of 1.00×10^5 Pa to a pressure of 5.00×10^5 Pa. The volume of the gas at A is 3.00×10^{-3} m^3.

a) Calculate the volume at B.

The work done by the gas during the compression AB is -410 J.

b) State the change in the internal energy of the gas during the compression AB.

The internal energy of the gas increases by a further 1390 J during the isobaric expansion BC.

c) Calculate, for the change BC:

   (i) the work done by the gas

   (ii) the energy transferred to the gas.

d) (i) Explain why the energy transferred from the gas during the change CA is less than the value calculated in part (c)(ii).

   (ii) Determine the efficiency of the cycle.

e) The minimum temperature of the gas during the cycle is 300 K and the maximum temperature is 1500 K. Calculate the efficiency of the Carnot cycle operating between these temperatures.

B.5 Current and circuits

You must know:

✔ that a flow of charge carriers is responsible for an electric current I and that $I = \frac{\Delta q}{\Delta t}$

✔ that direct current (dc) is a flow of charge carriers in one direction in an electrical circuit

✔ that electrical potential difference V is the work done per unit positive charge when the positive charge is moved and that $V = \frac{W}{q}$

✔ Ohm's law and the definitions of resistance and resistivity

✔ that resistors are subject to heating effects and that a metal conductor at a constant temperature is an ohmic device

✔ that chemical cells and solar cells provide an energy source for a circuit

✔ that an electrical cell is characterized by its emf ε and its internal resistance r

✔ that variable resistors can include devices such as thermistors, light-dependent resistors (LDRs) and potentiometers.

You should be able to:

✔ distinguish between emf and electrical potential difference (pd)

✔ draw and interpret circuit diagrams to represent the arrangement of electrical components in a circuit, including the use of cells, batteries, meters (ideal and non-ideal), variable and fixed resistors, and other devices

✔ describe the properties of electrical conductors and insulators in terms of the movement of charge carriers, and explain the microscopic origins of electrical resistance

✔ draw and interpret ohmic and non-ohmic behaviour using V–I characteristic graphs

✔ solve problems using the equations $R = \frac{V}{I}$, $\rho = \frac{RA}{L}$, $P = IV = I^2R = \frac{V^2}{R}$ and $\varepsilon = I(R + r)$ to calculate electrical quantities in circuits

✔ combine resistors arranged in series and parallel

✔ suggest the advantages and disadvantages of different sources of electrical energy.

Electrical current I and charge q

Electric currents can exist in solids, liquids and gases. The movement of charge in a conductor or through a vacuum is an electric current.

Conduction in solids is usually due to the movement of one type of charge carrier. Solid conductors, which include metals and materials such as carbon, contain fixed positive ions. These positive ions make up the bulk of the material. These ions release electrons to a "sea" of free electrons as part of the chemical bonding.

When charge flows in a circuit, kinetic energy is transferred to the electrons from the energy source (an electric chemical or solar cell, a battery or other power supply). The electrons then collide with the fixed positive ions. Because the charge carriers make repeated collisions with the ions, the carriers gain and lose kinetic energy as they travel.

Energy is transferred from the cell to the conductor when charge flows in the component. In a resistor, this appears as thermal energy.

The current depends on the mobility of charge carriers within the material. Conductors have a large mobility (many charge carriers each with a high random speed), whereas insulators have a small mobility (few charge carriers each moving slowly).

> The microscopic structure and behaviour of materials are described in Topics B.1 and B.3, and in Topics E.1 and E.3.
>
> An electric current can also cause magnetic effects (Topic D.3, Practice problem 1) and chemical effects as well as heating effects.

Assessment tip

Always remember that it is the charge that flows, **not** the current. There is a current (of flowing charge) in a conductor, just as the current in a river consists of the movement of flowing water.

> The mobility of charge carrier links to ideas of speed in Topic A.1. It is possible to show that the current is related to the number density of charge carriers, the charge of each one, the cross-sectional shape of the conductor and average speed of a charge carrier (called the **drift speed**).

> The unit of **electric current** is a fundamental unit in SI. It is the ampere (abbreviated A).
>
> **Electric charge** is defined in terms of the ampere. The current is **one ampere** when **one coulomb** of charge flows past a point in **one second**. The coulomb is abbreviated to C. In fundamental units, this is A s. In equation terms, $I = \dfrac{\Delta q}{\Delta t}$, where I is the electric current, q is the charge that flows and t is the time. When the current is constant, $q = It$.

Charge carriers move in an electric field that is created by the power supply.

Electric charge exists in two forms: **positive** and **negative**. An electric charge is associated with an electric field. An electric charge will accelerate in the presence of an electric field (Topic D.2). A moving electric charge will also accelerate in the presence of a magnetic field (Topic D.3).

As a result of the interaction between a charge and an electric field, **like charges repel** and **unlike charges attract**. An uncharged object can be attracted to a charged object because charge separation occurs in the uncharged object (Topic D.2).

As you will see in Theme D, the term "potential difference" can also be used in gravitation and electric field theory. However, in purely electrical theory the word "electric" in electric potential difference is usually dropped, and physicists and engineers simply talk about potential difference. Similarly, they will often abbreviate the term to "pd". You will see this usage throughout this book.

Electric potential difference V

When charges move, energy is transferred to them from the power supply. This energy is measured in terms of the energy transferred per unit of charge and is known as **electric potential difference**.

> The unit of **electric potential difference** V is the volt, abbreviated to V. 1 V is the potential difference that results in 1 J of work being done when 1 C of charge is moved between two points.
>
> Electric potential difference is often called **voltage** because of its unit. As an equation, $V = \dfrac{W}{q}$ and 1 V is the same as 1 J C^{-1}—the fundamental SI units are kg m^2 s^{-3} A^{-1}.

Voltmeters and ammeters are used to measure and record pd and current in a circuit. In theoretical work and examinations, meters are often assumed to be ideal and to require no energy from the circuit for their operation. In practice, voltmeters and ammeters are non-ideal and have resistance.

An **ideal voltmeter** has an infinite resistance so that no energy is transferred to it from the power supply. A voltmeter is placed in parallel with the component (or components) whose electrical potential difference is being measured.

An **ideal ammeter** has zero resistance, again so that no energy is transferred to it. An ammeter is placed in series with the component or components to measure the current flowing through them.

Real voltmeters always have a resistance less than an infinite value and real ammeters always have a small resistance.

Electrical resistance R

The electrical resistance of a component indicates the difficulty that charges have when moving through the component—or, alternatively, how easily energy can be transferred to the component from the charge carriers in it.

> **Resistance** $R = \dfrac{\text{potential difference across a component}}{\text{current in the component}} = \dfrac{V}{I}$
>
> Resistance is measured in ohms, abbreviated to Ω.
>
> 1 Ω is the resistance of a component that has a potential difference of 1 V across it when the current through it is 1 A. The resistance of a component depends on its size, shape and the material from which it is made. The resistance often varies with temperature.

Experiments show that resistance R of an object with a uniform cross-section is:
- proportional to length L across which the potential is applied
- proportional to the **resistivity** of the material ρ—resistivity is a **shape-independent** quantity
- inversely proportional to cross-sectional area A.

These lead to the equation for resistivity, $\rho = \dfrac{RA}{L}$.

The unit of resistivity is the ohm metre ($\Omega\,m$). All samples of the same pure material have the same resistivity at a given temperature.

> **Approaches to learning**
>
> Try to link themes and topics as you revise. For example, make a list of as many shape-independent quantities as you can from the whole course. Begin with electrical resistivity and density.

Example 1

The "lead" in a pencil is a conductor. It is made from graphite–clay mix that has a resistivity of $4.0\times10^{-3}\,\Omega\,m$.

Determine the resistance of a pencil "lead" of length 80 mm and diameter 1.4 mm.

Solution

$$R = \frac{\rho L}{A} = \frac{4.0\times10^{-3}\times 80\times 10^{-3}}{\pi\times(0.70\times 10^{-3})^2} = 0.21\,k\Omega \text{ (to 2 s.f.)}$$

Electrical power

Electrical power dissipated is the rate at which energy is transferred.

The potential difference V across the component is the energy W transferred per unit charge.

The power P dissipated in a component in time t is $\dfrac{W}{t}$. Also, $V = \dfrac{W}{q}$ and $q = It$.

Therefore, $V = \dfrac{W}{It}$ and $P = IV = \dfrac{W}{t}$.

Using the definition of resistance, $R = \dfrac{V}{I}$ leads to:

$$P = IV = I^2R = \frac{V^2}{R}$$

A metal conductor at a constant temperature obeys Ohm's law and is said to be an ohmic device or an ohmic resistor. Any component that obeys Ohm's law is called an ohmic device.

> **Ohm's law** states that the potential difference across a conductor is directly proportional to the current in the conductor provided that the physical conditions remain constant: $V \propto I$.

A metal at a constant temperature is considered to be an ohmic conductor.

Graphs of the variation of I with V for a component are called I–V characteristic graphs. They can be used to show whether a conductor is ohmic or non-ohmic.

Figure 1 shows the I–V characteristics for three different conductors.

> **Assessment tip**
>
> These equations are important—you should try to memorize them to save time looking them up during an examination.

> **Assessment tip**
>
> Take care with symbols. The IB Diploma Programme physics course uses W for electrical energy, but this is not to be confused with W (watts), the unit for power.

> **Assessment tip**
>
> The characteristics can be plotted with V on the x- or the y-axis. Read the axes labels carefully. Be prepared to draw either version.
>
> Remember to construct all four quadrants of the graph, even though the line continues through the origin without change in gradient for an ohmic conductor.

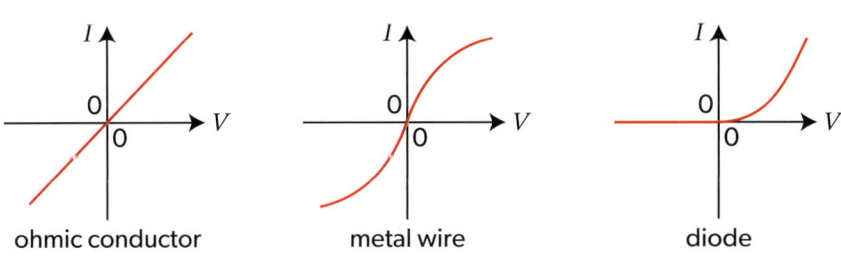

▲ Figure 1 I–V characteristic graphs

The temperature of a metal wire increases with current in the wire and this increases the resistance. The I–V characteristic shape for the wire changes from ohmic (straight-line) behaviour to a curve for which the value $\frac{V}{I}$ increases with increasing I.

A diode is an electrical component that only allows charge to flow through it in one direction. For negative values of V, the current is zero. Diodes only begin to conduct when V exceeds a certain value.

Example 2

The heating element of an electric heater is made of a metal wire that is an ohmic resistor. The heater is designed for an output power of 500 W when used with a 110 V supply.

a) Calculate:

   (i) the resistance of the wire

   (ii) the current in the wire.

b) The heater is accidentally connected to a 220 V supply. Calculate the power output of the heater.

Solution

a) (i) From $P = \frac{V^2}{R}$: $R = \frac{V^2}{P} = \frac{110^2}{500} = 24\,\Omega$

   (ii) $I = \frac{P}{V} = \frac{500}{110} = 4.5\,\text{A}$

b) The potential difference across the wire doubles, and since the wire has ohmic properties, its resistance remains unchanged. From $P = \frac{V^2}{R}$, the power increases by a factor four and is now 2000 W.

Assessment tip

The parallel equation for two resistors can also be written as $R = \frac{R_1 \times R_2}{R_1 + R_2}$.

When two identical resistors are placed in parallel, their combined resistance is half of the resistance of one of them.

Combining resistors

When resistors are joined in series or parallel, you can calculate the resistance of the combination using two rules.

In series: add the values:

$R = R_1 + R_2 + R_3 + \cdots$

In parallel: add the reciprocals of the values to form the reciprocal of the combined resistance:

$\frac{1}{R} = \frac{1}{R_1} + \frac{1}{R_2} + \frac{1}{R_3} + \cdots$

When there is a combination of series and parallel elements, break the network down into smaller sections that contain only series **or** parallel components. Then work out the resistance of each small section. Gradually combine the resistances of the sections until the whole circuit has been calculated.

Example 3

An electrical cable consists of eight parallel strands of copper wire, each of diameter 2.5 mm.

The resistivity of copper is $1.6 \times 10^{-8}\,\Omega\,\text{m}$.

The cable carries a current of 20 A.

a) Calculate:

   (i) the cross-sectional area of a **single strand** of copper wire

   (ii) the resistance of a 0.10 km length of the eight-strand cable.

b) Calculate the potential difference between the ends of the eight-strand cable.

c) State **one** advantage of using a stranded cable rather than a solid core cable with copper of the same total cross-sectional area.

Solution

a) (i) Cross-sectional area on one strand = $\pi \times (1.25 \times 10^{-3})^2 = 4.91 \times 10^{-6}\,m^2$

  (ii) Resistance of one strand = $\dfrac{1.6 \times 10^{-8} \times 100}{4.91 \times 10^{-6}} = 0.326\,\Omega$

  The resistance of the cable is one-eighth of this because there are eight strands in parallel, so $0.0407\,\Omega$.

b) $V = IR = 20 \times 0.0407 = 0.81\,V$

c) Possible answers include: the flexibility of a cable compared with one strand; the cable will still conduct even if one strand breaks; and the larger surface area gives a better heat dissipation.

Nature of science

Is Ohm's law a law at all? The term "law" can be used when scientists predict the behaviour of phenomena without explaining them. Clearly, Ohm's law is only partially true.

Example 4

Three ohmic resistors $20\,\Omega$, $60\,\Omega$ and $100\,\Omega$ are placed as shown in a circuit with a cell that supplies a potential difference of 12 V.

a) Calculate:

  (i) the overall resistance of the circuit

  (ii) the total power dissipated in the circuit

  (iii) the reading of the ammeter, assuming that it is ideal.

b) Explain how your answer in part (a)(iii) will change when the ammeter has a constant non-zero resistance.

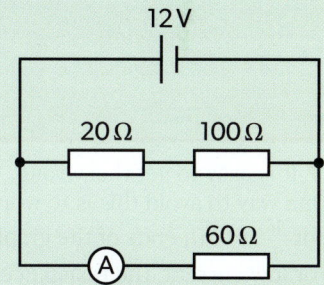

Solution

a) (i) The $20\,\Omega$ and $100\,\Omega$ resistors are in series with each other and are equivalent to a single resistor of $20\,\Omega + 100\,\Omega = 120\,\Omega$, which in turn is connected in parallel with the $60\,\Omega$ resistor. The combined resistance R of the circuit can be calculated from:

$$\frac{1}{R} = \frac{1}{120} + \frac{1}{60} = \frac{3}{120}$$

$$R = \frac{120}{3} = 40\,\Omega$$

  (ii) $P = \dfrac{V^2}{R} = \dfrac{12^2}{40} = 3.6\,W$

  (iii) The current in the ammeter is the same as in the $60\,\Omega$ resistor. With an ideal ammeter, the potential difference across the resistor is equal to the emf of the cell, so 12 V. The ammeter shows:

$$I = \frac{V}{R} = \frac{12}{60} = 0.20\,A$$

b) There is now a non-zero potential difference across the ammeter; hence the potential difference across the resistor will be less than 12 V (so that the combined potential difference across the series connection of the ammeter and the resistor is still equal to the emf of the cell). The reading of the ammeter will be less than 0.20 A.

Potentiometer circuits (sometimes called potential dividers) have advantages over variable resistors for controlling current. Figure 2 shows both arrangements used to determine the I–V characteristic for a metal wire.

B The particulate nature of matter

A potentiometer will usually vary the potential difference across a component X from zero to a maximum. A variable resistor will usually vary the potential difference from a maximum down to a non-zero minimum. The values of the maximum and minimum depend on the ratio of the resistances of the component and the variable resistor.

> **Assessment tip**
>
> Be prepared to answer both qualitative and quantitative questions about potential dividers. You need to be able to show that, for the circuit below, $V_1 = \dfrac{R_1}{R_1 + R_2} V_{supply}$ and $V_2 = \dfrac{R_2}{R_1 + R_2} V_{supply}$.
>
>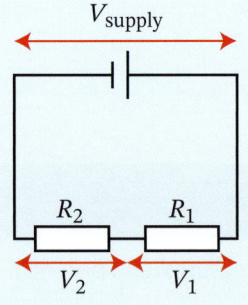

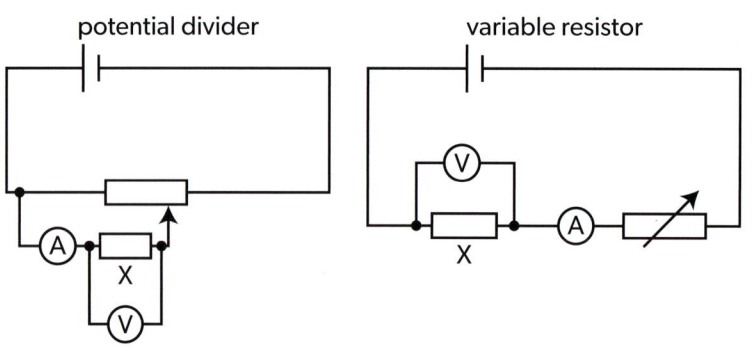

▲ Figure 2 Determining the I–V characteristic for component X

Sample student answer

a) The graph shows how current I varies with potential difference V for a resistor R and a non-ohmic component T.

(i) State how the resistance of T varies with the current going through T. [1]

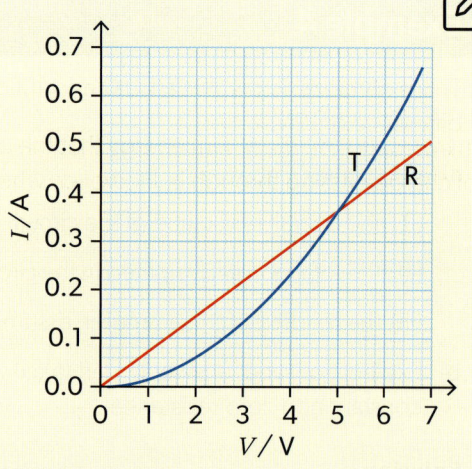

This answer could have achieved 0/1 marks:

The resistance of T increases at a decreasing rate with the current.

▼ It is easy to get this wrong. One way to avoid this is to work out $\dfrac{V}{I}$ at both ends of the graph. For low current, the resistance is about 50 Ω. At high current, it is about 10 Ω. This would only take a few seconds on a calculator to work out.

(ii) Deduce, without a numerical calculation, whether R or T has the greater resistance at $I = 0.40$ A. [2]

This answer could have achieved 2/2 marks:

R has greater resistant at $I = 0.40$ A. As shown on the graph, R has greater voltage than T when $I = 0.40$ A. According to the formula, resistance = $\dfrac{V}{I}$ so when V is greater it will have a greater resistance, and hence R has greater resistance.

▲ This time, a reference is made to $\dfrac{V}{I}$. The fact that current is the same for both components, but V is greater for one component, gives the answer directly.

b) Components R and T are placed in a circuit. Both meters are ideal. Slider Z of the potentiometer is moved from Y to X.

(i) State what happens to the magnitude of the current in the ammeter. [1]

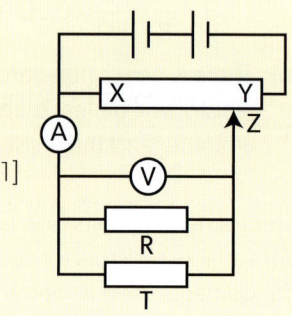

This answer could have achieved 0/1 marks:

The sum of R and T's current.

▼ The focus should be on the current in the ammeter. This is a potential divider arrangement, so moving the slider to X reduces the potential difference at Z. This will also reduce the current in the ammeter (because $V = IR$, and the resistance of R and T does not change).

(ii) Estimate, with an explanation, the voltmeter reading when the ammeter reads 0.20 A. [2]

This answer could have achieved 0/2 marks:

$V = IR \quad I = 0.20\ A \quad \dfrac{1}{0.3} + \dfrac{1}{0.7} = \dfrac{1}{R}$

$R = \dfrac{0.06}{0.2} = 0.3\ (\Omega)$

$V = 0.06\ (T) \quad R = 0.21 \quad V = 0.21 \times 0.2 = 0.042\ (V)$

$V = 0.14\ (R) \quad R = \dfrac{0.14}{0.2} = 0.7\ (\Omega)$

▼ This approach is going to be difficult (although not impossible). Answering this question is straightforward when you remember that the current of 0.2 A must be the sum of the currents in R and T. Because they are in parallel, the potential difference across each of them is the same.

Looking at the graph, you can see that, when $V = 2.0\ V$, the two currents are about 0.06 A and 0.14 A, respectively. These add to 0.20 A and, therefore, 2.0 V is the answer.

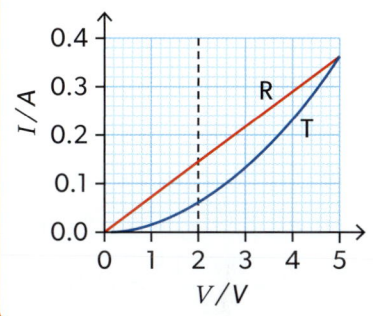

Chemical and solar cells

Society needs portable and compact power supplies—this drives research into innovative chemical cells, for example, modern lithium batteries (a battery is a collection of cells).

When a charged cell is not transferring energy, the potential difference across the cell terminals is equivalent to the electromotive force (emf) ε of the cell.

ε represents the maximum energy that the cell can deliver for each coulomb of charge passing through it. However, when a real cell is transferring energy it will always give a lower potential difference reading V across its terminals (called the terminal potential difference) because energy is required to move charge through the cell itself. This loss is due to the internal resistance r of the cell.

A cell with internal resistance can be represented as the ideal cell plus a resistance enclosed in dotted lines (see Figure 3). For this circuit, $V = \varepsilon - Ir$ or $\varepsilon = I(R + r)$.

The equation $V = \varepsilon - Ir$ suggests a graphical method for determining both r (from the negative of the gradient of the V–I graph) and ε (from the intercept on the V-axis) for a real cell. Figure 3 shows a circuit in which a variable resistor is used to control the current. As current increases, the terminal potential difference across the cell drops because a larger current requires more energy to drive charge through the cell.

Approaches to learning

You need to take care when assigning the direction of **conventional current**. Early scientists labelled the charge carriers in metals as positive. We now know that charge carriers in metals are electrons and are negative. Conventional current is due to the flow of **positive** charges in the external parts of the circuit from the positive terminal to the negative terminal of the power supply. Direction rules in Topics D.3 and D.4 rely on this. Unless you are told otherwise, always assume that "current" refers to the conventional current.

Topic B.5 involves only **direct current** (dc). The direct current is always in one, unchanging direction.

Another type of current alternates—that is, it cycles between charge flowing in one direction and the other. Alternating current arises when a conducting coil is rotated in a magnetic field (Topic D.4—AHL only).

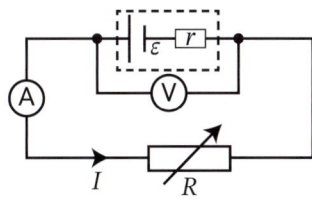

▲ Figure 3 A circuit to determine the internal resistance r and the emf ε of a practical cell

Assessment tip

The terminal potential difference is $V - IR$. This is often required in calculations.

B The particulate nature of matter

Example 5

The circuit in Figure 3 is used to determine the variation of V with I in the circuit. The results are shown in the table.

Voltmeter reading / V	Ammeter reading / A
1.25	0.10
0.70	0.20
0.43	0.25
0.15	0.30

a) Plot a suitable graph from these data.

b) Use your graph to determine:

   (i) the emf of the cell

   (ii) the internal resistance of the cell.

Solution

a) See the graph.

b) (i) Extrapolating the line to the y-axis gives the emf (1.8 V).

   (ii) r is the negative of the gradient:

   $$r = -\frac{0.7 - 1.8}{0.2 - 0} = 5.5\,\Omega$$

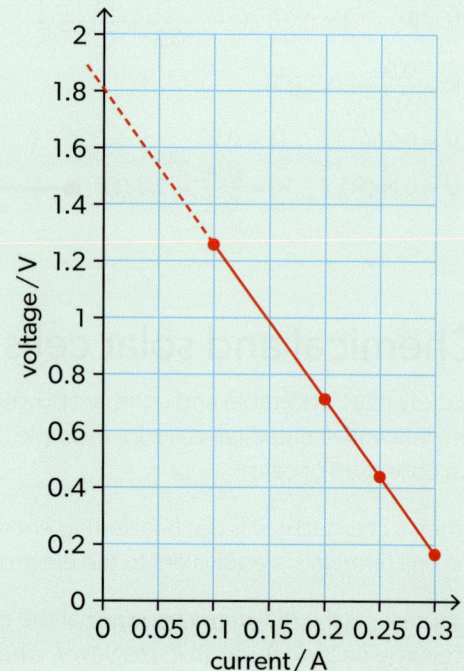

Sample student answer

A student adjusts the variable resistor and takes readings from the ammeter and voltmeter. The graph shows the variation of the voltmeter reading V with the ammeter reading I.

Use the graph to determine:

a) the electromotive force (emf) of the cell [1]

This answer could have achieved 0/1 marks:

> 9V

▼ You must clearly show the examiner what you are doing when the command is "determine" (ideally, explain this in words). In this example, the only clue is a dot drawn at the point where the student read the graph. The clue is that there are two lines for the answer—if the examiner simply required the answer, there would have been only one.

b) the internal resistance of the cell. [2]

This answer could have achieved 1/2 marks:

> Negative of the slope as $V = \varepsilon - IR$.
> $\frac{9 - 5.6}{4.2} = 0.81\,\Omega$

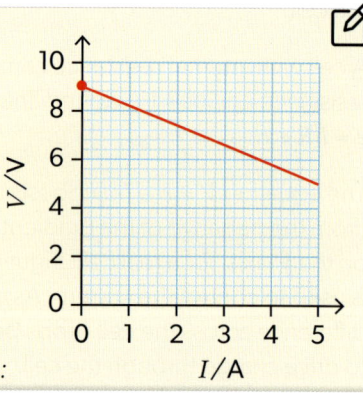

▼ This is not a perfect answer. The student makes it clear that the negative of the (negative) slope is r, that is, a positive value. The gradient calculation should be $\frac{5.6 - 9.0}{4.2}$ and the fact that the gradient is negative must be taken into account.

Practice problems

Problem 1
Three identical lamps are connected in parallel to a cell of emf 9.0 V and negligible internal resistance. Each lamp develops a power of 4.5 W.

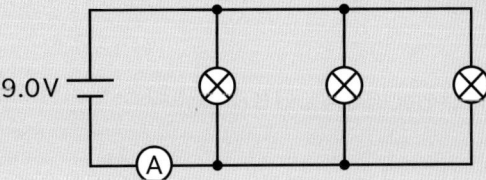

a) Calculate:

   (i) the resistance of each lamp

   (ii) the total resistance of the circuit

   (iii) the reading on the ammeter.

b) The three lamps are now connected in series to the same cell. Explain, without any further calculation, how the combined power developed by the lamps changes compared with the parallel connection.

Problem 2
Three resistors 25 Ω, 40 Ω and 100 Ω are connected as shown to a cell of emf 9.0 V and negligible internal resistance.

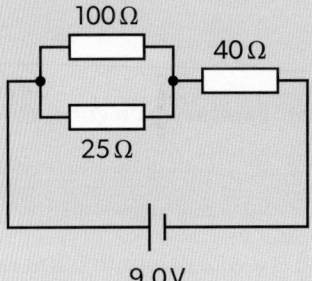

Calculate:

a) the total resistance of the circuit

b) the potential difference across the 100 Ω resistor

c) the current in the 40 Ω resistor.

Problem 3
A variable resistor is connected to a cell as shown. The ammeter and the voltmeter are ideal.

A student investigates how the reading V of the voltmeter varies with the resistance R of the variable resistor. The table shows some of the data collected by the student.

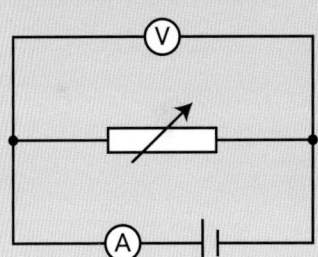

R/Ω	V/V
10.0	5.22
20.0	5.58

a) Calculate the reading of the ammeter when $R = 20.0\,\Omega$.

b) Show that the internal resistance of the cell is about 1.5 Ω.

c) Determine the emf of the cell.

d) The student suggests that the relationship between V and R is linear. Comment on the student's suggestion.

Problem 4
A bicycle is powered by an electric motor that transfers energy from a rechargeable battery. When fully charged, the 12 V battery can deliver a current of 14 A for 30 minutes before full discharge.

a) Determine the charge flowing through the battery during one discharge cycle.

b) Calculate the energy available from the battery.

When the cyclist goes uphill, the driving force on the bicycle is entirely provided by the motor. The mass of the cyclist and the bicycle is 75 kg. The overall efficiency of energy transfer in this system is 0.55.

c) Calculate the maximum height that the cyclist can climb before the battery needs recharging.

Problem 5

A 6.0 V cell and a 30 Ω resistor are connected in series. The cell has negligible internal resistance.

a) Calculate the current in the cell.

b) An arrangement of a 30 Ω resistor and a 60 Ω resistor in parallel is connected in series with the original 30 Ω resistor. Calculate the current in the cell.

Problem 6

The I–V characteristic graph for two conductors A and B is shown.

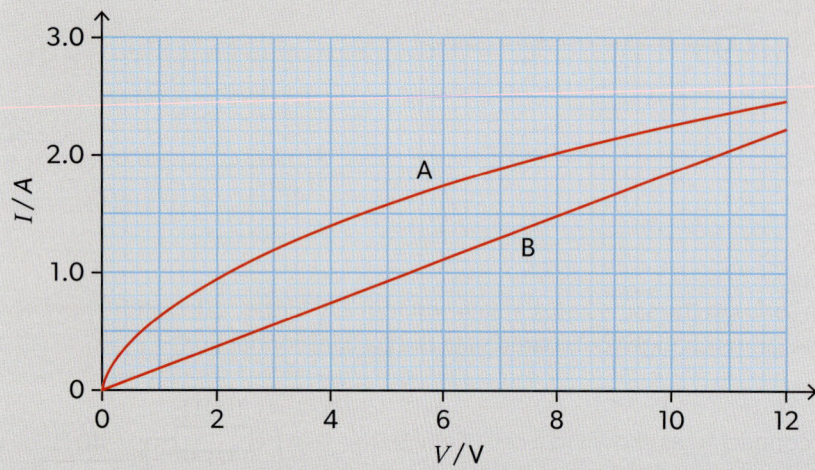

a) Explain which conductor is ohmic.

b) (i) Calculate the resistance of the conductor A when $V = 1$ V and $V = 10$ V.

   (ii) A is a lamp filament. Explain why the values of resistance in part (i) are different.

B is a wire of length 0.8 m with a uniform cross-sectional area of 6.8×10^{-8} m².

c) Determine the resistivity of B.

C Wave behaviour

C.1 Simple harmonic motion

You must know:
- what is meant by an oscillation
- the conditions for simple harmonic motion (shm)
- the defining equation for shm ($a = -\omega^2 x$)
- the definitions of time period, angular frequency, amplitude, displacement and phase difference

Additional higher level:
- the meaning of phase angle when expressed in radian measure.

You should be able to:
- sketch and interpret graphs for simple harmonic motion of displacement–time, velocity–time, acceleration–time and acceleration–displacement
- solve problems using the equation for the time period of a mass–spring system
- solve problems using the equation for the time period of a simple pendulum
- describe the energy changes that take place in one cycle of an oscillation

Additional higher level:
- solve problems involving the displacement, speed, acceleration, total energy and potential energy of a system that is undergoing simple harmonic motion.

A pendulum (a mass swinging at the end of a light string) is an example of an oscillating system. A cycle for this system is the movement of the mass from the rest position at one end of the swing, through to the opposite side and back to the original rest position. Motion from the maximum on one side to the maximum on the other is half a cycle.

Another example of an oscillating system is a mass moving up and down at the end of a spring. A cycle that begins when the mass is at the top of the motion ends when the mass has returned to the top for the first time.

The rest position (which is the middle of the swing for a pendulum and the rest position for the mass–spring) is also known as the **equilibrium position**. This is the position of the system when it is not oscillating.

Simple harmonic motion is an important oscillation to study because other types of oscillation can be modelled as combinations of multiple harmonic oscillators.

Nature of science
The observations of simple harmonic motion go back to at least the time of Galileo Galilei. He is said to have used his pulse to time the slow swings of candelabra in a cathedral. Making accurate, appropriate and repeatable measurements in science and then applying these to develop and validate a model is an essential part of the nature of science.

For **simple harmonic motion** (shm) oscillations:

- the acceleration a of the object is directly proportional to its displacement x: $a \propto -x$
- the defining equation for simple harmonic motion is $a = -\omega^2 x$, where ω is the constant of proportionality known as the **angular frequency**—the quantity ω^2 is always positive as ω is squared
- the vector direction of acceleration is opposite to the displacement x—this is why there is a minus sign in the equation.

Figure 1 shows the relationship between acceleration and displacement.

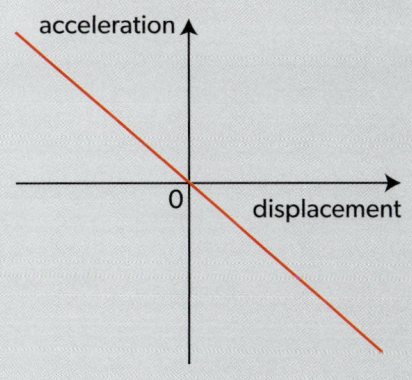

▲ Figure 1 The acceleration–displacement graph for simple harmonic motion

Approaches to learning
When learning about this topic and then answering questions about it, remember that there are two parts to the definition of shm: the direction and magnitude of the acceleration or the restoring force. The direction of the force is always towards the equilibrium position (motion centre) and the magnitude is directly proportional to the distance from the equilibrium position.

C Wave behaviour

Time period T is the time for one cycle of the oscillation. The unit is the second (s).

Frequency f is the number of cycles of the oscillation in one second. The unit is the hertz (Hz or s^{-1}).

Time period and frequency are connected by $f = \dfrac{1}{T}$.

Angular frequency $\omega = \dfrac{2\pi}{T}$, where the system goes through one cycle (2π rad) in time T. The unit is the hertz (Hz or s^{-1}) or rad s^{-1}.

Amplitude x_0 is the maximum displacement of the oscillating object from its equilibrium position. This can be expressed as an angle (in radians) or as a distance.

The **equilibrium position** is the position to which the system returns when it is not oscillating.

The **displacement** x is the distance between the equilibrium position and an instantaneous position and, as displacement is a vector that requires direction, can be positive or negative.

Phase difference is the difference, in degrees or radians, between two oscillations at the same instant in time.

> **Approaches to learning**
>
> Angular frequency ω links to the angular speed covered in Topics A.2 and A.4. In those topics it is called angular speed or angular velocity.

> **Approaches to learning**
>
> Use the links between definitions in wave motion and definitions in kinematics and mechanics to help your understanding of both areas of the subject.

Example 1

A mass, with an equilibrium position at **O** is displaced to point **X** and released from rest. Its motion is simple harmonic.

Identify where the acceleration of the mass is greatest.

Solution

Because acceleration is proportional to the negative of displacement, the greater the distance from **O**, the greater the magnitude of the acceleration. So, acceleration is greatest at **X**.

> **Approaches to learning**
>
> You can help your understanding of both Topic A.2 and Topic C.1 by comparing simple harmonic motion to motion in a circle. Imagine that the shadow of an object moving in a horizontal circle at constant speed is projected onto a vertical plane. The motion of the shadow is close to simple harmonic. You can see similarities in the mathematics of shm and circular motion too.

Example 2

A mass hanging on a spring is pulled vertically downwards through a distance of 0.15 m from its equilibrium position and released. The mass returns to the equilibrium position 0.75 s after release.

a) State the amplitude of the oscillation.

b) Calculate:

   (i) the time period

   (ii) the angular frequency of the oscillation.

Solution

a) The amplitude is the distance from the equilibrium position to the maximum displacement. This is 0.15 m.

b) (i) The mass has travelled one-quarter of a cycle when it reaches the equilibrium position for the first time after release.

   Time period $T = 4 \times 0.75 = 3.0$ s

   (ii) Angular frequency $= \dfrac{2\pi}{T} = \dfrac{2\pi}{3.0} = 2.1$ rad s^{-1}

Figure 2 shows graphs of the variation with time of displacement, velocity and acceleration in simple harmonic motion.

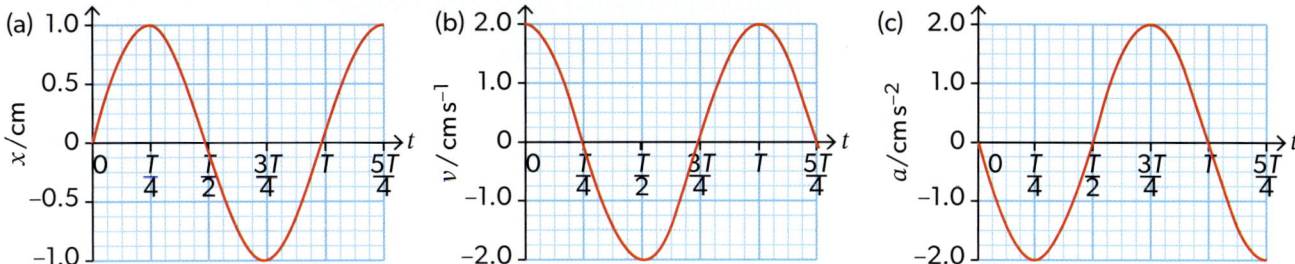

▲ Figure 2 The graphs of x–t, v–t and a–t for simple harmonic motion, where the motion starts at the equilibrium position

Instantaneous velocity at a particular time is equal to the gradient of the displacement–time graph.

Instantaneous acceleration is equal to the gradient of the velocity–time graph.

Mathematically, because $v = \frac{\Delta x}{\Delta t}$ and $a = \frac{\Delta v}{\Delta t}$, where Δ means "change in", this makes the a–t graph the inversion of the x–t graph. This is expected, since $a \propto -x$ (see Figure 1).

- The v–t graph lags the x–t graph by a quarter of one time period. This is equivalent to a 90° phase difference (v–t is 90° out of phase and reaches the maximum $\frac{\pi}{2}$ rad before x–t).
- The a–t graph is 180° out of phase with the x–t graph.

Energy transfers occur throughout the oscillator cycle. For a mass–spring system oscillating on a horizontal (frictionless) surface, the kinetic energy of the mass transfers to and from elastic potential energy in the spring. This energy transfer sustains the oscillation indefinitely when no friction acts.

When a mass–spring system is oscillating vertically, the energy transfers are more complicated because now gravitational potential energy must be taken into account.

Figure 3 shows the energy variations plotted against time for simple harmonic motion.

▲ Figure 3 Energy variation with time for simple harmonic motion. E_k is the kinetic energy and E_p is the potential energy of the system. The total energy is constant. The dashed (blue) curve shows the displacement against time for the system—it takes both positive and negative values unlike the energy curves

> **The terms phase and phase angle** are used in this theme. There are also links to Topics A.2 and A.4, where radian measure and degrees are used: 90° can be written as $\frac{\pi}{2}$ rad and 180° as π rad.

> **Assessment tip**
>
> Phase angle is the term used in the IB Diploma Programme physics course, but you may see the term "phase difference" used in other books. Phase difference is short for "the difference between the phase angles of two oscillations or waves".
>
> For example, when timing an oscillation, a time of half a cycle (half of a time period) is equivalent to a phase difference of π rad (180°).
>
> When a question asks for phase difference, then answers in either radians or degrees are accepted.
>
> An understanding of phase angle is only required for additional higher level.

> **Assessment tip**
>
> Although for standard level you only need to describe energy changes in shm qualitatively, you may find Figure 3 a good way to remember the links between kinetic and potential energies.

C Wave behaviour

The kinetic energy and the potential energy cycles:

- have double the frequency of the motion
- are never negative (because $E_k \propto v^2$)
- have different shapes from the displacement–time curves in Figure 3.

The total energy is constant with time when energy losses are zero. The variation with time of the stored elastic potential energy E_p has the same frequency as the E_k–time graph but is π out of phase with it.

> Spring constant was defined as the force per unit extension of a spring in Topic A.2, page 10.

Two contexts for simple harmonic motion are the **simple pendulum** and the **mass–spring system**.

Simple pendulum

This system consists of a mass suspended by a string of negligible weight and length l that can oscillate in a vertical plane. The gravitational force acting on the mass restores it from a displaced position back towards the equilibrium position where the string is vertical.

This is an example of approximate or **inexact simple harmonic motion** because some assumptions are required for the harmonic motion to be described correctly in mathematical terms.

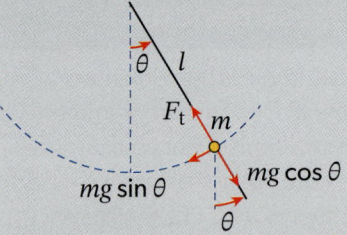

The mass is known as a **pendulum bob**. In this simple case, it is assumed to be a point object.

The force that acts on the bob to return it to the centre (equilibrium position) is $mg \sin\theta$ and this must be equal to ma. As with the mass–spring system, a negative sign is required. θ is measured as positive in the anticlockwise direction (see diagram), whereas the component of the gravitational force is acting clockwise.

Thus $ma = -mg \sin\theta$ and $a = -g \sin\theta$. However, this equation does not yet satisfy the shm conditions because $\sin\theta$ is not proportional to the displacement from the equilibrium position.

However, when θ is small (<10°, so that $\sin\theta \approx \theta$ in radians), you can write $\theta = \dfrac{x}{l}$, where x is the distance of the bob from the equilibrium point. This leads to $a = -\dfrac{g}{l}x$. The comparison with $a = -\omega^2 x$ shows that $\omega = \sqrt{\dfrac{g}{l}}$ and $T = 2\pi\sqrt{\dfrac{l}{g}}$. This is the equation for the time period of a simple pendulum.

Mass–spring system

Providing that a spring is elastic and that its extension is always directly proportional to the force acting on it, a mass–spring system performs **exact simple harmonic motion**.

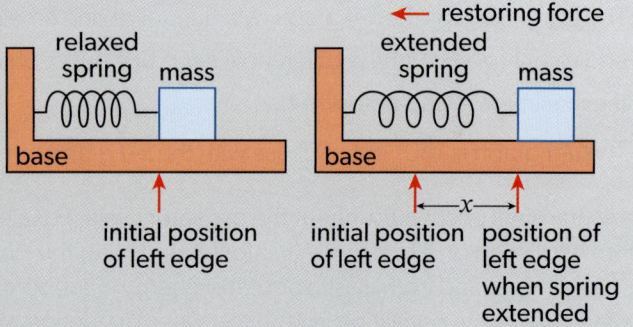

In Topic A.2, the force F_H that acts on a spring is shown to be directly proportional to its extension x and $F_H = -kx$, where k is the spring constant. The negative sign shows that when the spring is extended (for example, to the right in the diagram) a force acts to return the mass to its equilibrium position (towards the left in the diagram).

So, if we equate F_H to the acceleration a and mass m at the end of the spring, we get $ma = -kx$, with the acceleration positive to the right and the force acting to the left.

Rearranging this expression gives $a = -\left(\dfrac{k}{m}\right)x$, which is in the form $a = -\omega^2 x$, where $\omega = \sqrt{\dfrac{k}{m}}$. The motion of the mass–spring system therefore satisfies the conditions for simple harmonic motion.

Therefore, $T = 2\pi\sqrt{\dfrac{m}{k}}$ because $T = \dfrac{2\pi}{\omega}$. This equation equates the time period of oscillation to the properties of the spring and the mass.

Notice that it is the system that oscillates, not the spring or mass alone. Both spring and mass are required for the oscillation to occur.

Look carefully at these two equations for the time periods and you will see that they have the general form: $T = 2\pi \sqrt{\dfrac{\text{inertial (mass) term}}{\text{elastic (spring) term}}}$.
All practical oscillators operate with the transfer of energy between two (or more) forms: between a mass-type form and an elastic-type form (e.g. between kinetic energy of the mass and stored elastic potential energy in the mass–spring system). This can help you to analyse other forms of harmonic oscillator.

Example 3

A simple pendulum oscillates with frequency 1.4 Hz.

Determine the length of the pendulum.

Solution

Period of the pendulum $T = \dfrac{1}{f} = \dfrac{1}{1.4} = 0.71\,\text{s}$

The period only depends on the length l:

$T = 2\pi \sqrt{\dfrac{l}{g}}$

$l = \left(\dfrac{T}{2\pi}\right)^2 g = \left(\dfrac{0.71}{2\pi}\right)^2 \times 9.8 = 0.13\,\text{m}$

The pendulum is 13 cm long.

Example 4

A 120 g mass is attached to a spring of spring constant 5.6 N m⁻¹ and undergoes simple harmonic oscillations.

a) Calculate the frequency of the oscillations.

The same mass is now attached to a different spring. The frequency of the oscillations is halved compared with the value calculated in part (a).

b) Determine the spring constant of the second spring.

Solution

a) Period $= 2\pi \sqrt{\dfrac{0.12}{5.6}} = 0.92\,\text{s}$, so frequency $= \dfrac{1}{0.92} = 1.1\,\text{Hz}$

b) The square of the period of the mass–spring system is inversely proportional to the spring constant $\left(T^2 \propto \dfrac{1}{k}\right)$. The period has doubled and so the spring constant of the second spring is smaller than that of the first spring by a factor of four.

Spring constant of second spring $= \dfrac{5.6}{4} = 1.4\,\text{N m}^{-1}$

C Wave behaviour

> **Sample student answer**
>
> A mass oscillates horizontally at the end of a horizontal spring. The mass moves through a total distance of 8.0 cm from one end of the oscillation to the other.
>
> a) State the amplitude of the oscillation. [1]
>
> *This answer could have achieved 0/1 marks:*
>
> 8.0 cm
>
> b) Outline the conditions that the system must obey for the motion to be simple harmonic. [2]
>
> *This answer could have achieved 2/2 marks:*
>
> a must be in the opposite direction to x, and a must be proportional to x.

▼ The student has not visualized the arrangement. The mass is moving from one extreme to the other and passes through the equilibrium position halfway through this distance. The amplitude is 4.0 cm.

▲ These are the two conditions that are required. However, it would be better to define the symbols x and a for clarity.

Nature of science

Simple harmonic motion can be used to model other oscillations too. A car suspension, the oscillations of a diving board, and a bungee jump are all examples of motion that can be approximated as simple harmonic. Modelling of this type is all part of the nature of science.

Assessment tip

Notice that the displacement and acceleration equations satisfy the original shm definition because $x = x_0 \sin \omega t$ and therefore $a = -x_0 \omega^2 \sin \omega t$ so that $a = -\omega^2 x$, as required.

The equation $a = -\omega^2 x$ can be solved. Its solution is $x = x_0 \sin \omega t$ when the object begins its motion at the centre at time $t = 0$. When the object is released from rest at one extreme of the motion, then $x = x_0 \cos \omega t$.

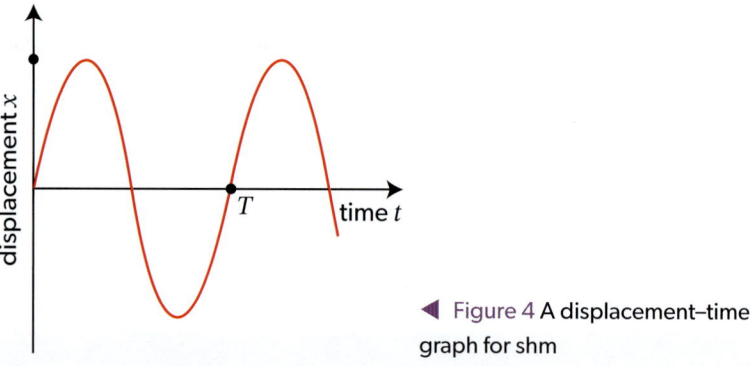

◀ **Figure 4** A displacement–time graph for shm

Figure 4 shows the graph of x against t. It is a sine wave and the oscillation begins at the equilibrium position for $t = 0$. When the motion begins at one extreme, the graph must begin at the maximum displacement (amplitude x_0) to continue as a cosine graph.

The equation $v = \pm \omega \sqrt{(x_0^2 - x^2)}$ gives the variation with displacement rather than time. It has a ± sign because, at any single position between the ends of the motion, the object can be moving either towards or away from the centre.

Knowing how displacement varies with time allows you to draw velocity–time and acceleration–time graphs. These are connected through the gradients of the respective displacement and velocity graphs, as shown in Table 1.

Displacement in shm can be modelled by $x = x_0 \sin(\omega t = \phi)$ where ϕ is the **phase angle**. The expression for the velocity is then $v = \omega x_0 \cos(\omega t = \phi)$. The phase angle determines the initial condition of the motion at time $t = 0$. When $\phi = 0$ or π, the object is initially at the centre of the motion, moving towards positive or negative displacement. When $\phi = \frac{\pi}{2}$, the object starts from rest at the maximum positive displacement ($x_0 \sin\left(\omega t + \frac{\pi}{2}\right)$ is equivalent to $x_0 \cos \omega t$). When $\phi = -\frac{\pi}{2}$ or $\frac{3\pi}{2}$, the object is initially at rest at the maximum negative displacement (because $x_0 \sin\left(\omega t - \frac{\pi}{2}\right)$ is equivalent to $-x_0 \cos \omega t$). Other values of ϕ represent intermediate positions of the object at $t = 0$.

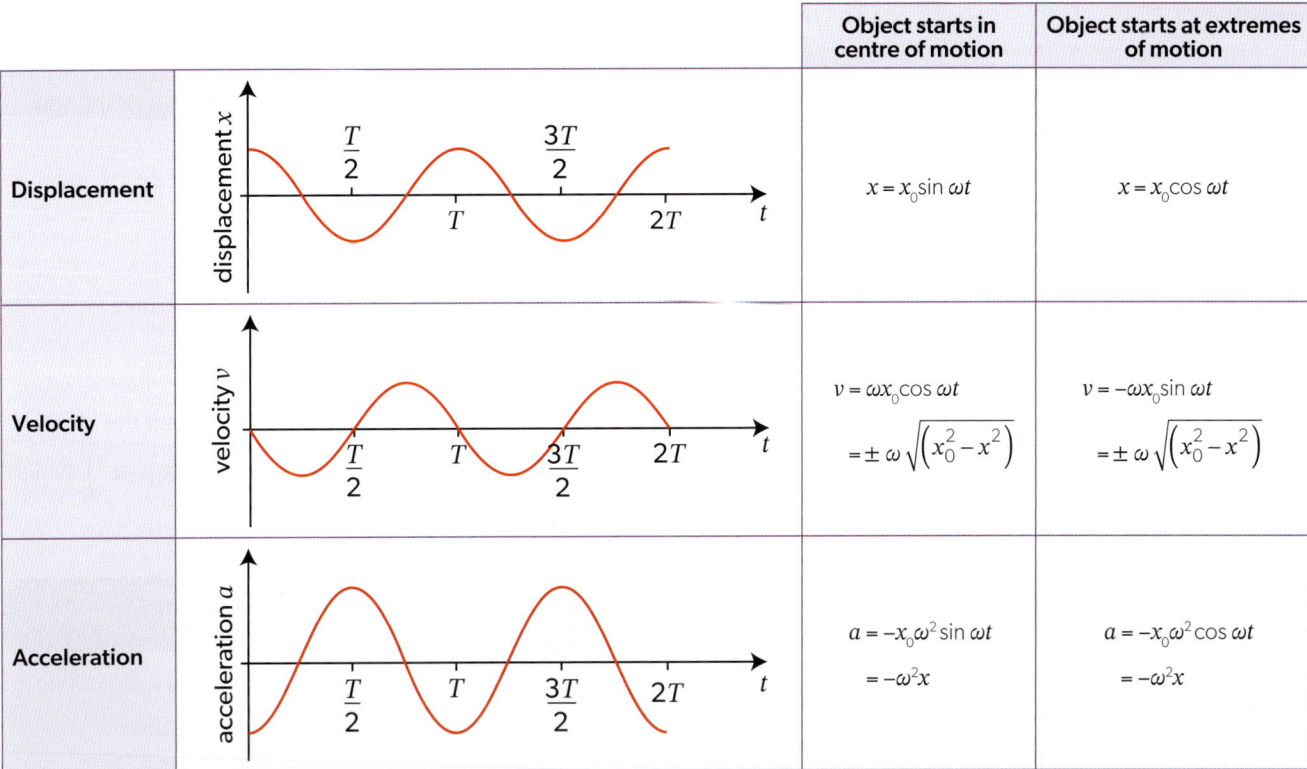

Table 1 Relationship between displacement, velocity and acceleration

Example 5

Two mass–spring systems undergo simple harmonic oscillations.

The displacement x_1, in cm, of System 1 is given by $x_1 = 6.0 \times \sin(2.1t)$, where t is time in s.

The displacement x_2, in cm, of System 2 is given by $x_2 = 2.0 \times \sin\left(2.1t - \frac{\pi}{2}\right)$.

a) Explain why the systems oscillate with the same time period T.

b) Calculate T.

c) Compare the relative motion of the systems.

d) For System 2, calculate:

   (i) the maximum velocity

   (ii) the velocity at $t = 0.50$ s.

Solution

a) The angular frequency, $\omega = 2.1$ rad s^{-1}, is the same for both systems. Hence, the time period must also be the same.

b) $T = \frac{2\pi}{2.1} = 3.0$ s

c) System 1 oscillates with a greater amplitude than System 2 (6.0 cm, compared with 2.0 cm).

   The phase difference between the systems is $\frac{\pi}{2}$, which corresponds to a quarter of oscillation. System 2 lags behind system 1 by $\frac{3.0}{4} = 0.75$ s. For example, System 2 passes the equilibrium position 0.75 s after System 1.

d) (i) Maximum velocity $\omega x_0 = 2.1 \times 2.0 = 4.2$ cm s^{-1}

   (ii) For System 2 at time t, velocity $v = 4.2 \cos\left(2.1t - \frac{\pi}{2}\right)$

   When $t = 0.50$ s, $v = 4.2 \times \cos\left(2.1 \times 0.50 - \frac{\pi}{2}\right) = 3.6$ cm s^{-1}

The maximum kinetic energy E_k for a simple harmonic system must be when the system is passing through its equilibrium position, when $x = 0$. The maximum E_k of $\frac{1}{2}m\left(\omega\sqrt{(x_0^2 - x^2)}\right)^2$ becomes $E_k = \frac{1}{2}m\omega^2 x_0^2$. At this point, the potential energy E_p must be zero.

The total energy in the oscillation does not change.

- In true shm, no energy is lost due to friction or other dissipation.
- The energy simply transfers continuously between the kinetic and potential forms.

So, the total energy E_T in the oscillations is also constant at $E_T = \frac{1}{2}m\omega^2 x_0^2$ as this is the maximum E_k value. To find the potential energy, subtract the kinetic energy from the total energy: $E_P = E_T - E_k = \frac{1}{2}m\omega^2 x_0^2 - \frac{1}{2}m\omega^2(x_0^2 - x^2)$, which simplifies to $E_P = \frac{1}{2}m\omega^2 x^2$.

Example 6

A particle of mass m executes simple harmonic motion in a straight line with amplitude A and frequency f. Calculate the total energy of the particle.

Solution

This can be approached in a number of ways. The total energy of the particle is equal to its maximum kinetic energy, which occurs at the centre of the motion, where $x = 0$.

Maximum velocity of particle $v_{max} = A\omega$

Since $\omega = 2\pi f$ and $v_{max} = 2\pi A f$, maximum kinetic energy $= \frac{1}{2}mv^2 = \frac{1}{2}m 4\pi^2 A^2 f^2 = 2\pi^2 m A^2 f^2$.

Example 7

Particle P moves with simple harmonic motion. This graph shows the variation of the displacement x of P in the medium with time t.

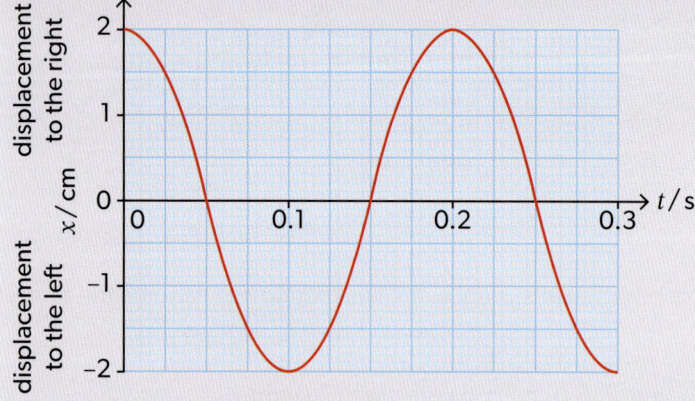

a) Calculate the magnitude of the maximum acceleration of P.
b) Calculate its speed at $t = 0.12$ s.
c) State its direction of motion at $t = 0.12$ s.

Solution

a) Period $T = 0.20$ s. $a_{max} = \omega^2 x_0 = \left(\frac{2\pi}{0.20}\right)^2 \times 2.0 \times 10^{-2} = 19.7 \approx 20 \text{ m s}^{-2}$

b) The displacement at $t = 0.12$ s is -1.62 cm. $v = \omega\sqrt{x_0^2 - x^2} = \frac{2\pi}{0.2}\sqrt{(2.0 \times 10^{-2})^2 - (1.62 \times 10^{-2})^2} = 0.37 \text{ m s}^{-1}$

c) To the right.

Example 8

The graph shows the variation of total potential energy with time for a mass–spring system. The mass is 0.32 kg and the maximum kinetic energy in the system is 20 mJ.

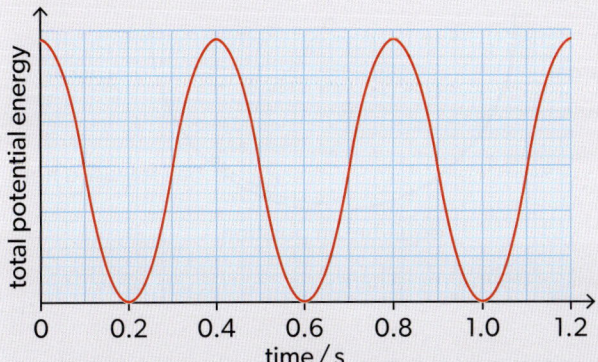

a) State, with a reason, the time period of oscillation of the mass on the spring.

b) Calculate the spring constant k of the spring used.

c) Determine the amplitude of the oscillation.

Solution

a) The potential energy cycles twice for one time period of the system. The time period of the mass on the spring is 0.80 s.

b) $T = 2\pi\sqrt{\dfrac{0.32}{k}}$

$k = \dfrac{4\pi^2 \times 0.32}{(0.80)^2} = 20\,\text{N m}^{-1}$

c) $2.0 \times 10^{-2} = \dfrac{1}{2}m\omega^2 x_0^2 = 0.5 \times 0.32 \times \left(\dfrac{2\pi}{0.80}\right)^2 x_0^2$

$x_0 = 4.5\,\text{cm}$

Example 9

Which of the following statements is true for an object performing simple harmonic motion about an equilibrium position O?

A. The acceleration is always away from O.

B. The acceleration and velocity are always in opposite directions.

C. The acceleration and the displacement from O are always in the same direction.

D. The graph of acceleration against displacement is a straight line.

Solution

The acceleration can be in the same or the opposite direction to velocity, so B is incorrect. The negative sign in the definition of shm means that acceleration and displacement are always opposed, so A and C cannot be correct. Response D is the alternative way to express this.

The correct answer is D.

Assessment tip

Multiple-choice questions demand care. The incorrect responses in Example 9 test the relationships between acceleration, velocity and displacement. The negative sign in $a = -\omega^2 x$ means that, because the acceleration and velocity are 90° out of phase, the acceleration and velocity can be in the same direction or opposite.

C Wave behaviour

Sample student answer

A small ball of mass *m* is moving on the inside surface of a frictionless hemispherical bowl.

a) The ball is placed a small distance *x* from the bottom of the bowl and released from rest.

The magnitude of the force on the ball towards the equilibrium position is given by $\frac{mgx}{R}$, where R is the radius of the bowl.

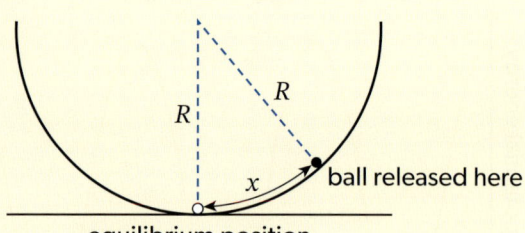

Outline why the ball will perform simple harmonic oscillations about the equilibrium position. [1]

This answer could have achieved 1/1 marks:

> The acceleration of the ball is proportional to displacement and is directed towards the equilibrium position.

▲ This answer captures both essential points about the definition of shm: the magnitude and the direction.

b) The radius of the bowl is 8.0 m.

Show that the period of oscillation of the ball is about 6 s. [2]

This answer could have achieved 2/2 marks:

> $F = \frac{mgx}{R} \quad a = \frac{gx}{R} \quad \frac{g}{R} = \omega^2$
>
> $\omega = \sqrt{\frac{g}{R}} = 1.1 \text{ s}^{-1}$
>
> $T = \frac{2\pi}{\omega} = \frac{2\pi}{1.1 \text{ s}^{-1}} = 5.75 \approx 6 \text{ s}$

▲ This is a good answer. The acceleration of the sphere is deduced (and is shown as proportional to *x*) and thus identifies the constant of proportionality. This is ω^2 and leads to a correct calculation.

Practice problems

Problem 1

An object performs simple harmonic motion with amplitude x_0 and time period T.

a. Identify the phase difference between velocity and displacement for this motion.

The object passes the equilibrium position at time $t = 0$.

b. Sketch a graph to show the variation of kinetic energy with time for the object for time T. Explain your answer.

Problem 2

A particle oscillates with simple harmonic motion without any loss of energy. What is true about the acceleration of the particle?

A. It is always in the opposite direction to its velocity.

B. It is least when the speed is greatest.

C. It is proportional to the frequency.

D. It decreases as the potential energy increases.

Problem 3

A crane moves a load of mass 15×10^3 kg that is supported by a vertical cable. The mass of the cable is negligible compared with the mass of the load. When the crane stops, the load and cable behave like a simple pendulum. The oscillation of the load has an amplitude of 2.0 m and period of 8.0 s.

a) Calculate the length of the cable that supports the load.

b) (i) Determine the maximum acceleration of the load as it swings.

(ii) Calculate the force on the load that produces the acceleration in part (b)(i).

Problem 4

The acceleration a, in m s^{-2}, of a particle varies with the displacement x, in m, according to the equation $a = -25x$.

a) Outline why the particle performs simple harmonic motion.

b) Calculate:

(i) the angular frequency

(ii) the period of the oscillations.

Problem 5

A particle undergoes simple harmonic motion with period 60 ms and amplitude 8.0 mm.

a) Determine the maximum speed of the particle.

b) Calculate, when the displacement of the particle is 2.0 mm:

(i) the speed of the particle

(ii) the acceleration of the particle.

Problem 6

A mass of 0.25 kg is attached to a weightless spring of spring constant 9.6 N m^{-1}. The mass performs simple harmonic oscillations with amplitude 0.050 m.

a) Calculate the total energy of the system.

b) Calculate:

(i) the maximum speed of the mass

(ii) the maximum acceleration of the mass.

The displacement x, in m, of the mass is given by $x = 0.050 \sin(\omega t)$, where t is the time in s.

c) Calculate the value of ω.

d) Determine the velocity of the mass when $t = 0.20$ s.

C.2 Wave model

You must know:

- what is meant by wavelength, frequency, time period and wave speed when applied to wave motion
- the difference between transverse waves and longitudinal waves
- the nature of electromagnetic waves
- the nature of sound waves
- the differences between mechanical waves and electromagnetic waves.

You should be able to:

- solve problems using the equation $v = f\lambda = \dfrac{\lambda}{T}$
- explain the motion of particles in a medium that leads to transverse and longitudinal waves
- sketch and interpret displacement–time and displacement–distance graphs for transverse and longitudinal waves
- use the *Physics data booklet* to find the order of magnitude of wavelengths for the principal regions of the electromagnetic spectrum.

Waves transfer energy without any overall change in the medium through which they pass.

For **transverse waves**, particles of the medium oscillate at 90° to the direction of energy propagation (Figure 1).

For **longitudinal waves**, particles oscillate parallel to the direction of energy propagation (Figure 2).

C Wave behaviour

Wavelength λ is the distance between the two nearest points on the wave with the same phase.

Wave speed c is the speed at which the wave moves in the medium.

Quantities in the wave model link to the definitions of distance and speed in Topic A.1 and quantities used in Topics A.2 and A.4 for circular motion.

Radian measure is defined in Topic A.2.

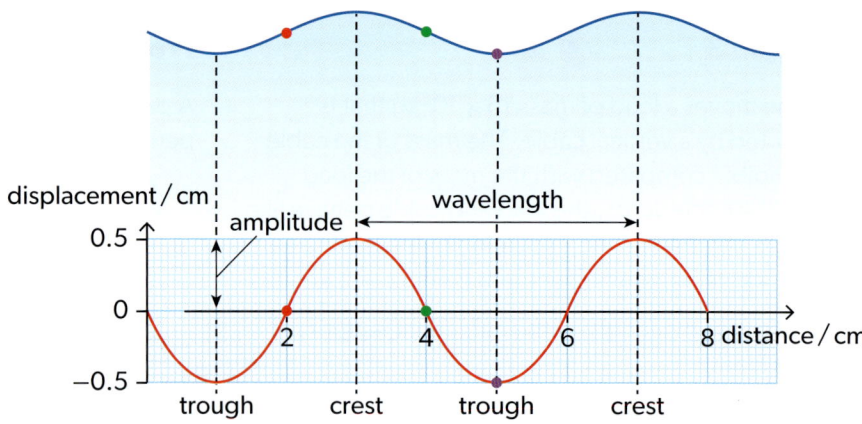

▲ **Figure 1** The displacement–distance graph for a transverse wave. The top diagram shows the position of three particles as the wave moves past them. This is a snapshot of the wave at one instant of time

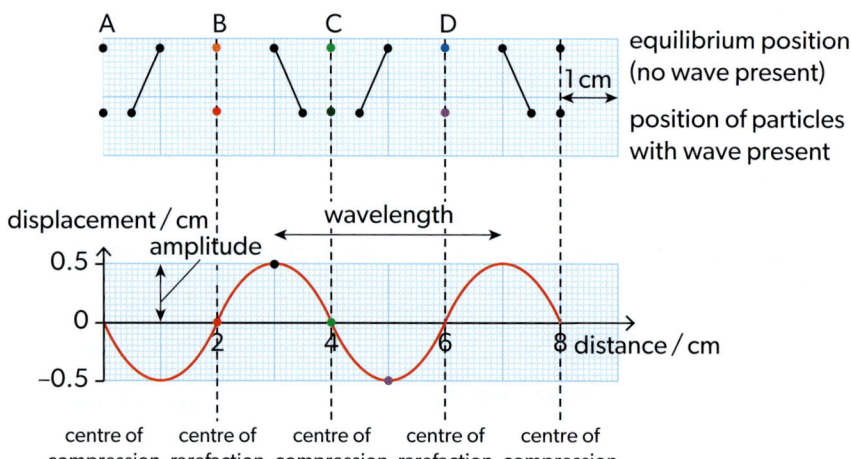

▲ **Figure 2** The displacement–distance graph for a longitudinal wave. The top diagram shows the movement of particles as the wave moves past them. This is a snapshot of the wave at one instant of time

A representation of the motion of an individual point on the wave is given by a displacement–time graph (Figure 3). This may look similar to the graphs in Figures 1 and 2, but this graph gives a direct value for the time period of the wave, rather than its wavelength.

Assessment tip

You need to be able to derive the equation for the speed of a wave: $c = f\lambda$.

Remember that a displacement–distance graph shows that the wave moves forward by λ in one cycle, whereas a displacement–time graph gives the time T that one particle takes to go through one cycle. Therefore $c = \dfrac{\lambda}{T}$.

As $f = \dfrac{1}{T}$, $c = f\lambda$.

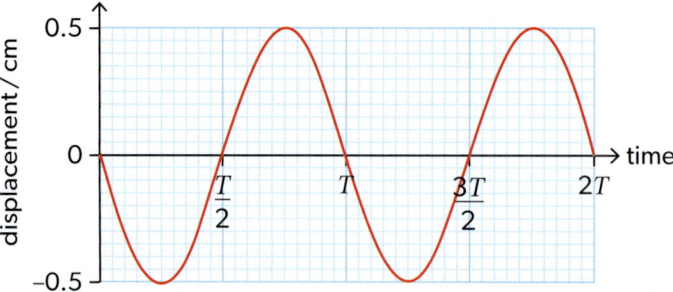

▲ **Figure 3** A displacement–time graph for an individual particle in a wave. This graph can apply to both a transverse and a longitudinal wave

Assessment tip

Always check the graph axes carefully for both the quantity and the unit.

Example 1

The shortest distance between two points on a progressive transverse wave which have a phase difference of $\frac{\pi}{3}$ is 0.050 m.

The frequency of the wave is 500 Hz.

Determine the speed of the wave.

Solution

$\frac{\pi}{3}$ rad is 60°, which is $\frac{1}{6}$ of a cycle. Therefore,

wavelength $\lambda = 6 \times 0.050 = 0.30$ m

The speed of the wave is $c = f\lambda = 500 \times 0.30 = 150$ m s^{-1}.

Sound is transmitted through a fluid (gas or liquid) as a longitudinal wave. The molecules of the fluid are the medium in this case and they move as the wave passes through them.

- There are areas of low pressure (rarefactions: below atmospheric pressure) and high pressure (compressions: above atmospheric pressure).
- The positions of maximum and minimum **pressure** are at the points where the displacement is zero.
- The positions of maximum and minimum **displacement** are the places where the pressure is atmospheric (in other words, at the average gas pressure).

Sound waves show the common properties of waves: reflection, refraction, diffraction and interference (see Topic C.3).

Example 2

a) Outline the difference between a longitudinal wave and a transverse wave.

b) State an example of:

  (i) a transverse wave

  (ii) a longitudinal wave.

c) Sound with a frequency of 860 Hz travels through steel with a speed of 4.2 km s^{-1}.

  Calculate the wavelength of the sound wave.

Solution

a) In a transverse wave, the vibrations are perpendicular to the direction of energy propagation.

  In a longitudinal wave, the vibrations of the particles are in the same direction as the energy propagation.

b) (i) A wave on the surface of water.

  (ii) A sound wave in a gas.

c) $c = f\lambda$

  $\lambda = \frac{c}{f} = \frac{4200}{860} = 4.9$ m

Approaches to learning

All real waves are caused by disturbances in the medium that occur at a particular place. The disturbance must start and stop rather than go on forever. The wave that results from this disturbance therefore has an overall length. This idea links Topics A.1 and C.2.

When the disturbance that creates a wave undergoes n cycles, then the time for the disturbance is nT. When the wavelength of the wave is λ, the overall length of the wave is $n\lambda$.

Assessment tip

Notice that the answer to Example 2, part (b)(ii) specifies "sound wave **in a gas**". A solid can transmit both transverse and longitudinal waves. For example, waves can propagate on a solid surface like a "ripple" on a liquid surface.

C Wave behaviour

Electromagnetic waves:
- do not need a medium and can travel through a vacuum
- travel at the same speed in a vacuum irrespective of frequency, in other words, at 3×10^8 m s^{-1}
- have decreased speeds in matter—this speed depends on the frequency.

Figure 4 shows the principal regions in the electromagnetic spectrum.

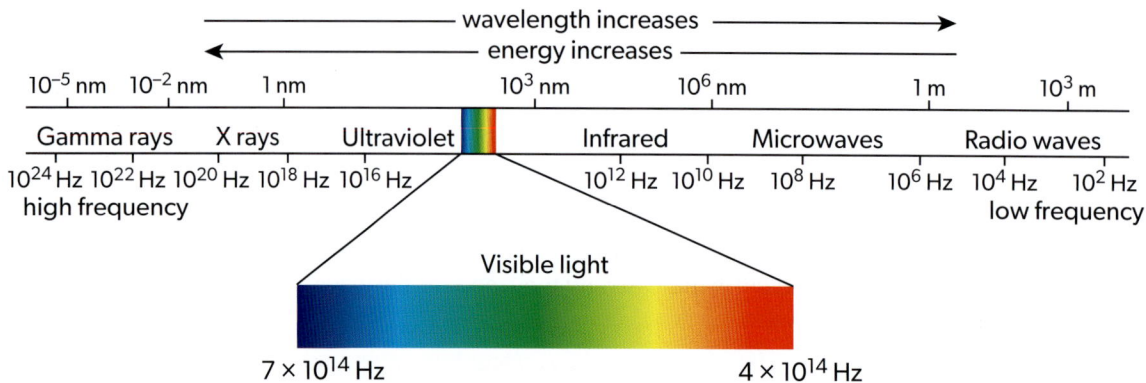

▲ Figure 4 The principal regions of the electromagnetic spectrum and their wavelengths

Assessment tip

There is a similar diagram to Figure 4 in the *Physics data booklet* and you can obtain approximate wavelengths from there.

You should know a use for each wave. For some parts of the spectrum, the advantages of using them has to be balanced against their disadvantages.

Sample student answer

A longitudinal wave is travelling in a medium from left to right. The graph shows the variation with distance x of the displacement y of the particles in the medium. The solid line and the dotted line show the displacement at $t = 0$ and $t = 0.882$ ms, respectively.

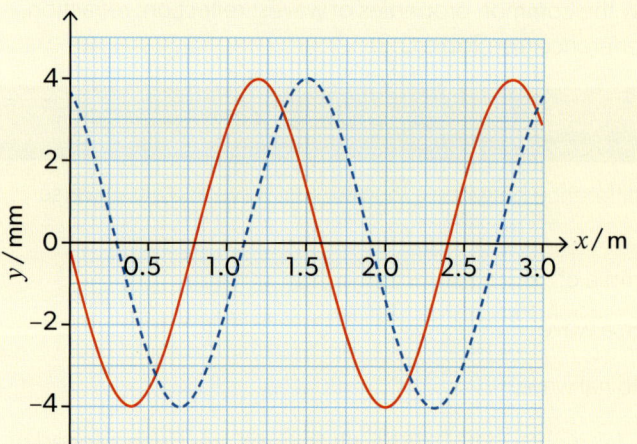

The period of the wave is greater than 0.882 ms. A displacement to the right of the equilibrium position is positive.

a) State what is meant by a longitudinal travelling wave. [1]

This answer could have achieved 1/1 marks:

> A wave where the energy is oscillating in the same direction as the motion of the particles.

▲ The student has conveyed the sense that this direction is parallel to the displacement of the particles in the wave (although it would be better to talk about the "direction of propagation of energy").

b) Calculate the speed of this wave. [2]

This answer could have achieved 0/2 marks:

> $v = f\lambda$ $\lambda = 1.6$ (m) $f = \frac{\lambda}{T} < 1.13$ Hz
>
> 0.3 m in 0.882
>
> $v < 1.81$ m s^{-1}

▼ The graphs show the motion of two particles separated by 0.3 m. The time taken for the maximum amplitude to travel from the particle represented by the solid line to that represented by the dashed line is 0.882 ms. $\frac{\text{displacement}}{\text{time}}$ gives the speed of the wave. There is no credit for this approach which tries to use $\frac{1}{0.882}$ as the frequency. Note that there is an additional power of ten error as the time was quoted in milliseconds but used here in seconds.

Practice problems

Problem 1

For a sound wave travelling through air, explain what is meant by "particle displacement", "amplitude" and "wavelength".

Problem 2

A transverse wave travels along a string. Each point in the string moves with simple harmonic motion.

a) Sketch a graph showing:

(i) the variation of displacement for a point on the wave with time

(ii) the variation of displacement with distance along the wave.

Label your graphs and axes clearly.

b) Label, where possible, on your graphs:

(i) the amplitude x_0 of the wave

(ii) the period T of the vibrations

(iii) the wavelength λ of the wave

(iv) points P and Q, which have a phase difference of $\frac{\pi}{2}$.

Problem 3

The frequency range that a child can hear is from 30 Hz to 16 500 Hz. The speed of sound in air is 330 m s^{-1}. Calculate the shortest wavelength of sound in air that the child can hear.

Problem 4

A longitudinal wave of frequency 1200 Hz travels in a railway rail. There are 25 full wavelengths of the wave in a length of 100 m of the rail. Calculate the speed of the wave.

Problem 5

a) State **two** differences between sound waves and electromagnetic waves.

b) A home Wi-Fi network uses electromagnetic waves of frequency 5.0 GHz to transmit information between connected devices.

(i) Calculate the wavelength of these waves.

(ii) State the region of the electromagnetic spectrum to which these waves belong.

Problem 6

A wave travels in a medium at a speed 180 m s^{-1}. The diagram shows how the displacement of two particles P and Q in the medium varies with time t.

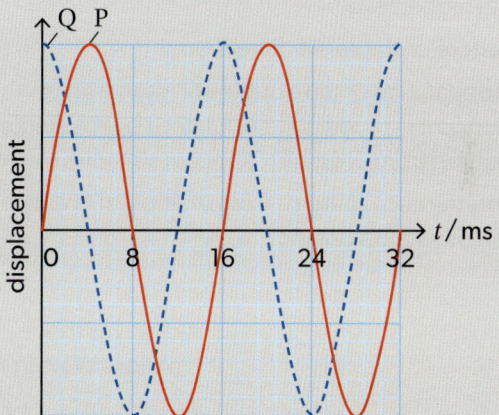

a) Calculate the frequency of the wave.

b) Determine the minimum possible distance between the equilibrium positions of P and Q.

C.3 Wave phenomena

You must know:
- the concept of wavefronts and rays for waves travelling in two or three dimensions
- what is meant by reflection and refraction
- what is meant by diffraction
- that superposition can occur when two or more waves are added together
- what is meant by coherence for double-source interference
- what is meant by constructive and destructive interference and the conditions under which they can arise

Additional higher level:
- the nature of single-slit diffraction
- the appearance of the double-slit interference pattern as it is modified (modulated) by the single-slit pattern from each slit
- the interference patterns from multiple slits and from a diffraction grating.

You should be able to:
- solve problems involving reflection and refraction
- solve problems including Snell's law, critical angle and total internal reflection
- sketch and interpret the superposition of pulses and waves
- describe the diffraction of waves around objects and through apertures
- calculate the resultant of two waves or pulses using algebra or graphs
- solve problems involving path difference for constructive and destructive interference
- solve problems involving Young's double-slit interference

Additional higher level:
- solve problems involving the wave wavelength and the slit width for single-slit diffraction
- solve problems involving interference patterns from multiple slits and diffraction gratings.

As waves move, they can change both their shape and their direction of motion. Wavefronts and rays are used to visualize these changes (Figure 1).

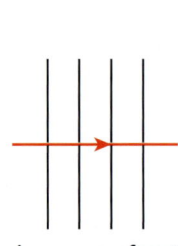
plane wavefronts

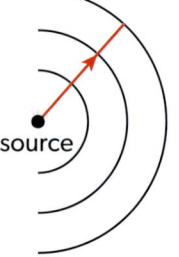
source
circular wavefronts

▲ Figure 1 Plane and circular wavefronts and rays

A **wavefront** shows the shape of the wave at one instant. A series of successive wavefronts makes it possible to deduce the origin and history of the wave. A common assumption in drawing wavefronts is that they are one wavelength apart.

A **ray** is a line at 90° to a wavefront as it crosses the wavefront. An arrow on the ray shows the direction of wave movement.

The concept of intensity links to many themes in the course. It is used, for example, in Topics B.1, B.2 and E.5 in connection with black-body and stellar radiation. There are also links to the intensity of gamma photons in Topic E.3.

The quantities I and P in this topic are the same as the quantities b and L that occur in Topic B.1. The terms apparent brightness b and luminosity L are mainly used when discussing astronomical sources of radiation. Outside that field of study, intensity and power are the terms we normally use.

The rate of transfer of wave energy is measured using the quantity intensity I. A point source of a wave spreads out through space. The source radiates power P (as usual, this is the energy transferred by the source in one second).

At a distance r from the source, this energy spreads over a sphere of radius r, which has an area equal to $4\pi r^2$. The intensity is given by $I = \dfrac{P}{4\pi r^2}$.

The **intensity** I of a wave is related directly to the square of its amplitude A: $I \propto A^2$. When a wave has its amplitude doubled, the intensity increases by a factor of four, so four times as much energy falls on a given area every second.

Intensity is the amount of energy that a wave can transfer to an area of one square metre in one second.

The unit of intensity is W m^{-2}. In fundamental units, this is kg s^{-3}.

Example 1

A bicycle lamp and a floodlight deliver their output energy into a cone that covers 0.20 of the area of a sphere.

The output power of the bicycle lamp is 1.5 W.

The output power of the floodlight is 300 W.

a) Calculate the intensity of the bicycle lamp when viewed by an observer from 20 m away.

b) The observer's eye has a pupil diameter of 5.0 mm. Determine the light power entering the eye from the bicycle lamp.

c) The bicycle lamp and the floodlight are equally bright to the observer. Deduce the distance of the floodlight from the observer.

Solution

a) 1.5 W of light is delivered to 0.20 of a sphere of radius 20 m.
$$I = \frac{P}{0.20 \times 4\pi r^2} = \frac{1.5}{0.20 \times 4\pi \times 20^2} = 1.5 \times 10^{-3} \text{ W m}^{-2}$$

b) Area of pupil = $\pi \times (2.5 \times 10^{-3})^2 = 1.96 \times 10^{-5} \text{ m}^2$
Power entering eye = $1.5 \times 10^{-3} \times 1.96 \times 10^{-5} = 29$ nW

c) The intensity entering the eye must be the same for both lamps.
$$\frac{P_b}{r_b^2} = \frac{P_f}{r_f^2}; \text{ therefore } \frac{r_f^2}{r_b^2} = \frac{P_f}{P_b} \text{ and } \frac{r_f}{r_b} = \sqrt{\frac{P_f}{P_b}}.$$
As the power ratio is 300 : 1.5 = 200 : 1, the distance ratio is 14 : 1 and the floodlight is 20 × 14 = 280 m away.

Approaches to learning

You will meet the inverse-square law on many occasions in this course. The intensity due to a point source of radiation is directly proportional to $\frac{1}{r^2}$, where r is the distance from the point source. Doubling the distance from a source means that the intensity decreases to one-quarter of its original value.

Reflection and refraction

When waves are incident on the interface between two different media, changes in direction of the wave are observed.

- **Reflection**: when the wave continues to travel in the original medium.
- **Refraction**: when the wave travels from the original medium to a new medium.

The angles of incidence, reflection and refraction are always measured from the normal to the interface between the media. Incident, reflected and refracted rays all lie in the same plane. These quantities are defined in Figure 2.

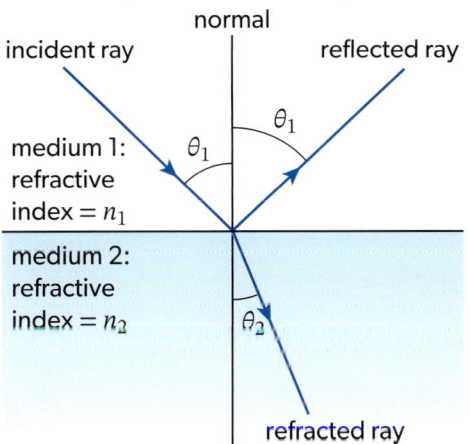

◀ Figure 2 **Rays at a boundary between two media**

The **reflected angle** is always equal to the incidence angle θ_1 when measured from the normal to the interface between the media.

The **refraction angle** θ_2 is related to the incidence angle by $\frac{\sin \theta_1}{\sin \theta_2} = \frac{n_2}{n_1} = \frac{v_1}{v_2}$. This is **Snell's law**. This expression is given in the *Physics data booklet*.

C Wave behaviour

The ratio $\frac{n_2}{n_1}$ which is equal to $\frac{v_1}{v_2}$, where v_1 is the speed in medium 1 and v_2 is the speed in medium 2, is the (relative) **refractive index** going from medium 1 to medium 2 and is written as $_1n_2$. The **absolute refractive index** is

$$\frac{\text{speed of light in a vacuum}}{\text{speed of light in the medium}}.$$

As the speed of light in air is close to the speed in a vacuum, the relative refractive index going from air to the medium ($_{air}n_{medium}$) is used in practice.

Media with large optical densities have large values of *n* and correspondingly small wave speeds.

Rays are reversible: the waves can be reversed, so that their rays trace out the same path but in the opposite direction.

When waves travel between media from an optically denser medium to a less dense medium, the angle of refraction is greater than the angle of incidence (Figure 3). As the angle of incidence increases from a small value, there comes a point where the angle of refraction grazes the interface (ray 4 in the diagram) — the incident angle when this happens is known as the **critical angle**.

- For incident angles less than the critical angle, a weak reflected ray is observed.
- For incident angles greater than the critical angle, there is no refraction; only a strong total internal reflection is observed.

The **critical angle** *c* occurs when incident ray θ_1 gives rise to a refracted ray with $\theta_2 = 90°$. In Figure 3, the ray is travelling from medium 1 to medium 2 (from high refractive index to low refractive index), so the angle of incidence is θ_c and $\theta_2 = 90°$ is the angle of total internal reflection. Therefore, $\frac{\sin\theta_c}{\sin 90°} = {_1n_2} = \frac{1}{_2n_1}$, and so $\sin\theta_c = \frac{1}{_2n_1}$.

Notice that $_2n_1$ is the relative refractive index when medium 2 is air or a vacuum. When you have measured *n* for a ray travelling **from** air **to** an optically denser substance (e.g. water or glass), then the critical angle for rays going **from** the substance **to** air will be given by $\sin\theta_c = \frac{1}{n}$.

Total internal reflection occurs when waves travel from a medium with high refractive index to one with a lower refractive index and the angle of incidence is greater than the critical angle.

When a wave goes from medium 1 through medium 2 to medium 3, then $_1n_3 = {_1n_2} \times {_2n_3}$.

▲ Figure 3 Critical angle and total internal reflection

Example 2

A plane mirror consists of a parallel-sided glass sheet coated with a reflector. A ray of light travelling in air enters the glass at an angle of 30° to the glass surface.

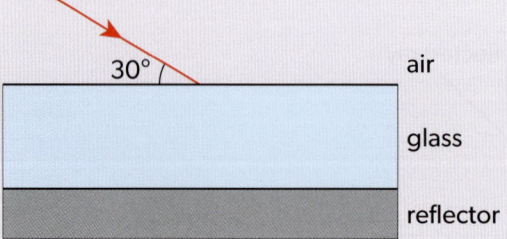

The refractive index of the glass is 1.5.

a) Determine the angle of reflection at the reflector.

b) Calculate the critical angle for the glass–air interface.

C.3 Wave phenomena

Solution

a) The angle of incidence at the air–glass interface is 60° and Snell's law states that $\sin r = \dfrac{\sin 60°}{1.5}$. Hence, the angle of refraction r is 35°. Because the glass has parallel sides, the angle of incidence at the glass–reflector interface and the angle of reflection are also both 35°.

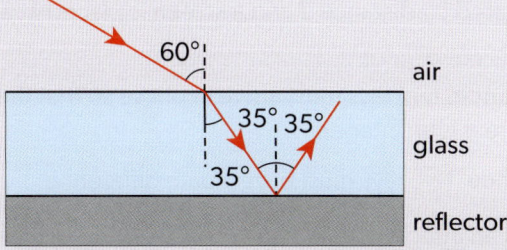

b) Using $\sin\theta_c = \dfrac{1}{1.5}$ gives $\theta_c = 42°$.

Wave pulses and continuous waves can superpose. Figure 4 shows what happens when two pulses on a stretched rope meet. Because the pulses are of equal amplitude but in opposite directions, the two disturbances cancel out to give zero displacement when they meet.

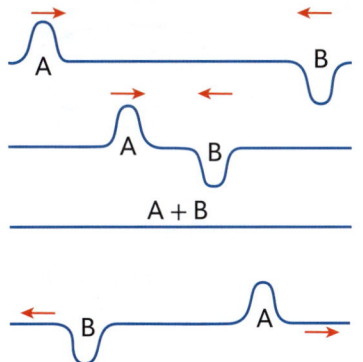

◀ Figure 4 Two pulses, identical in size but in opposite directions, approach each other on a rope. At one instant there is no disturbance in the rope as the two pulses cancel out.

> **Superposition** is when two (or more) waves add. This addition is vectorial, so you must take account of the sign of the displacement. Two waves that have the same magnitude of displacement but opposite signs give zero displacement overall.

For the **interference** of two light waves to be observable, the waves must be **coherent**. The phase relationship between the waves must be constant over a sufficiently long time for the observation to be made. In practice, this means that the waves must:

- have an unchanging phase relationship
- which implies that they must also have the same wavelength
- arrive at the same place in space simultaneously for the observation to be made.

A model for **double-source interference** is shown in Figure 5. A_1 and A_2 are two wave sources. C is the point midway between A_1 and A_2. O and B are points on a screen that is a distance D from the wave sources.

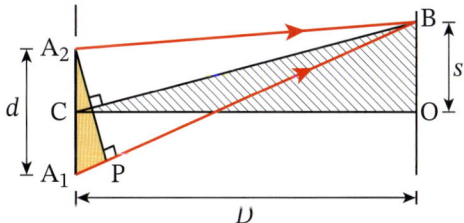

▲ Figure 5 A model for double-source interference

> **Approaches to learning**
>
> A travelling electromagnetic wave is a self-propagating pair of electric and magnetic fields. When two or more waves of these fields meet, the disturbances add to give the resultant sum of the fields at the position and time of the meeting. This is the effect called superposition when applied to electromagnetic waves.

Both waves will have travelled the same distance to reach O, the centre of the interference pattern, so there is a maximum here because the waves arrive in phase.

C Wave behaviour

> **Assessment tip**
>
> You should be able to show that Young's double-slit interference equation is true. When $D \gg d$, the angle at P is close to a right angle (because A_1B and A_2B are effectively parallel) and when this is true, the two triangles shown ($\Delta A_1 A_2 P$ and ΔOBC) are similar, so $\dfrac{BO}{CO} = \dfrac{A_1 P}{A_2 P}$ and $\dfrac{s}{D} = \dfrac{\lambda}{d}$.

To reach point B, the wave from A_1 must travel a distance PA_1 further than the wave from A_2. For the first maximum, which is at B, a distance s from O, the distance PA_1 will be one wavelength, so the two waves arrive in phase.

The **Young's double-slit interference** equation $s = \dfrac{\lambda D}{d}$ gives the separation s of successive bright fringes when light of wavelength λ is incident on two slits separated by distance d and a screen is placed at a distance D from the slits.

For **constructive interference**, two rays must arrive in phase so that the path difference is $n\lambda$, where n is an integer.

For **destructive interference**, two rays must have a path difference $\left(n + \dfrac{1}{2}\right)\lambda$, again where n is an integer.

These expressions, together with Young's equation, are provided in the *Physics data booklet*.

Example 3

The diagram shows two identical loudspeakers placed 0.75 m apart. The loudspeakers are in phase and emit sound of frequency 2.0 kHz.

Y is midway between the loudspeakers and 5.0 m away from the line that joins them. Y is a position of maximum sound intensity.

Positions X and Z are equidistant from Y and have zero sound intensity. There are further maxima and minima beyond X and Z.

The speed of sound in air is 330 m s⁻¹.

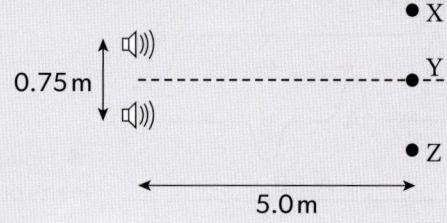

not to scale

a) Explain why the sound is:

   (i) a maximum at Y

   (ii) a minimum at X and Z.

b) Calculate:

   (i) the wavelength of the sound

   (ii) the distance XY.

Solution

a) (i) Position Y is equidistant from the loudspeakers, so both waves travel an equal distance to Y. When the waves superpose, they are in phase. This produces constructive interference and an intensity maximum.

   (ii) At X and Z, the waves arrive 180° out of phase.

   Path difference $= \left(n + \dfrac{1}{2}\right)\lambda$

   They interfere destructively to produce complete cancellation of the superposed waves.

b) (i) $\lambda = \dfrac{330}{2 \times 10^3} = 0.165$ m

   (ii) Separation between maxima $= \dfrac{\lambda D}{d} = \dfrac{0.165 \times 5.0}{0.75} = 1.10$ m

   X is halfway between the central maximum at Y and the next maximum; therefore,

   $XY = \dfrac{1.10}{2} = 0.55$ m

C.3 Wave phenomena

Nature of science

The work in Theme C assumes that electromagnetic radiation is a wave phenomenon. This is how it was regarded until the end of the 19th century. However, the modern view has changed. Electromagnetic radiation has particle-like and wave-like properties. Its photon properties are a discussion for Theme E. The history of this topic explores many of the issues arising from the nature of science.

Assessment tip

Take care with the phrase "out of phase". This describes two waves that have **any** phase difference other than zero and it is ambiguous. Always quote the magnitude of the phase difference where possible—for example, write "$\frac{\pi}{2}$ rad out of phase".

Two terms that are equivalent to a π phase difference are "completely out of phase" and "in antiphase".

Example 4

Monochromatic light of wavelength 520 nm is incident at a double slit. An interference pattern is formed on a screen that is 1.6 m from the slits.

15 bright fringes are observed in a length of 5.0 cm of the interference pattern.

a) Determine the distance between the slits.

When a different monochromatic source of light is used, the fringe separation on the screen increases by 25% compared with the first experiment.

b) Calculate the wavelength of the second light source.

Solution

a) Fringe separation on the screen $s = \frac{0.050}{15} = 3.33 \times 10^{-3}$ m

Slit separation $d = \frac{\lambda D}{s} = \frac{520 \times 10^{-9} \times 1.6}{3.33 \times 10^{-3}} = 2.5 \times 10^{-4}$ m $= 0.25$ mm

b) From $s = \frac{\lambda D}{d}$, the fringe separation is directly proportional to the wavelength; hence the wavelength must have been 25% longer than in the first experiment.

Wavelength = $520 \times 1.25 = 650$ nm

Assessment tip

When light of a single wavelength (**monochromatic**, meaning one colour) is diffracted by a single slit, the red light is diffracted more than the blue light. This is the opposite to what happens when white light is refracted. In refraction, red light is refracted less than the blue because the red light travels faster in most media than blue.

Interference and diffraction

Waves incident on a slit or an obstacle undergo **diffraction**. The wave "spreads out" as it interacts with the aperture (Figure 6).

The effect is most obvious when the aperture dimensions are of the same order as, or smaller than, the wavelength. Diffraction is easily demonstrated with light but is observed for all waves. For example, long-wavelength radio waves are diffracted by hills to allow radio reception in what is apparently a shadow area.

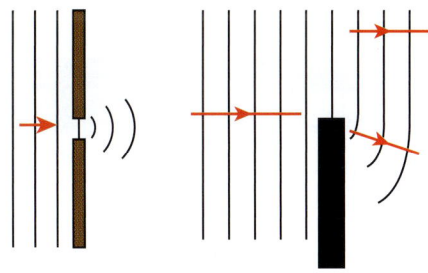

▲ Figure 6 Diffraction at a single aperture and at an edge

Example 5

Monochromatic light is incident on a narrow slit.

Sketch a graph showing the variation of light intensity with distance from the centre of the diffraction pattern.

Solution

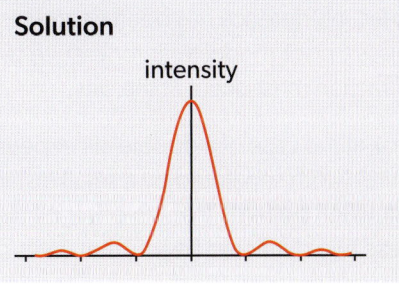

Assessment tip

Take care not to confuse the words **refraction** and **diffraction**. It is easy to write one instead of the other when under exam pressure, so be careful to avoid this error.

C Wave behaviour

> **Assessment tip**
>
> Monochromatic light, diffracted by a slit, gives the instantly recognizable diffraction pattern of Example 5. You should be able to sketch diffraction patterns with precision and confidence.
>
> Take care drawing the single-slit diffraction pattern. Remember:
> - the central maximum is about nine times more intense than the first maximum
> - maxima further from the central position have even smaller intensities than the first maximum
> - individual maxima are not symmetrical.
>
> When comparing the diffraction patterns produced by single slits of different widths, the intensity of the central maximum of the pattern is proportional to the (slit width)^2 because intensity $\propto$ amplitude^2. The diagram shows two diffraction patterns drawn to scale with one pattern displaced vertically upwards for clarity. The monochromatic light is the same for both patterns. The upper (red) pattern is formed by a slit that is three times wider than the slit for the lower (blue) pattern.

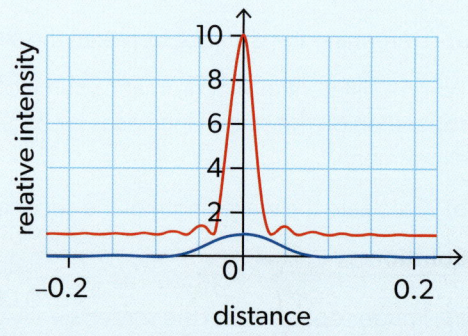

> **Assessment tip**
>
> You should be able to carry out straightforward calculations of the position of the first minimum position.
>
> θ must be in radians when using $\theta = \dfrac{\lambda}{b}$. When you have to calculate θ you can express your answer in radians or in degrees.
>
> All wave types demonstrate diffraction. You can be asked questions that involve sound waves or microwaves as well as light. The basic physics is the same.

The first minimum position of a single-slit diffraction pattern is given by $\theta = \dfrac{\lambda}{b}$, where θ is the angle between the central maximum and the first minimum, λ is the wavelength and b is the width of the single slit. The proof for this equation uses the approximation $\sin\theta \approx \theta$, so that the equation is only true for small θ.

Example 6

Sound waves of wavelength 35 cm are incident on a gap in a fence of width 2.7 m.

a) The first minimum in the intensity of the sound is at an angle of θ from the central maximum. Calculate θ.

b) The frequency of the sound is reduced without changing its amplitude. State and explain how this will affect the position of the first minimum.

Solution

a) $\theta = \dfrac{\lambda}{b} = \dfrac{0.35}{2.7} = 0.13$ rad, which is 7.4°.

b) A reduced frequency means a greater wavelength, so the sound will be diffracted through a larger angle.

So far, interference has been treated as occurring at slits that are infinitely thin. This is a poor model for interference by multiple slits as both interference and diffraction must occur at the slits.

Look at the case of a double-slit experiment that has two slits of the same (finite) width separated by a small distance. Both slits give rise to identical diffraction patterns that are offset by the slit separation. It is the interference of these two diffracted beams that gives rise to the final fringe pattern. (Figure 7).

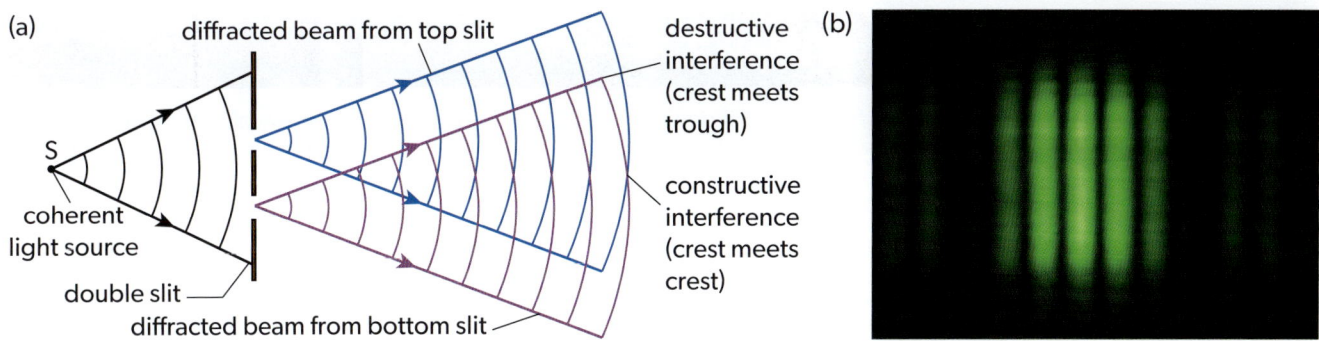

▲ Figure 7 (a) The arrangement to produce two diffracted beams from double slits that interfere.
(b) The appearance of a double-source interference pattern with diffraction

The Young's double-source interference pattern (each fringe of equal intensity) is **modulated** by a single-slit diffraction pattern (Figure 7(b)). The diffraction pattern acts as an envelope for the interference fringe pattern, setting a maximum limit on the intensity at any position (Figure 8).

As the number of slits increases to three or more, with the individual slit width and spacing unchanged, other effects appear (Figure 9). The fringes become sharper. Subsidiary maxima appear between the fringes but, as the number of slits increases, these become relatively less intense than the main peaks between them.

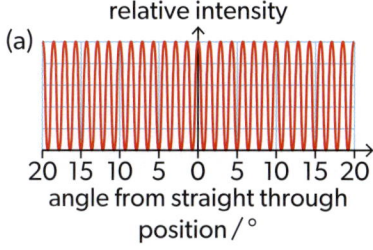
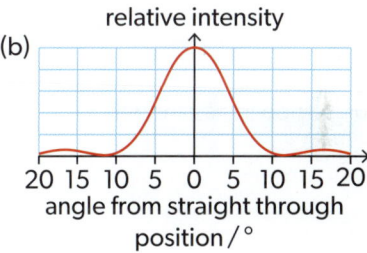
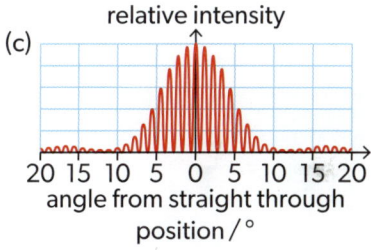

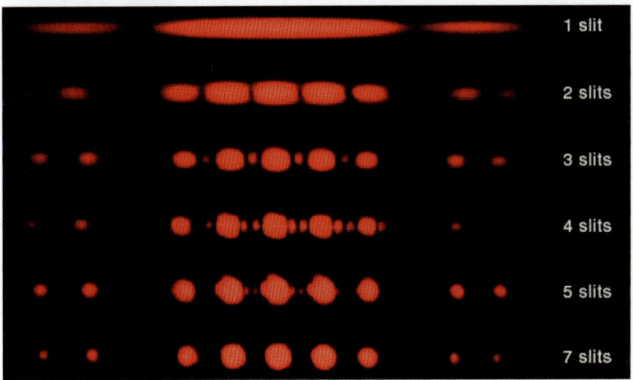

▲ Figure 9 As the number of interference slits increases, the fringes become sharper and more intense

▲ Figure 8 The overall effects of diffraction and interference for double-source interference through finite slits. (a) The interference pattern. (b) The diffraction pattern of one slit alone. (c) The combination of the diffraction and the interference

When the number of slits becomes very large, the arrangement becomes that of a **diffraction grating**. The pattern for monochromatic light becomes a bright central maximum surrounded by intense sharp lines with darkness between them. These lines are called **orders** and are given integer labels counting out from the centre, which is assigned the value zero—leading to central or zeroth order, first order, second order, …). When there are many wavelengths in the light, individual lines may merge to form a spectrum—for light, this would be a continuous red to violet band (Figure 10).

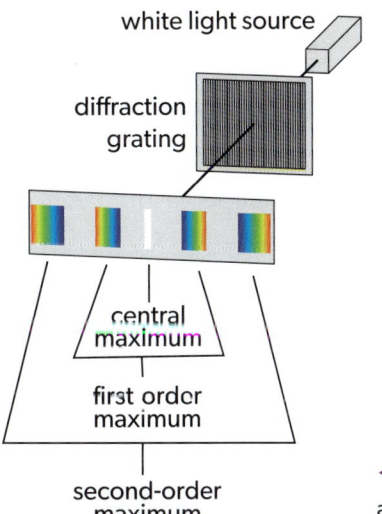

◀ Figure 10 The diffraction-grating arrangement

For a diffraction grating, the angle θ between the straight-on beam and a line is given by $n\lambda = d \sin \theta$, where n is the **order** of the spectrum, λ is the wavelength of a line in the spectrum and d is distance between adjacent slits or the slit separation.

The **number of slits per metre** is one way in which the closeness of the lines in a diffraction grating can be specified—it is equal to $\frac{1}{d}$.

C Wave behaviour

Assessment tip

There are several occasions in the IB Diploma Programme physics course where an answer can only be an integer. The order of a diffraction-grating maximum is one of these.

Whenever an answer must be an integer, take care to quote the answer appropriately. It may not, for example, be correct to round an answer such as 12.7 revolutions up to 13 if you are asked for the number of **complete** revolutions made by a wheel.

Example 7

A diffraction grating has 4.5×10^5 lines m^{-1}. Light of wavelength 486 nm is normally incident on the grating. Determine the highest order diffracted image that can be produced for this wavelength by the diffraction grating.

Solution

Distance between adjacent lines of the diffraction grating $d = \dfrac{1}{4.5 \times 10^5}$
$= 2.22 \times 10^{-6}$ m

The highest order occurs at $\theta = 90°$. Using this value in the equation gives

$$n = \frac{d \sin \theta}{\lambda} = \frac{2.22 \times 10^{-6} \times \sin 90°}{4.86 \times 10^{-7}} = 4.6$$

This means that the fourth order is the highest order that can be observed.

Sample student answer

A student investigates how light can be used to measure the speed of a toy train.

Light from a laser is incident on a double slit. The light from the slits is detected by a light sensor attached to the train.

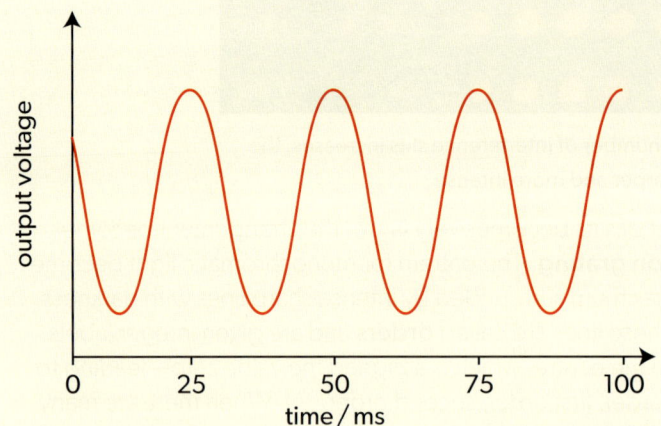

The graph shows the variation with time of the output voltage from the light sensor as the train moves parallel to the slits. The output voltage is proportional to the intensity of light incident on the sensor.

a) Explain, with reference to the light passing through the slits, why a series of voltage peaks occurs. [3]

This answer could have achieved 3/3 marks:

> Light hits the two slits and diffracts into two radially propagating waves. If the path difference at the sensor is a multiple of the wavelength of the light then, by superposition, a maximum occurs (constructive interference). If an $\left(n + \dfrac{1}{2}\right)$ multiple occurs, then destructive interference occurs (the troughs). A series of peaks are created because they are at multiples of the wavelength. Because the voltage is proportional to intensity which is proportional to (amplitude)^2, the constructive interference forms voltage peaks.

▲ This is a complete answer from a student who has read the question thoroughly. The process of superposition is well described and the distinction between constructive and destructive interference is clear. The answer also links intensity of the fringe pattern to the voltage peaks, which is also a clear requirement of the question.

C.3 Wave phenomena

This answer could have achieved 0/3 marks:

> As the light is passing through the slits, maxima and minima will appear as the light sensor is in the different positions.

▼ Although there is a recognition that there are maxima and minima, this really just restates the details of the graph in the question. There is no attempt to explain how the fringes form or how their intensity variation relates to changes in voltage shown on the graph.

b) The slits are separated by 1.5 mm and the laser light has a wavelength of 6.3×10^{-7} m. The slits are 5.0 m from the train track. Calculate the separation between two adjacent positions of the train when the output voltage is at a maximum. [1]

This answer could have achieved 1/1 marks:

> $s = \dfrac{\lambda D}{d} = \dfrac{(6.3 \times 10^{-7})(5)}{(1.5 \times 10^{-3})} = 2.1 \times 10^{-3}$ m = 2.1 mm

▲ A careful, legible answer with a clear substitution and evaluation.

Practice problems

Problem 1

A ray of light in air is incident on a rectangular glass block. The ray enters the block at point A, at an angle of 45.0° with the normal. The refracted ray reaches the surface of the block at point B.

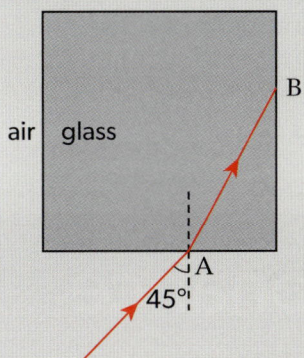

The refractive index of the glass is 1.48.

a) Calculate:

(i) the angle of refraction at A

(ii) the angle of incidence at B.

b) (i) Calculate the critical angle for light entering air from the glass.

(ii) State, with a reason, whether the ray of light will enter air at B.

Problem 2

A plane sound wave travels from air to water. The diagram shows some of the wavefronts of the sound wave. The wavefronts in the air make an angle of 12° with the water surface and those in the water make an angle of 66° with the surface.

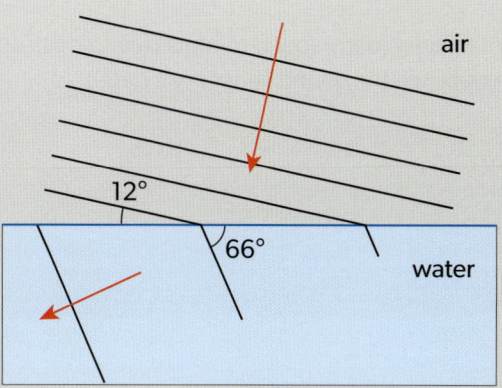

a) Calculate the relative refractive index for sound waves entering water from air.

b) The speed of the wave in air is 340 m s^{-1}. Calculate the speed of the wave in water.

c) The frequency of the wave is 800 Hz. Calculate the smallest distance between the wavefronts in the water.

Problem 3

In a double-slit experiment with monochromatic light, the slits are separated by 0.15 mm and the interference pattern is viewed on a screen that is 2.0 m from the slits. The fringe separation on the screen is 6.0 mm.

a) Calculate the wavelength of light.

b) The position of the screen is changed and the bright fringes are now separated by 1.0 cm. Calculate the distance from the slits to the screen after the change.

C Wave behaviour

Problem 4

Two loudspeakers A and B are driven in phase and emit sound of frequency 1700 Hz. The loudspeakers are 2.0 m apart. A sound sensor moves away from loudspeaker A, in a direction that is perpendicular to the line joining the loudspeakers. The sound sensor is initially at point P, 4.8 m from loudspeaker A. The angle BAP is 90°. The speed of sound in air is 340 m s^{-1}.

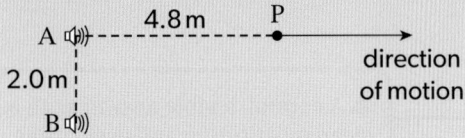

a) Calculate the wavelength of the sound emitted by the loudspeakers.

b) Determine the nature of the interference at P.

c) Explain how many sound minima due to destructive interference the sound sensor will detect.

Problem 5

Laser light with wavelength 6.2×10^{-7} m is incident on a single slit of width 0.15 mm.

a) Calculate, in degrees, the angle between the central maximum and the first minimum in the diffraction pattern.

b) Describe and explain the change in the appearance of the pattern when the monochromatic laser light is replaced by white light.

Problem 6

A diffraction grating with 10 000 lines m^{-1} is used to analyse light emitted by a source.

The source emits a range of wavelengths from 500 nm to 700 nm.

a) Calculate the angle between the central maximum and the first-order maximum for the 500 nm wavelength.

b) Calculate the angular width of the first-order spectrum.

c) A detector is positioned a distance of 2.0 m from the grating to detect the maxima. Calculate the distance between the extreme ends of the first-order spectrum in this position.

C.4 Standing waves and resonance

You must know:

- the nature of standing waves and how they form from the superposition of two travelling waves
- what is meant by a node and an antinode
- how boundary conditions in strings and air in pipes influence the node–antinode pattern in a standing wave
- what is meant by the term harmonic
- what is meant by the natural frequency of a vibration
- what is meant by resonance, in response to a periodic stimulus, and by driving frequency
- what is meant by damping.

You should be able to:

- describe the nature and formation of standing waves in terms of superposition
- distinguish between standing and travelling waves
- sketch and interpret the standing wave patterns that form in strings and the air in pipes
- solve problems involving the frequency of a harmonic, the wavelength of the standing wave and the wave speed
- describe and interpret examples of light, heavy and critical damping
- describe, using a graph, how the amplitude of vibration varies with driving frequency for an object close to its natural frequency of vibration
- describe both useful and destructive effects of resonance.

Standing waves

Standing waves are constant in space but vary with time. They arise when two or more travelling waves interact. The simplest case is that of two identical travelling waves moving in opposite directions in the same medium.

Permanent zero positions on a standing wave are **nodes** and the positions of peak amplitude are **antinodes**.

The table shows comparisons between standing and travelling waves.

	Standing wave	Travelling wave
Amplitude	Zero at nodes and maximum at antinodes equal to $2x_0$.	Same for all particles in wave. Amplitude is x_0.
Energy	No energy transfer but there is energy associated with the motion of individual particles.	Energy transfer along wave.
Frequency	Same for all particles except those at nodes (where particles are at rest).	Same for all particles.
Wavelength	Twice the distance between any pair of adjacent nodes or antinodes.	Distance between nearest particles with the same phase.
Phase	Phase for all particles between adjacent nodes is the same. Phase difference of π rad between one internodal segment and the next.	All particles within one wavelength have different phases (the difference varies from 0 to 2π rad).

The formation of a standing wave is closely linked to resonance, which is described later in this topic.

There are links to oscillations of all types, in particular, to simple harmonic oscillation (Topic C.1).

A standing-wave condition can be used to visualize the orbitals in the Bohr model of the hydrogen atom (Topic E.1, AHL only).

Standing waves form when a wave is reflected at a boundary, and this reflected wave and the original wave superpose. Figure 1 shows the cases when a pulse travelling along a string meets either a fixed end or a free end.

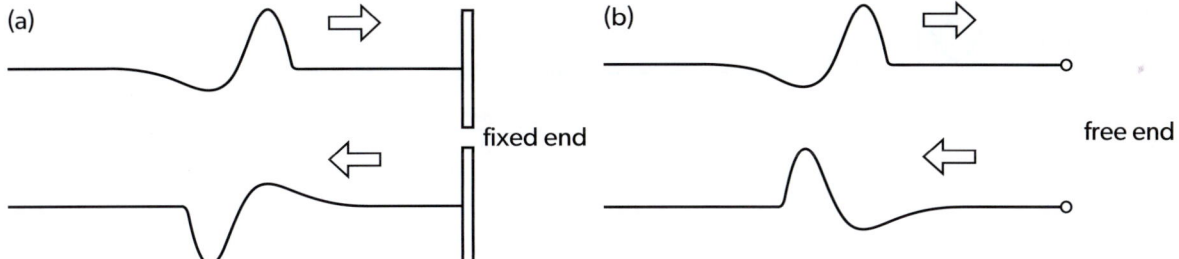

▲ Figure 1 (a) A pulse reflected at a fixed end. (b) A pulse reflected at a free end

- When the string has a fixed end, the reflected pulse is inverted.
- When the string has a free end, the reflected pulse is the original amplitude non-inverted.

The boundaries (free or fixed) determine the standing-wave shape for a specific frequency.

Each end of a string can be fixed or free. At the points where a string is fixed there must be displacement nodes where the reflection of the waves occurs. Midway between nodes (assuming the string is uniform) there is an antinode.

Nature of science

There are many laboratory demonstrations of the motion of standing waves. Some are based on real apparatus, some on computer simulations. When viewing a model that involves a computer-created image, you should always ask: to what extent is this a real model (i.e. based on the strengths and weaknesses of the underlying physics)? Or is it simply an animation?

C Wave behaviour

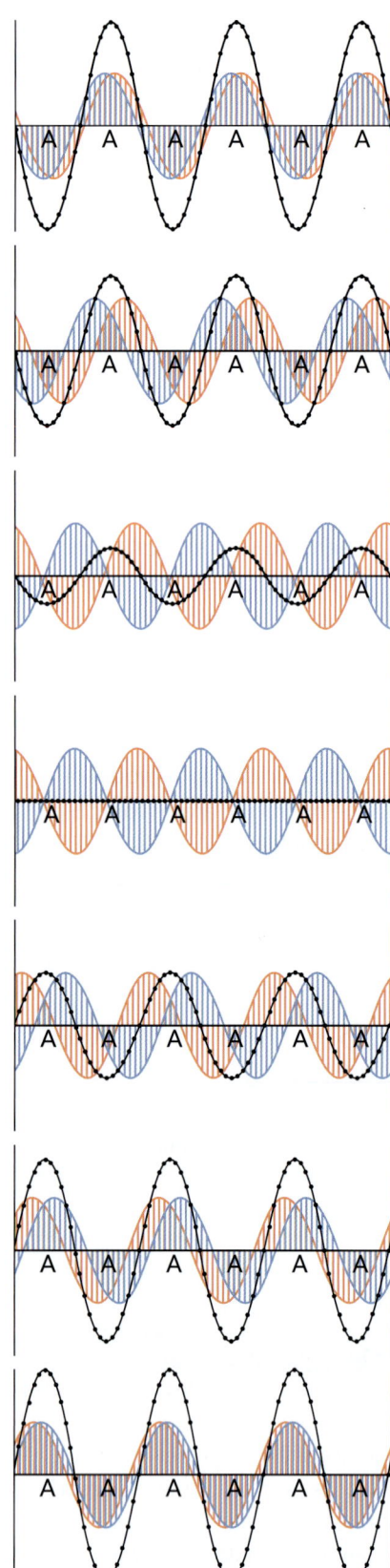

▲ Figure 2 A standing wave (black) forms when two waves (red and blue) with the same wavelength, travelling in opposite directions, superpose. Label A marks the position of the antinodes

The formation of a standing wave on a string can be seen in the time-sequence of Figure 2. The blue wave is travelling to the left. The red wave is travelling to the right. The black wave shows the superposition of the red and blue waves. It does not move in space but oscillates with the same frequency as the two travelling waves.

When a string is stretched between two fixed points, there will be a smallest frequency f_1 for which a standing-wave forms, with a peak amplitude in the centre of the string. This is the **first harmonic.** It is an important oscillation mode for many stringed instruments.

At greater frequencies, further harmonics form with more nodes along the string. These are other members of the **harmonic series** (second harmonic, third harmonic, ...).

(a)
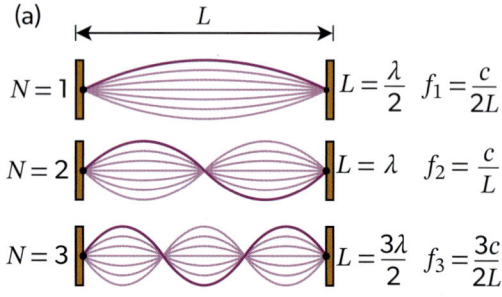

(b)
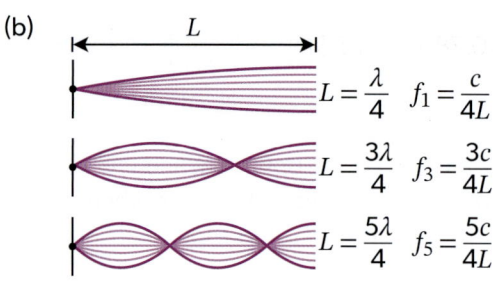

(c)
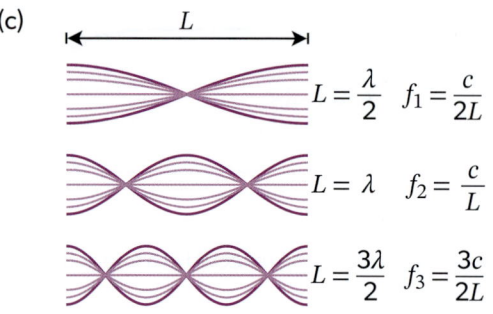

◀ Figure 3 The first three shapes of the harmonic series for (a) a string fixed at both ends, (b) a string fixed at one end and free at the other and (c) a string free at both ends

Figure 3(a) shows the first three standing waves for a string of length L fixed at both ends. The first harmonic corresponds to a wavelength of $2L$ and therefore $f_1 = \dfrac{c}{2L}$, where c is the speed of the wave on the string. Figure 3(a) also shows the next two standing waves and their frequencies f_2 and f_3.

Assessment tip

The physics of the standing waves formed in the gas in a pipe is complicated by the fact that it is possible to discuss the situation either in terms of displacement of the air molecules or in terms of the pressure variations in the gas. In the IB Diploma Programme physics course, the descriptions are always in terms of the displacement nodes and antinodes, **never** the pressure variations. This is why the word "displacement" appears in the text in brackets.

You may see discussions of pressure variations in other books.

Figures 3(b) gives the first three standing waves and their frequencies for a string fixed at one end, and 3(c) gives the waves and frequencies for a string free at both ends.

To observe the "two free ends" case in practice, it is necessary to clamp a stiff metal wire at a nodal point, the position of which will vary for different harmonics.

Example 1

A standing wave oscillates with three loops on a string of length 0.78 m, fixed at both ends. The frequency of oscillation is 140 Hz.

a) Calculate the speed of transverse waves along the string.

b) Calculate the first harmonic frequency of the string.

Solution

a) This is the third harmonic standing wave:

wavelength $= \frac{2}{3} \times 0.78 = 0.52$ m

As the frequency is 140 Hz,

wave speed $= f\lambda = 140 \times 0.52 = 73$ m s^{-1}

b) First harmonic wavelength $= 0.78 \times 2 = 1.56$ m

$f_1 = \frac{73}{1.56} = 47$ Hz

This is one-third of the frequency of the standing wave in part (a).

Assessment tip

Take care with the harmonic notation. For the "both fixed" and "both free" cases, the harmonics are numbered in integer values: first harmonic, second harmonic, third harmonic, first order, second order, ...

For the case where the string is fixed at one end and free at the other, there are no even harmonics. The first harmonic is one-quarter of the wavelength, the next harmonic is three-quarters of the wavelength and so is called the third harmonic, the one after that is the fifth harmonic (five-quarters of the wavelength), and so on.

Standing waves also form in the air in a pipe. These lead to the series of harmonics shown in Figure 4.

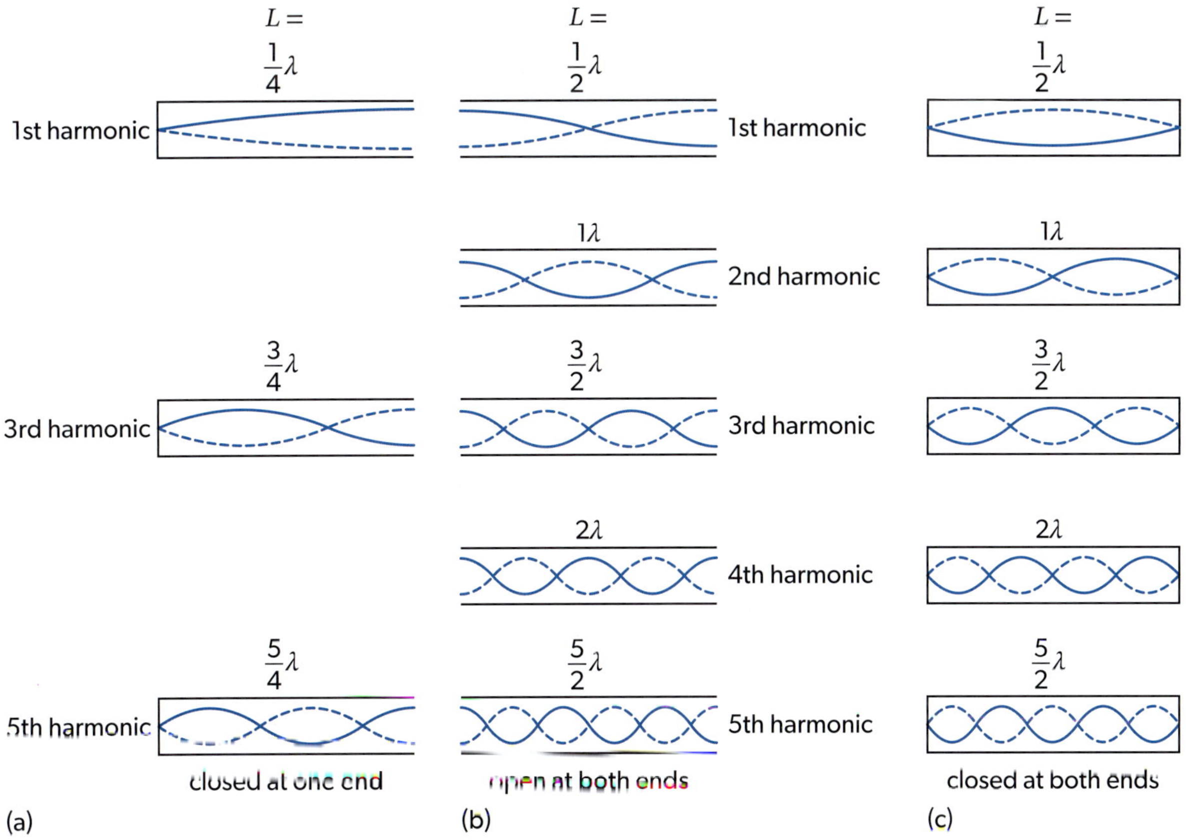

▲ Figure 4 The standing waves formed in a pipe where (a) one end is open and the other closed, (b) both ends are open and (c) both ends are closed

C Wave behaviour

Look again at Figure 2 in Topic C.2 on page 108. Notice that the positions of 2 cm and 6 cm correspond to points where the pressure is a minimum (i.e. below average pressure) but the displacement is zero. In other words, a displacement node but a pressure antinode. Convince yourself that at displacement antinodes (e.g. at 3 cm) the pressure must be atmospheric (because all wave particles are displaced by the same amount) so this is a pressure node.

In pipes, the nodes are formed at points of zero displacement. The antinodes form halfway between the nodes, as for strings.

A closed end of a pipe forces a (displacement) node to form. An open end must correspond to a displacement antinode.

The phase of the reflections at the ends of a pipe arises from the nature of the boundary and the way in which longitudinal waves travel in a gas. Air molecules at a closed end cannot move parallel to the pipe axis. Therefore, an incident wave is forced to be reflected as an inversion of the original wave, otherwise the molecules would not remain stationary along the central axis at the fixed end.

At an open end, the pressure in the tube must always equal atmospheric pressure. This forces the open end to correspond to a displacement maximum. There must be a maximum amplitude of the movement of air molecules at an open end.

Example 2

A horizontal glass tube contains fine powder. A loudspeaker at one end of the tube emits a single frequency, and a standing wave forms in the tube. Powder heaps occur at nodes.

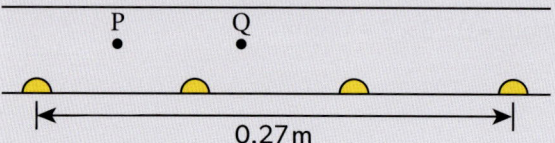

Speed of sound waves in air = $340 \,\text{m s}^{-1}$

a) Identify the type of wave formed in the tube.

b) Determine, for the sound in the tube:

  (i) its wavelength

  (ii) its frequency.

P and Q are points in the tube.

c) Compare the frequency, amplitude and phase of air particles at P with those at Q.

Solution

a) Longitudinal because this is a sound wave in a gas.

b) (i) $\lambda = \dfrac{0.27}{3} \times 2 = 0.18 \,\text{m}$

  (ii) $f\left(=\dfrac{c}{\lambda}\right) = \dfrac{340}{0.18} = 1.9 \,\text{kHz}$

c) The frequencies are the same. P is midway between two nodes so must be an antinode.

Q is between the antinode and node, so the amplitude of P is greater than the amplitude of Q.

P and Q are in adjacent node–node segments. This means that the wave displacements are in opposite directions, so the phase difference is π rad.

C.4 Standing waves and resonance

Example 3

An organ pipe of length L is closed at one end.

a) Sketch the displacement of the air in the pipe when the emitted frequency of sound is:

   (i) the lowest possible f_a

   (ii) the next harmonic above the lowest possible f_b.

b) Deduce $\dfrac{f_a}{f_b}$.

Solution

a) (i) and (ii) See Figure 4(a), first and third harmonics.

b) L is the length of the pipe. So,

$$\lambda_a = 4L \qquad \lambda_b = \frac{4L}{3}$$

Using $c = f\lambda$,

$$f_a = \frac{c}{4L} \qquad f_b = \frac{3c}{4L}$$

$$\frac{f_a}{f_b} = \frac{1}{3}$$

Sample student answer

A toy train travels away from a loudspeaker that is emitting sound waves of constant amplitude and frequency. The train is moving at constant speed towards a reflecting barrier.

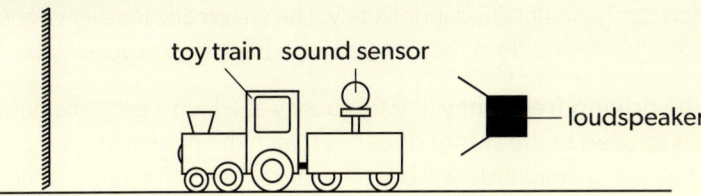

A sound sensor is mounted on the train. The graph of the variation with time of the output voltage from the sensor is shown.

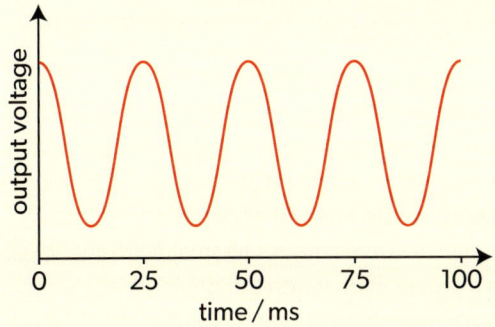

Explain the shape of the graph. [3]

This answer could have achieved 1/3 marks:

> The sound wave goes from the loudspeaker and is reflected at the barrier. This means that there is a standing wave in the space where the train is moving. The sensor hears the wave and gives a large voltage when the sound is loud. When the train goes through a node then the sound is loud.

▲ The key to answering this question is to recognize that a standing wave forms in the space between the barrier and the loudspeaker. A reference to the reflection at the barrier is required and this point is made by the student.

▼ The answer needs to go on to describe the formation of the standing wave and its sequence of nodes and antinodes. This part of the answer is absent. The student makes matters worse by associating the node of the standing wave with a maximum intensity rather than a minimum (or zero).

C Wave behaviour

Simple harmonic motion is covered in Topic C.1.

- **Lightly damped** oscillators lose energy gradually and take many oscillations to come to rest. In light damping, the amplitude of the oscillation decreases exponentially. This implies that the time to lose half of the amplitude (and three-quarters of the energy) is constant for the system.
- **Critically damped** oscillators stop moving (and therefore lose their total kinetic energy) in the shortest time possible.
- **Heavily damped** oscillators stop moving in a longer time than the critically damped case.

Assessment tip

Figure 6 shows four amounts of damping and the variation of the maximum amplitude. As damping increases, the point of maximum amplitude drifts to lower frequencies. The curves are not symmetrical about the maximum amplitude, and you should always ensure that you draw them asymmetrically as here.

Forced vibrations and resonance

Simple harmonic motion has, by definition, a constant amplitude and constant total energy—this was an implicit assumption of earlier topics. In real cases, a freely oscillating system gradually loses energy through resistive losses. The transfer of energy out of the oscillating system by frictional forces is known as **damping**.

Damping can be light, critical or heavy (Figure 5).

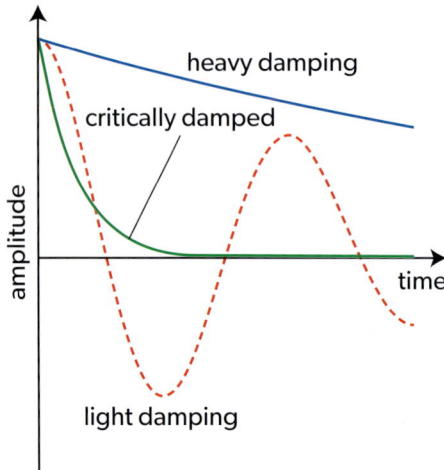

▲ Figure 5 An amplitude–time graph for oscillators with various degrees of damping

Suppose that a child on a swing is given a push to begin the pendulum-like oscillation. Without intervention, the swing oscillates at its **natural frequency**. The swing will eventually stop moving because of energy transfers to the bearings of the swing and the surroundings.

However, if we add energy to the swing system in a systematic way, the oscillation can be maintained indefinitely. The systematic transfer of energy to the swing is usually made at the same point in the swing's cycle.

When the **driving frequency** (the frequency at which a periodic pulse of energy is applied to the swing) does not match the natural frequency of the swing, the swing amplitude will become larger or smaller depending on how close the driving frequency is to the natural frequency. The driven system is said to undergo forced oscillations. After a short time, the swing will reach an equilibrium. This is where the energy input from the driver in one cycle is equal to the energy transferred out of the system in one cycle through resistive losses.

Figure 6 shows how the amplitude of the driven system varies with frequency of the driver. An important feature of these curves is that there is a maximum amplitude for each level of damping—this occurs at the **resonance** condition. As the amount of damping increases, the maximum in the resonance curves moves to smaller frequencies. Amplitude resonance is the condition when the driving frequency produces a maximum amplitude in the driven system.

The maximum amplitude at resonance is also reduced as damping increases. With very heavy damping it may not be possible to observe a maximum in the curve at all, and the amplitude simply increases as the driving frequency is reduced.

In the limit of very light damping, the resonant frequency is the same as the natural frequency of the driven system.

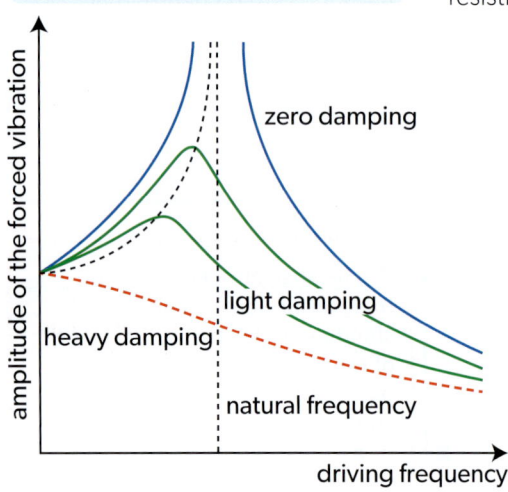

▲ Figure 6 Resonance curves for different degrees of damping

Figure 7 shows the phase relationship between the driver system and the driven system. When the driver frequency is greater than the driven frequency, the driver is 180° out of phase with the driven system. When the driver frequency is smaller than the natural frequency, the driver is in phase. At resonance, the driver and the driven system are 90° apart, with the driver leading.

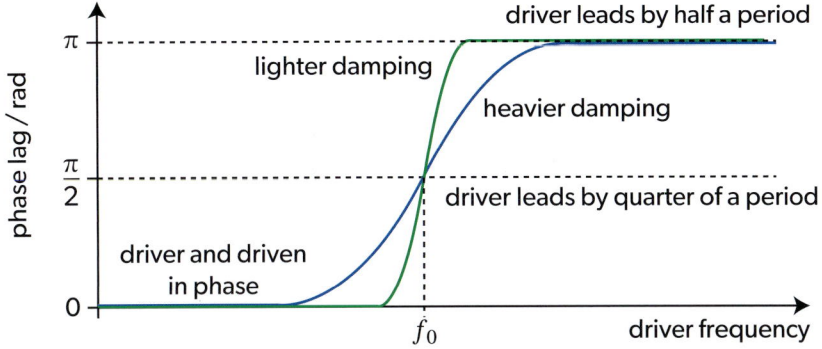

▲ Figure 7 Phase relationships between driver frequency and the natural frequency for different amounts of damping

Approaches to learning

It is always helpful to relate graphs such as Figure 7 to reality. Think about pushing a child on a swing. Not only must the rate at which the push is supplied be correct, but it has to be supplied at the right moment. The maximum push has to be given when the child is moving through the equilibrium position (maximum speed). The push (the driver) has to be one-quarter cycle ahead of the swing itself.

Resonance is an important phenomenon. Resonating tuning circuits are used in radios to select the received signal at its transmitted frequency. Tides are caused by resonances in the oceans driven by the gravitational attraction of the Sun and Moon.

Disadvantages of resonance include large and potentially destructive amplitudes in mechanical systems when an oscillating system is driven close to its natural frequency by a varying driving system.

Sample student answer

A driven system is lightly damped. The graph shows the variation with driving frequency f of the amplitude A of oscillation.

On the graph, sketch a curve to show the variation with driving frequency of the amplitude when the damping of the system increases. [2]

This answer could have achieved 2/2 marks:

▼ The endpoints are not clear: on the left, the curves appear to cross. Try to avoid this ambiguity. There was no penalty, however.

▲ The lower curve is reasonably drawn and incorporates the features that examiners were looking for. The peak is lower than the original because the damping is greater. The peak is shifted to lower frequencies, which is also correct.

C Wave behaviour

Practice problems

Problem 1

A standing wave of frequency 260 Hz is formed on a string of length 0.80 m that is fixed at both ends. The standing wave has two antinodes.

a) State the wavelength of the standing wave.

b) Calculate the speed of the waves in the string.

c) Calculate the frequency of the first harmonic mode.

Problem 2

A stiff wire of length 1.1 m is clamped at one end and has the other end free. Two successive harmonic frequencies of standing waves in the wire are 1.0 kHz and 1.4 kHz.

a) Determine the frequency of the first harmonic mode.

b) Calculate the wave speed.

c) State the number of antinodes of the 1.4 kHz harmonic.

Problem 3

A pipe of length 0.45 m is open at both ends.

a) Calculate the first two harmonic frequencies that can be produced in this pipe.
The speed of sound in air is 340 m s^{-1}.

b) Compare the nth harmonic standing wave in this pipe with the nth harmonic standing wave in a pipe of the same length that is closed at both ends.

Problem 4

a) Outline a condition necessary for resonance to occur in an oscillating system.

One of the wheels of a car is out of balance and provides a periodic force on the suspension of the car. The period of this force is equal to the period of rotation of the wheel. The radius of the wheel is 0.32 m.

The graph shows how the amplitude A of oscillations of the car varies with frequency f_D of the driving force acting on the car.

b) Determine the speed of the car at which it will have the greatest amplitude of oscillation.

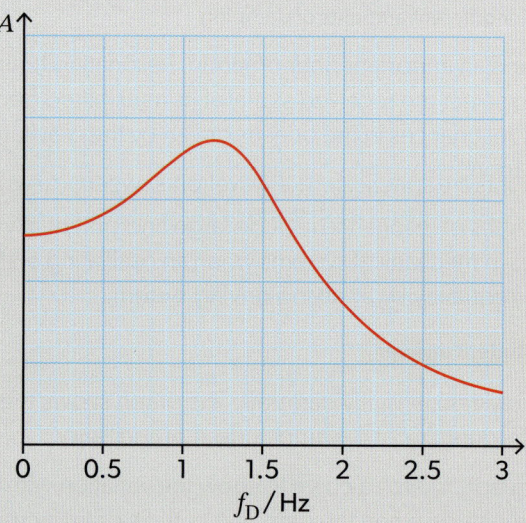

C.5 Doppler effect

You must know:
✔ that shifts in frequency are observed when there is relative motion between a source and an observer
✔ that there are differences between the Doppler effect as observed in light waves and in sound waves
✔ that spectral-line shifts give information about the relative motion of astronomical bodies
✔ that the Doppler effect has applications in medical physics and radar techniques.

You should be able to:
✔ sketch and interpret diagrams showing how the Doppler effect occurs when either the source or the observer is moving
✔ solve problems for the relative change in frequency or wavelength for light when the speed of light is much larger than the relative speed between source and observer

Additional higher level:
✔ solve problems for sound and mechanical waves involving the Doppler effect.

When a source S of a sound wave or electromagnetic radiation is at rest relative to an observer O, both S and O agree about the frequency of the wave. Figure 1 shows that the wavefronts emitted by a point source S spread outwards. At one instant the waves form concentric circles in two dimensions (in three dimensions the wavefronts are spherical). The wavefronts sweep across O at the same rate at which they were emitted by S. S and O agree about the frequency.

A Doppler shift occurs in the observed frequency when the source and observer move relative to each other. The cases for sound and electromagnetic radiation need to be treated differently.

For sound waves, when S is moving relative to the transmission medium (taken to be air here) and O is stationary, the effect is easy to imagine (Figure 2(a)). The wavefronts from the source are emitted at a constant rate (the source frequency) but the source "catches up" with its own waves. Once the waves have been emitted, they continue to move in the medium at the wave speed. They cross O more quickly than they were emitted, so a higher frequency is detected by O.

For S moving away from O, the distances between waves are "stretched" compared with the stationary case and the observer detects a lower frequency than that emitted (Figure 2(b)).

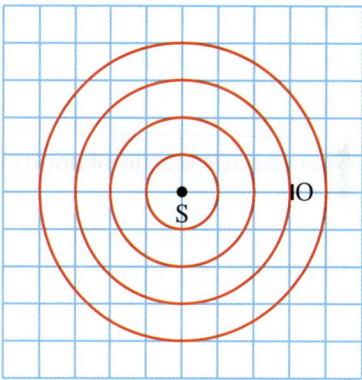

▲ Figure 1 When a wave source and an observer are stationary relative to each other, there is no Doppler shift observed

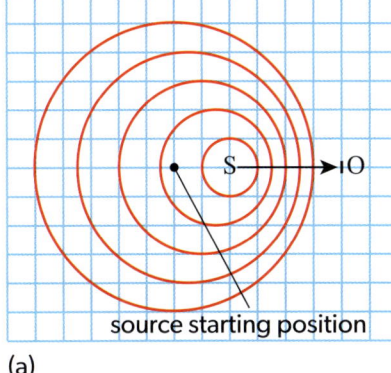
(a)

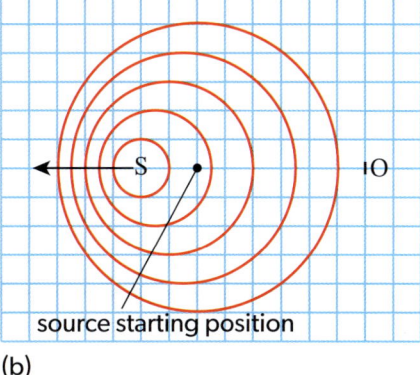
(b)

▲ Figure 2 (a) O is stationary and S is moving towards O at half a square per time interval. (b) S is moving away from O at the same speed as in (a)

When the observer O moves relative to a stationary source S, the emitted wavefronts are not distorted by source movement and are concentric circles (spheres in three dimensions) centred on the source. However, the observer crosses the wavefronts more quickly (when approaching) or more slowly (when receding), so a Doppler frequency shift is detected again.

Nature of science

The mechanisms that lead to wave characteristics in light and sound are very different. Light reflection involves absorption and re-emission of photons, whereas sound reflection involves compressions and rarefactions at a solid boundary. Electromagnetic radiation in a vacuum has a universal speed. These differences lead to further differences between the behaviour of light and sound waves in the Doppler effect.

It is part of the nature of science to seek to identify the similarities and differences between phenomena such as this.

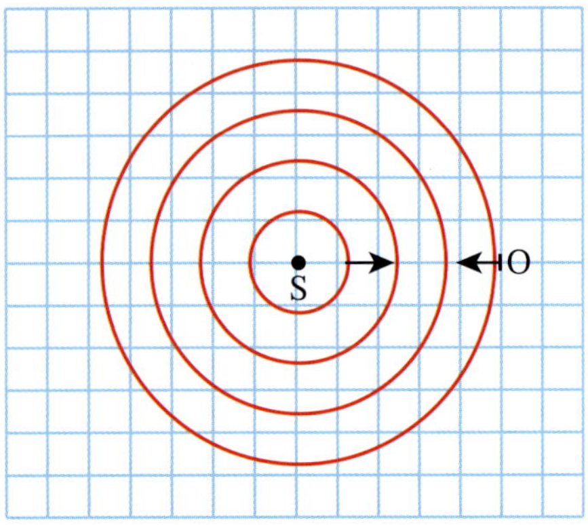

C Wave behaviour

▲ Figure 3 O is moving towards S. O crosses the wavefronts more often than they are emitted, giving a higher frequency

When the source S is stationary relative to a moving observer O, the wavefronts move symmetrically but they are detected by O more or less frequently than they were emitted. Figure 3 shows O moving towards S.

These diagrams are drawn on the basis that the source and observer are moving along the line between their centres. When this is not true, the speed components along the line between centres must be used.

The Doppler effect with electromagnetic waves needs care, so the IB Diploma Programme physics course only considers cases where the observer and source speeds are much less than the speed of light ($v \ll c$). Using this assumption, $\frac{\Delta f}{f} = \frac{\Delta \lambda}{\lambda} \approx \frac{v}{c}$, where Δf is the change in frequency due to the Doppler effect and $\Delta \lambda$ is the change in wavelength. Because the changes are small, f and λ can be the source or observed frequency, or the average of the two.

At higher speeds, the equation is more complicated.

When the source and observer are moving away from each other, the observed wavelength is longer than that emitted, so the frequency is reduced. This is called a **redshift**. When the source and observer are moving closer together, there is a **blueshift**.

For a reflection, the frequency shift $\Delta f = 2f \frac{v}{c}$.

> The Doppler effect has many applications. It is used in radar to determine the speeds of moving objects and in weather forecasting. In radar astronomy, microwaves are reflected from objects, for example the Moon, to detect their distance from Earth and their speed.
>
> The effect is used in medical applications to determine blood flow and other flow measurements. It has the advantage that the measurements are non-invasive for the patient.

Example 1

A stationary source emits a spectral line of wavelength 396.8 nm. The same line observed in a spectrum of a distant galaxy has wavelength 399.7 nm.

a) State the direction of motion of the galaxy relative to Earth.

b) Calculate the relative speed of the galaxy.

Solution

a) The observed wavelength is longer than the emitted wavelength, so the galaxy is moving away from Earth.

b) Shift in wavelength $\Delta \lambda = 399.7 - 396.8 = 2.9$ nm

Relative speed $= \frac{\Delta \lambda}{\lambda} c = \frac{2.9}{396.8} \times 3.0 \times 10^8 = 2.2 \times 10^6 \, \text{m s}^{-1}$

> Spectral lines arise from transitions between atomic energy levels and their wavelengths are specific to each chemical element. This is discussed in Topic E.1.

The *Physics data booklet* gives two separate equations for the Doppler effect when applied to a sound wave or a mechanical wave:

- for a moving source, $f' = f\left(\frac{v}{v \pm u_s}\right)$, where u_s is the source speed

- for a moving observer, $f' = f\left(\frac{v \pm u_o}{v}\right)$, where u_o is the observer speed.

In both equations, f is the emitted (source) frequency and f' is the observed frequency, v is the speed of the wave, and u_s and u_o are the speeds of the source and observer, respectively.

Example 2

A whistle emits sound of frequency of 1000 Hz and is attached to a string. The whistle is rotated in a horizontal circle at a linear speed of 30 m s^{-1}. The speed of sound in air is 330 m s^{-1}.

An observer is standing a long way from the whistle on the same horizontal plane.

a) Explain why the sound heard by the observer changes regularly.

b) Determine the maximum frequency of the sound heard by the observer.

Solution

a) The whistle (source) moves first towards the observer and then away. When the whistle is moving towards the observer, the wavefronts are compressed and the rate at which they cross the observer increases compared with a stationary whistle. The frequency is increased. When source and observer are moving apart, the wavefronts are further apart and the frequency decreases. As the relative speed is constantly changing, the frequency heard changes regularly too.

b) The maximum frequency occurs when the source and observer are approaching.

$$f' = 1000 \times \frac{330}{330-30} = 1100 \text{ Hz}$$

Assessment tip

It is important to get the signs correct with these equations. Work from first principles each time. Imagine you are standing by the roadside and a car goes along the road at speed, sounding its horn. The sound of the horn appears to go from a high to a low frequency, changing as the car draws level with you. In the moving source equation, the minus sign must be used for the approaching car and the positive sign used for the receding car.

An alternative way to remember this is to memorise the full expression (it is not in the *Physics data booklet*), but only for the approaching case, this is

$f' = f\left(\dfrac{v + u_O}{v - u_S}\right)$. When either u_S or u_O is zero, that particular quantity disappears from the equation.

When the objects are moving apart, simply reverse the signs.

Sample student answer

Police use radar to detect speeding cars. A police officer stands at the side of the road and points a radar device at an approaching car. The device emits microwaves which reflect off the car and return to the device. A change in frequency between the emitted and received microwaves is measured at the radar device.
The frequency change Δf is given by

$$\Delta f = \frac{2fv}{c},$$

where f is the transmitter frequency, v is the speed of the car and c is the wave speed.

The following data are available.

Transmitter frequency $f = 40$ GHz

$\Delta f = 9.5$ kHz

Maximum speed allowed = 28 m s^{-1}

a) Explain the reason for the frequency change. [3]

This answer could have achieved 0/3 marks:

> As the car moves towards the police officer, the device's waves have an increased perceived frequency due to the relative motion of the car and the police officer.
>
> $f' = f\left(\dfrac{v}{v \pm u_s}\right)$

C Wave behaviour

As the reflected microwaves are reflected by the car, they travel back as if they had the additional speed of the car. As a result, the waves reach the device faster than they would have if they were reflected off a non-moving source, and therefore there is a frequency change.

This answer could have achieved 2/3 marks:

This case is a Doppler effect where there is a moving source of waves and a stationary observer (receiver). Because the source is moving, the distance that the waves travel is reduced, which causes the same waves to be 'squeezed' into a smaller distance. The wavelength is then decreased while the wavespeed remains the same. Therefore, frequency increases ($c = f\lambda$).

b) Suggest why there is a factor of 2 in the frequency-change equation. [1]

This answer could have achieved 0/1 marks:

Because the distance travelled by the wave is twice the distance between the source and the radar (car).

▼ The answer begins well with a discussion of relative motion. However, towards the end there is a statement that the "waves reach the device faster than they would have". The waves are travelling at the speed of light and this does not change depending on the observer's speed. This major error in physics disqualifies the whole answer.

▼ However, the idea that the distance the waves travel is reduced is incorrect—better to say that the wavefronts are closer because the source has moved since the emission of the previous wavefront.

▲ There is clear identification of the reason for the frequency shift. There is also the recognition that the frequency increases (this is not given in the question and is always worth stating if not).

▼ The factor of 2 is because of the reflection from the moving object which acts first as observer and then as source. The echo distance idea here is wrong.

Practice problems

Problem 1
Barnard's star is one of the nearest stars. It moves towards the solar system and its velocity component along the line of sight is 1.1×10^5 m s^{-1}.

a) Explain, with reference to wavefronts, why spectral lines observed in light from Barnard's star have different wavelengths from those observed in light from a stationary source.

b) A line of wavelength 612.5 nm is present in a laboratory spectrum of a stationary source. Calculate the wavelength of the same line in light from Barnard's star.

Problem 2
An airport radar emits a beam of microwaves of frequency 2.7 GHz towards an approaching aircraft. The frequency of the reflected beam is shifted by 3.0 kHz relative to the emitted beam.

a) Outline why the frequency shift is approximately given by $\Delta f = \dfrac{2vf}{c}$, where v is the component of the velocity of the aircraft towards the radar and f is the emitted frequency.

b) Calculate the velocity of the aircraft in the direction of the radar. State the answer in km h^{-1}.

Problem 3
A train horn emits a sound with frequency f. An observer moves towards the stationary train at constant speed and measures the frequency of the sound to be f'.

Explain, using a diagram, the difference between f' and f.

Problem 4
The frequency f emitted by the train horn in Problem 3 is 300 Hz. The speed of the observer is 15 m s^{-1}. The speed of sound in air is 330 m s^{-1}.

Calculate the frequency f' measured by the observer.

Problem 5
An ambulance moves towards a stationary observer P and away from another stationary observer Q. The siren of the ambulance emits a sound of frequency 1840 Hz. Observer P hears a frequency of 2000 Hz.

a) Calculate the ratio $\dfrac{\text{speed of the ambulance}}{\text{speed of sound in air}}$.

b) Determine, to the nearest hertz, the frequency heard by observer Q.

D Fields

D.1 Gravitational fields

You must know:
- Kepler's three laws of orbital motion
- Newton's universal law of gravitation
- the conditions for which extended bodes can be treated as point masses
- that the gravitational field strength at a point is the gravitational force per unit mass experienced by a small point mass placed at that point
- that a gravitational field line indicates the direction of a gravitational field and that the density of field lines indicates the strength of the field

Additional higher level:
- what is meant by gravitational potential and gravitational potential energy
- that the gravitational field strength is the negative of the gravitational potential gradient
- what is meant by an equipotential surface
- the effect of a small viscous atmospheric drag on the mechanics of an orbiting body.

✔ You should be able to:
- solve problems involving Kepler's three laws of orbital motion when applied to circular orbits
- apply Newton's law of gravitation to the motion of an object in a circular orbit around a point mass

Additional higher level:
- solve problems involving gravitational potential and the gradient of gravitational potential with distance
- solve problems involving the gravitational potential energy for a two-body system
- solve problems involving escape speed for a gravitational field and orbital speed for a small body orbiting a large mass.

The term **field** is used when two separated objects exert a force on each other.

A gravitational field is said to exist in the region where a gravitational force acts "at a distance" on a mass.

Newton's law of gravitation relates the gravitational force between two objects to their masses and separation. It states that the gravitational force F between two point objects of mass m_1 and m_2 is related to the distance r between them by the relationship $F \propto -\dfrac{m_1 \times m_2}{r^2}$.

The negative sign indicates that the force is attractive (although you can ignore the sign for the purposes of the IB Diploma Programme physics course).

In the IB Diploma Programme physics course, you will meet gravitational, electrostatic and magnetic fields, and their effects. Gravitational effects arise through interactions involving mass. The remaining two involve the interaction of stationary or moving charges (Topic D.2).

Newton's law refers to point masses, but it can be extended to spherically symmetric masses (such as Figure 1) of uniform density as a special case. In these cases, r is the distance between the centres of the masses.

Gravitational field strength is the force per unit mass that acts on a mass in a gravitational field.

The constant of proportionality G is given by $G = \dfrac{F \times r^2}{m_1 \times m_2}$.

G has magnitude 6.67×10^{-11} N m² kg⁻² (or, in fundamental units, m³ s⁻² kg⁻¹).

For an object on the Earth's surface, a mass m is gravitationally attracted to the centre of Earth, mass m_E. Newton's law of gravitation becomes $F = \dfrac{Gm \times m_E}{r_E^2}$, where r_E is the radius of the Earth.

This is called "the force of gravity". The weight of the object is written as mg, where g is the gravitational field strength at the surface, which leads to $g = \dfrac{G \times m_E}{r_E^2}$. This is also the acceleration of free fall for an object near the surface of the Earth.

Assessment tip

Newton's law of gravitation is the first of two inverse-square laws that you study in depth in the IB Diploma Programme physics course. Inverse square relates to the dependence on the reciprocal of the distance squared, $\frac{1}{r^2}$. The other inverse-square law is Coulomb's law in Topic D.2.

Learning and revising these topics together will help you to understand both areas of work better.

Gravitational field lines show the direction of the force acting on a mass in a gravitational field. Figure 1 shows a planet with its radial field lines radiating from the centre of the planet.

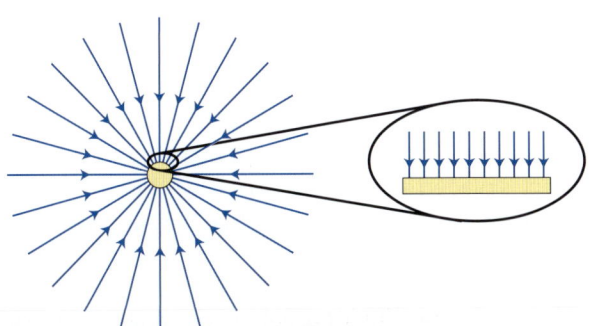

▲ Figure 1 The radial gravitational field lines around a planet. The inset shows how the field appears uniform when close to the surface

The Earth has an (approximately) radial gravitational field but, because we normally stay close to the surface, the divergence of the lines is not apparent to us—broadly speaking, we live in a uniform gravitational field.

In a uniform gravitational field, the gravitational field lines will be parallel and equally spaced. This is described in more detail in Topic D.2 when dealing with electric fields, but the physical description is the same for gravitational force.

Gravitational field strength,

$$g = \frac{\text{force acting on a small test object}}{\text{mass of the test object}} = \frac{F}{m}$$

The symbol g as it is used here does not refer specifically to the Earth's surface but can be anywhere in a gravitational field.

Two units can be used for g:

- N kg^{-1} (when dealing with gravitational field strength)

- m s^{-2} (when dealing with acceleration of free fall).

Test objects are required because the object itself contributes to the gravitational field where it is placed. The mass of the test objects must be much smaller than that of the mass producing the field. This is usual when defining any field strengths (see Topic D.2).

Example 1

Knowledge of the acceleration of free fall g_E at the Earth's surface is one way to determine the mass of the Earth m_E.

a) Show that $m_E = \dfrac{g_E r_E^2}{G}$, where r_E is the radius of the Earth.

The gravitational field strength of the Earth at its surface is 6.1 times that of the Moon at the surface of the Moon.

b) Calculate the mass of the Moon as a fraction of the mass of the Earth.

Radius of Moon = 1.7×10^6 m

Radius of Earth = 6.4×10^6 m

Solution

a) $mg_E = \dfrac{Gmm_E}{r_E^2}$

Cancelling and rearranging gives $r_E^2 g_E = Gm_E$ and hence the result.

b) The equation also applies to the Moon:

$$m_E = \frac{g_E r_E^2}{G} \qquad m_M = \frac{g_M r_M^2}{G}$$

Dividing gives:

$$\frac{m_M}{m_E} = \frac{g_M r_M^2}{g_E r_E^2} = \frac{1}{6.1} \times \left(\frac{1.7}{6.4}\right)^2 = 0.012$$

The mass of the Moon is only about 1.2% of the mass of Earth.

D Fields

Nature of science

Physics distinguishes between centre of mass and centre of gravity. They are at the same point when an extended object is in a uniform gravitational field. But when the field lines are not parallel, the centre of gravity is the effective point in an object where all its weight acts. The centre of mass is a point in the object where the mass distribution is symmetrically arranged.

Johannes Kepler analysed planetary data provided by Tycho Brahe to arrive at his laws. This makes the laws **empirical** in nature. Later scientists, such as Newton, used Kepler's laws to arrive at the rules for gravitational force using **theory**.

You can also contrast empirical and theoretical approaches with the kinetic theory of gases in Topic B.3.

Topic A.2 links to this topic to describe the mathematics of satellite orbits.

The gravitational force of a planet provides the centripetal force for the satellite. For a circular orbit (assumed here), the force between satellite and planet is at right angles to the direction of motion of the satellite.

Equating centripetal force (from Topic A.2) and gravitational forces, using $v = r\omega$ and $\omega = \frac{2\pi}{T}$ leads to $v = \sqrt{\frac{GM}{r}}$, $\omega = \sqrt{\frac{GM}{r^3}}$ and $T = 2\pi\sqrt{\frac{r^3}{GM}}$ (this is an algebraic form of Kepler's third law of planetary motion).

Example 2

The gravitational field strength at the surface of planet **P** is 17 N kg^{-1}. Planet **Q** has twice the diameter of **P**. The masses of the planets are the same.

Calculate the gravitational field strength at the surface of **Q**.

Solution

From Newton's law of gravitation, $g \propto \frac{1}{r^2}$.

$$\frac{g_Q}{g_P} = \frac{r_P^2}{r_Q^2}$$

$$g_Q = g_P \times \frac{r_P^2}{r_Q^2} = \frac{17}{2^2} = 4.3 \text{ N kg}^{-1}$$

The real objects that we need to deal with in gravitational theory are known as extended bodies. When two extended bodies interact gravitationally, they can be treated (approximately) as point masses by assuming that the whole of the mass of one object is placed at its centre of gravity. This also assumes that the objects are in a uniform gravitational field or that they are well away from any other mass.

When an object is in a non-uniform gravitational field, the system cannot be treated as a series of point masses. More complex mathematics is required, as in the case of ocean tides on Earth.

Kepler's laws

Johannes Kepler devised three laws of orbital motion in the early 17th century.

Kepler's laws are applied here in the context of a planet orbiting its Sun. However, the laws apply equally to other orbital situations such as a moon orbiting its planet.

These are the laws when written in terms of circular orbits (the only orbits examined in the IB Diploma Programme physics course). Kepler's original laws were written in terms of the more general elliptical orbits. The orbits of the planets in the solar system are close to circular (except for the dwarf planet Pluto).

- Planets move in circular (elliptical) orbits with the Sun at one focus—**Kepler's first law** of planetary motion.
- The line connecting a planet to the Sun sweeps out equal areas in equal times—**Kepler's second law** of planetary motion. (This law is self-evident with the constant angular speed of a circular orbit.)
- The square of the periodic orbital time T of a planet is directly proportional to the cube of the orbital radius r: $T^2 \propto r^3$—**Kepler's third law** of planetary motion.

Example 3

A satellite orbits above the equator so that it stays over the same point on Earth.

a) Calculate the angular speed ω of the satellite.

b) Show that the radius r of the orbit of the satellite satisfies the equation $r^3 = \dfrac{Gm_E}{\omega^2}$, where the mass of the Earth $m_E = 6.0 \times 10^{24}$ kg.

c) Calculate r.

Solution

a) The satellite has a periodic time of 24 hours ≡ 86 400 s.

$$T = \frac{2\pi}{\omega}$$

$$\omega = \frac{2\pi}{T} = 7.3 \times 10^{-5} \text{ rad s}^{-1}$$

b) The centripetal acceleration of the satellite is equal to the gravitational field strength at the orbit.

$\omega^2 r = \dfrac{Gm_E}{r^2}$; therefore $\omega^2 r^3 = Gm_E$, which leads to the result.

c) $r = \sqrt[3]{\dfrac{Gm_E}{\omega^2}} = \left(\dfrac{6.67 \times 10^{-11} \times 6.0 \times 10^{24}}{(7.3 \times 10^{-5})^2}\right)^{\frac{1}{3}} = 42$ Mm

Assessment tip

Distances in the solar system can be expressed in metres or in astronomical units (AU). The astronomical unit is the mean distance between the centre of Earth and the centre of the Sun. 1 AU = 1.50×10^{11} m (this value is given in the *Physics data booklet*).

When solving problems about planetary orbits, you can assume that the orbit of Earth is a circle of radius 1 AU.

The symbol ≡ means "is equivalent to".

Nature of science

Kepler's laws were seen to be extremely successful both in his own time and later. However, he had pushed his theory well beyond the validity of the data he used. Kepler's law ought to include the mass of both Sun and planet. However, because the Sun's mass is so much greater than that of a planet, Kepler's results were still valid within the limits of observation of his time. A healthy skepticism about the validity of data is important for all scientists.

Example 4

The orbit of Earth is approximately a circle of radius 1 AU = 1.50×10^{11} m.

Saturn orbits the Sun with a period of 29.4 years.

Calculate the orbital radius of Saturn. Give the answer in AU and in metres.

Solution

Kepler's third law can be applied to the orbits of Saturn and Earth, leading to:

$$\left(\frac{r_S}{r_E}\right)^3 = \left(\frac{T_S}{T_E}\right)^2$$

$$\frac{r_S}{r_E} = \left(\frac{T_S}{T_E}\right)^{\frac{2}{3}}$$

The orbital period T_E of Earth is 1 year, so $\dfrac{r_S}{r_E} = \left(\dfrac{29.4}{1}\right)^{\frac{2}{3}} = 9.53$.

The orbital radius of Saturn is $r_S = 9.53$ AU = 1.43×10^{12} m.

Nature of science

One of the major paradigm shifts in science was the revolution in our understanding of the nature of the Universe. Galileo, Copernicus, Kepler, Tycho Brahe, Newton and others all played a part in developing the new concept of a heliocentric (Sun-centred) solar system and the accompanying ideas of gravity and gravitational force. Theories and their development play a major role in what we mean by the nature of science.

D Fields

▲ The gravitational force acting between the Moon and Earth is correctly equated with the centripetal force on the Moon. This statement alone is worth one mark.

▼ The student attempts to rearrange the equation but fails to express the orbital speed in terms of the period, $v = \frac{2\pi R}{T}$, and does not reach the expected relationship.

▲ The student has rearranged the equation for M and substituted the data. The final value is consistent with the substitutions; hence partial marks are awarded despite the incorrect value for T.

▼ The orbital period is not converted to seconds, so the calculated answer is wrong. The student should have used $T = 27.3 \times 24 \times 3600 = 2.36 \times 10^6$ s.

Sample student answer

The radius of the Moon's orbit around Earth is R and the orbital period is T.

a) Show that $T^2 = \frac{4\pi^2 R^3}{GM}$, where M is the mass of Earth. [2]

This answer could have achieved 1/2 marks:

Let m be the mass of Moon.

$$\frac{GMm}{R^2} = \frac{mv^2}{R}$$

$$GM = v^2 R$$

b) The orbital radius of the Moon around Earth is 3.86×10^8 m and the orbital period is 27.3 days. Estimate the mass of Earth. [2]

This answer could have achieved 1/2 marks:

$$M = \frac{4\pi^2 R^3}{GT^2}$$

$$\frac{4\pi^2 (3.86 \times 10^8)^3}{6.67 \times 10^{-11} \times 27.3^2} = 4.6 \times 10^{34} \text{ kg}$$

Because a force acts on an object, work must be done to move the object in the field:

work done = force × distance

Energies involved in field theories are expressed using the concept of **potential** and **potential energy**. There is a linked quantity known as **potential difference**, which is covered in Topic B.5.

These new quantities are common to the descriptions of all fields and you meet them again in Topic D.2, which deals with electrostatic and magnetic fields.

Gravitational potential at a point V_g is the work done W in moving unit mass from infinity to the point.

Gravitational potential difference ΔV_g between two points is given by $\Delta V_g = \frac{W}{m}$, where m is the mass and W is the work done in moving the mass between the points.

The unit of gravitational potential is J kg^{-1}.

The concept of **infinity** is that of a "standard" place where potential is defined to be zero. In a gravitational field, force varies as $\frac{1}{(\text{distance})^2}$. At very large distances (at infinity), the force must be zero.

The gravitational potential at infinity is defined to be zero.

Assessment tip

Potentials are usually easier to handle than field strengths. This is because potential is a scalar quantity, so potentials are added using normal arithmetic and you only need to pay attention to the sign.

Field strengths are vectors and need to be added vectorially.

Gravitational potential is a quantity that is independent of the magnitude of mass of the object on which the gravitational force is acting.

The change in gravitational potential between two points is the gravitational potential difference.

Gravitational potential is always negative. This is because gravitational force is always attractive. To move a test object to infinity *from* somewhere close to the mass producing the field means that work must be done (to overcome the attractive force). Energy must be added to the gravitational system for the mass to be moved away. At infinity, the potential of the system will become zero (by definition). Therefore, the potential must have been negative before the transfer to infinity took place.

When a system has to be assembled from many components, the gravitational potential energy of the system is the work done to bring *every* component from infinity to the final position.

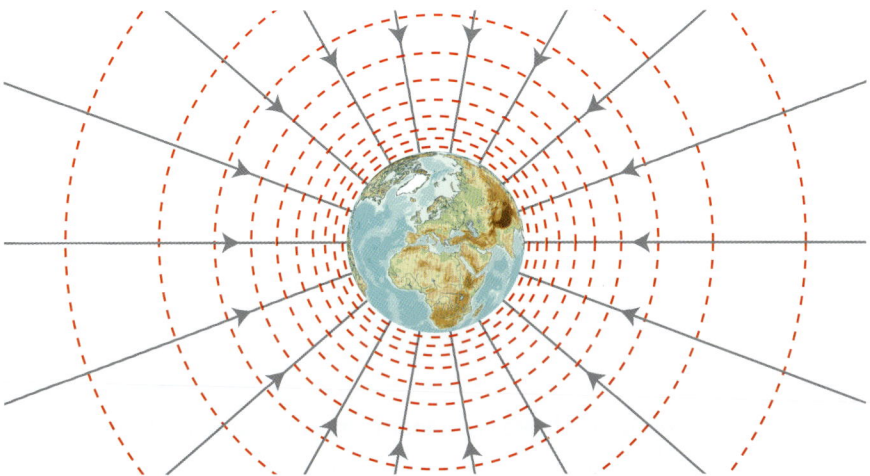

▲ Figure 2 The radial gravitational field lines (grey) and the spherical equipotential surfaces (red dashed lines) around the Earth

Figure 2 shows the gravitational field due to a spherical planet. The diagram has to be imagined in three dimensions (see also Figure 1). Points that are the same distance from the centre of the sphere have the same gravitational potential. When a mass moves on the red dashed surface (in three dimensions) no overall work is done. This gives the equipotential surface.

In a radial field for which the inverse-square law holds, potential depends on $\frac{1}{r}$, where r is the distance from the origin of the field to the point mass.

For a radial field:

Gravitational potential $V_g = -G\frac{M}{r}$, where M is the mass of the object that produces the radial field.

Gravitational potential energy (zero potential at infinity) for two objects of masses m_1 and m_2 is $E_p = -G\frac{m_1 m_2}{r}$, where r is the separation of the centres of mass of the two objects.

Another important, and universal, relationship between field strength and potential is:

field strength $= -\dfrac{\text{change in potential}}{\text{change in distance}} = -\dfrac{\Delta V_g}{\Delta r}$

In graphical terms, field strength at point x is the gradient of the potential against distance graph at point x. This is known as the **potential gradient**.

Nature of science

The idea that potential is zero at infinity—an unmeasurable point in space—is a curious one. But it works well because we can use it both in imagination and mathematically. It is the nature of science that a definition can rely on an agreement that such an unattainable position has a meaning. Are there other such definitions that you can identify?

Assessment tip

Potential is always a property of the field, not of individual objects moving in it.

Because work is done when a charge or mass moves along a field line, equipotentials must always meet field lines at 90°.

Mass moves on an equipotential surface **without** work being transferred to or from the system.

D Fields

> **Assessment tip**
>
> You need to understand the links between gravitational and electric fields and know the connections between field strength and potential.

Figure 3 shows the relationships between gravitational field strength and gravitational potential.

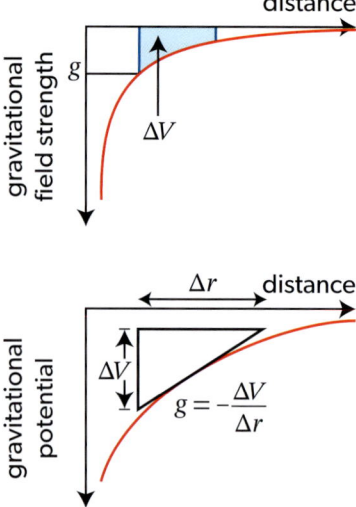

▲ **Figure 3** The graphical relationships between gravitational potential and gravitational field strength

> Close to a planet surface there is a uniform gravitational field strength given by $g = -\frac{\Delta V_g}{h}$, where h is the change in height of an object of mass m. So $\Delta V_g = -gh$ and the change in gravitational potential energy (E_p) is $m\Delta V_g = -mgh$. The change in E_p is therefore mgh, which links to Topic A.2.
>
> The minus sign arises because gravitational field strength is a vector quantity. When V_g is increasing in one direction, g is acting in the opposite direction.

> **Assessment tip**
>
> To help your understanding, link this way of describing energy change in a field to the meaning of potential difference in current electricity. This is explored in more detail in Topic B.5.

Example 5

An object of mass 2.4 kg is moved from a point where the gravitational potential is −800 J kg⁻¹ to a point where the gravitational potential is −780 J kg⁻¹. The distance between the points is 4.0 m.

Calculate:

a) the change in the gravitational potential energy of the object

b) the average component of the gravitational field strength in the direction of displacement of the object.

Solution

a) Change in gravitational potential = −780 − (−800) = 20 J kg⁻¹

The change in the gravitational potential energy is equal to the work done in moving the object.

This is equal to $m\Delta V_g = 2.4 \times 20 = 48$ J.

b) $g = -\frac{\Delta V_g}{\Delta r} = -\frac{20}{4.0} = -5.0$ N kg⁻¹

The object is moved in the direction in which the gravitational potential increases, so the field strength is in the opposite direction to the object's displacement. This is indicated by the minus sign in the answer.

The gravitational field inside a solid sphere is unusual in that, for any point inside the sphere, there are two contributions to the field: the contribution from the shell that is "outside" the point and the contribution from the material "inside" the point.

The net result of these two contributions is that the gravitational field strength *inside* the solid sphere varies directly with distance from the centre, whereas *outside* the sphere the variation follows the inverse-square law as usual.

Compare the case of the gravitational field inside the sphere with the case for the electric field strength and electric potential inside a solid or hollow conducting sphere, as described in Topic D.2.

Sample student answer

Explain what is meant by the gravitational potential at the surface of a planet. [2]

This answer could have achieved 2/2 marks:

> It is the work done per unit mass to bring a small test mass from a point at infinity (zero PE) to the surface of that planet (in the gravitational field).

▲ There are two marks for this question and two points to make—this answer has them both: work done per unit mass and the idea of taking the mass (it does not have to be "small" in a potential definition) from infinity to the surface.

Rockets and satellites that orbit a planet or star have both kinetic energy and stored gravitational potential energy when in orbit.

The gravitational potential energy is $E_p = -\dfrac{GM_p m_s}{r}$, where r is the orbital radius (from the centre of the planet and not from the surface), M_p is the mass of the planet and m_s is the mass of the satellite.

The kinetic energy of the satellite is $\dfrac{1}{2}m_s v^2$.

The gravitational force provides the centripetal force required to keep the satellite in its circular orbit. Therefore, $\dfrac{GM_p m_s}{r^2} = \dfrac{m_s v^2}{r}$, which leads to $\dfrac{1}{2}m_s v^2 = \dfrac{GM_p m_s}{2r}$.

The term on the left-hand side is the kinetic energy of the satellite and is half the magnitude of the gravitational potential energy.

> The total energy of a satellite in orbit is the sum of its gravitational potential energy and its kinetic energy: $E = E_p + E_k$.
>
> Adding the two contributions (remembering the signs) gives
> $E_K = \dfrac{GM_p m_s}{2r} \left(= -\dfrac{E_p}{2}\right)$ and $E_p = -\dfrac{GM_p m_s}{r}$, so the total energy E of the satellite at orbital radius r is $-\dfrac{GM_p m_s}{2r}$.
>
> Note the minus sign in this expression. It reminds you that the satellite is bound in its orbit and cannot escape from the planet.

What happens to a satellite when it moves to a lower orbit? The energy equations predict that, at a lower radius, the gravitational potential energy is more negative (that is, it has a larger magnitude in a negative direction) and the kinetic energy is larger (this time, more positive)—but remember that the kinetic energy is only half the magnitude of the potential energy. The total energy is also more negative because the potential energy magnitude increases more than the kinetic energy.

The **orbital speed** $v_{orbital}$ of the satellite increases as the orbital radius decreases even though, overall, it has lost energy. The total energy has become more negative—that is, more tightly bound to the system.

The equation $\dfrac{1}{2}m_s v^2 = \dfrac{GM_p m_s}{2r}$ also leads to $v^2 = \dfrac{GM_p}{r}$.

This shows that the orbital speed is $v_{orbital} = \sqrt{\dfrac{GM_p}{r}}$.

Notice that $v_{orbital}$ does not depend on the mass of the satellite but only on G, the orbital radius and the mass of the planet.

A **geosynchronous** orbit is one with an orbital period that matches the Earth's rotation on its axis.

A special case of this type of orbit is the **geostationary** orbit which is geosynchronous and also positioned above the equator. A satellite in geostationary orbit remains apparently fixed in position when viewed from Earth. This type of orbit is used for many communication satellites.

D Fields

> The **escape speed** v_{esc} for a system is the minimum speed required for an object to leave a gravitational field and (just) reach infinity—where the gravitational potential energy of the system will be zero.
>
> The initial kinetic energy for escape must provide an energy equal in magnitude to the gravitational potential energy at the point where the satellite is in the field. At a distance r from the centre of the spherical or point mass, $\frac{1}{2}m_s v_{esc}^2 = \frac{GM_p m_s}{r}$ and
>
> $v_{esc} = \sqrt{\frac{2GM_p}{r}}$.
>
> For a satellite on the surface of a planet of radius R, this is
>
> $v_{esc} = \sqrt{\frac{2GM_p}{R}}$.
>
> This is $\sqrt{2}\ \times$ the orbital speed **at the planet surface**.

Rockets can leave Earth's gravity completely as well as remaining bound in orbits. The escape speed is the speed required to do this.

A satellite in a low Earth orbit (about 160 km up to 1000 km from the surface) is subject to drag caused by collisions with the ions and molecules in the atmosphere (Topic A.2). This leads to a drag force acting in a direction at a tangent to the direction of motion of the satellite, in the opposite direction to the linear velocity of the satellite.

The atmospheric drag removes energy from the Earth–satellite system. The total energy of the system $E = -\frac{GM_E m_s}{2r}$ decreases so that r also decreases (because of the negative sign). Another way to put this is that the satellite becomes more bound to the system. As shown earlier, the consequence is that the orbital radius decreases but the kinetic energy $E_K = +\frac{GM_E m_s}{2r}$ increases and, therefore, the orbital speed of the satellite increases too.

> This discussion of the effect of atmospheric drag strictly only applies when the drag force is small. For cases where the drag force is large, the deceleration of the satellite will be so great and the kinetic energy will be lost so quickly that the satellite cannot be treated as being in orbit. It is now effectively a projectile that is subject to Earth's gravity, and it will obey the rules of mechanics and dynamics as described in Theme A, in a complicated way.

Example 6

The mass of Earth is 6.0×10^{24} kg and its radius is 6.4×10^6 m.

A satellite of mass 750 kg is raised from the surface of Earth to a circular orbit 300 km above the surface.

a) Calculate the speed of the satellite in its final orbit.

b) Determine:

   (i) the change in the potential energy of the satellite

   (ii) the total energy needed to put the satellite into orbit.

A viscous drag force due to the residual atmosphere acts on the satellite in the orbit.

c) Explain the effect of this force on:

   (i) the orbital radius

   (ii) the kinetic energy of the satellite.

Solution

a) Radius of orbit = $6.4 \times 10^6 + 3 \times 10^5 = 6.7 \times 10^6$ m

Orbital speed = $\sqrt{\frac{GM}{r}} = \sqrt{\frac{6.67 \times 10^{-11} \times 6.0 \times 10^{24}}{6.7 \times 10^6}} = 7.7 \times 10^3$ m s^{-1}

b) (i) When R is the Earth's radius and r is the orbital radius:

Change in potential = $-GM\left(\frac{1}{r} - \frac{1}{R}\right)$

$= 6.67 \times 10^{-11} \times 6.0 \times 10^{24}\left(\frac{1}{6.4 \times 10^6} - \frac{1}{6.7 \times 10^6}\right) = 2.8 \times 10^6$ J kg^{-1}

Potential energy change $\Delta E_p = \Delta V_g \times m_{satellite} = 2.8 \times 10^6 \times 750$
$= 2.1 \times 10^9$ J

> **Approaches to learning**
>
> Pay particular attention to the use of positive and negative signs in Example 6. Being aware of the signs of quantities has particular importance for your learning in field theory.

D.1 Gravitational fields

(ii) The kinetic energy of the satellite on the surface of Earth can be ignored compared with the kinetic energy in orbit, which is related to the speed calculated in part (a).

Kinetic energy change $\Delta E_K = \frac{1}{2} \times 750 \times (7.7 \times 10^3)^2 = 2.2 \times 10^{10}$ J

Energy required to put the satellite in orbit $= \Delta E_P + \Delta E_K$
$= 2.1 \times 10^9 + 2.2 \times 10^{10} = 2.4 \times 10^{10}$ J

c) (i) The drag force does work on the satellite, so its mechanical energy decreases (becomes more negative). From $E \propto -\frac{1}{r}$, the orbital radius decreases and the satellite moves closer to Earth.

(ii) From $v_{\text{orbital}} = \sqrt{\frac{GM}{r}}$, smaller r means greater orbital speed. The kinetic energy of the satellite increases.

> **Assessment tip**
>
> Much of the theory of satellite motion is still correct when a satellite has an elliptical orbit (the usual case). The assumption in the IB Diploma Programme physics course is that moons and satellites have circular orbits about their planets, and that a planet has a circular orbit around its star.

Sample student answer

The gravitational potential due to the Sun at its surface is -1.9×10^{11} J kg^{-1}. The following data are available.

Mass of Earth = 6×10^{24} kg

Distance from Earth to Sun = 1.5×10^{11} m

Radius of Sun = 7.0×10^8 m

a) Outline why the gravitational potential is negative. [2]

This answer could have achieved 0/2 marks:

> It works in the opposite direction to the gravitational force.

▼ Although there is some truth in the answer, it does not get us very far. This answer is really answering a question about the relationship between gravitational force and the defined positive direction.

This answer could have achieved 2/2 marks:

> Because zero of gravitational potential is defined at infinity. Since gravity attracts, work always has to be done on the object to get closer to infinity, which is the same thing as closer to zero potential, so potential is negative, since work need to be added to move away from the Sun's gravitational field.

▲ There are two marks and therefore two points need to be made. This answer scores both: (i) potential is zero at infinity, (ii) work has to be done to move a mass away from the Sun to infinity. The potential before the mass was moved must have been negative for it to have gained potential and still be zero at infinity.

The gravitational potential due to the Sun at a distance r from its centre is V_s.

b) Show that:

rV_s = constant [1]

This answer could have achieved 0/1 marks:

> $V_s = \frac{GM}{r}$ $rV_s = -GM$ ← mass of the Sun

▼ The first sentence is not very clear. It should be expressed the other way round for clarity. Fortunately the remainder of the answer makes up for this.

▼ You may be surprised that this answer scored zero. The key is in the command term: show that. The answer correctly rearranges the equation for gravitational potential and shows that the value of the constant is GM, but that is not enough. It should have gone on to say that G is constant. Therefore, $G \times M$ must be constant too.

D Fields

Sample student answer

A planet has radius R. At a distance h above the surface of the planet the gravitational field strength is g and the gravitational potential is V.

a) (i) State what is meant by gravitational field strength. [1]

This answer could have achieved 1/1 marks:

Gravitational field strength is the force per unit mass for a point mass in the field.

▲ The student gives a clear definition of g.

(ii) Show that $V = -g(R + h)$. [2]

This answer could have achieved 0/2 marks:

$g = -\dfrac{\Delta V g}{\Delta r}$ $g = -\dfrac{V}{R + h}$ $g(R + h) = -V$

$V = -g(R + h)$

▼ The student does not appear to know what to do here and rearranges the final equation several times. In fact, the student should have begun with the full equation for gravitational potential, recognizing that $g = G\dfrac{M}{(R + h)^2}$ and then cancelling terms such as M and G.

(iii) Draw a graph on the axes to show the variation of the gravitational potential V of the planet with height h above the surface of the planet. [2]

This answer could have achieved 2/2 marks:

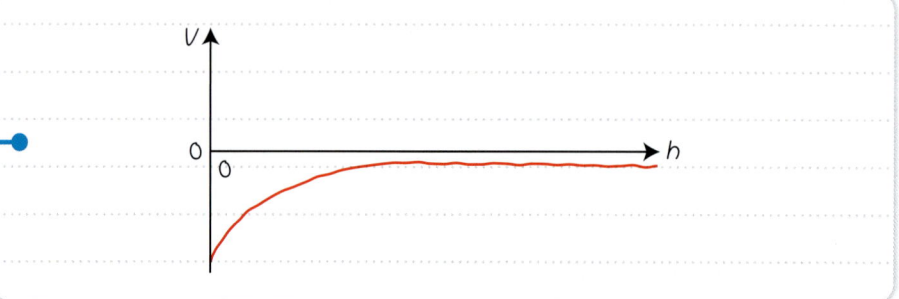

▲ The student has recognized that the origin on the graph corresponds to the surface, and has realized that the gravitational potential will have some finite value here and that the potential is always negative. The graph is asymptotic to 0 for large h but there is a suggestion that the line curves downwards. Had this been a definite downturn, then credit would have been lost.

A planet has a radius of 3.1×10^6 m. At a point P, a distance of 3.4×10^7 m above the surface of the planet, the gravitational field strength is 2.2 N kg⁻¹.

b) Calculate the gravitational potential at point P. Include an appropriate unit for your answer. [1]

This answer could have achieved 1/1 marks:

$V = -g(R + h) = -2.2 \text{ N kg}^{-1} \times (3.1 \times 10^6 + 2.4 \times 10^7)$

$= -59.62 \times 10^6$

$= -6 \times 10^7 \text{ J kg}^{-1}$

▲ A simple calculation that the student explains well.

The diagram shows the path of an asteroid as it moves past the planet. When the asteroid was far away from the planet it had negligible speed.

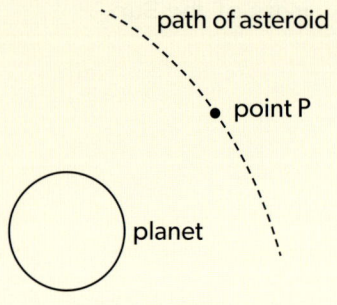

c) Estimate the speed of the asteroid at point **P** as defined in part (b). [3]

This answer could have achieved 3/3 marks:

$$\frac{1}{2}v^2 = -V \quad v^2 = -2V$$
$$v = \sqrt{-2V} = \sqrt{12.0 \times 10^7 \text{ J/kg}}$$
$$= 10.95 \times 10^3 \text{ m/s} = 11 \text{ km/s}$$

▲ Again, the student manipulates the energy equation successfully and arrives at the correct answer.

Practice problems

Problem 1

a) Outline what is meant by gravitational field strength.

The radius of Jupiter is 7.1×10^4 km and the gravitational field strength at the surface of Jupiter is 21 N kg^{-1}.

b) Estimate the mass of Jupiter.

Problem 2

An Earth satellite of mass 1.5×10^3 kg moves in a circular orbit. The gravitational field strength due to Earth at the orbital radius of the satellite is 7.4 m s^{-2}.

a) Calculate the gravitational force acting on the satellite.

b) Explain why the speed of the satellite is constant, even though a force acts on it.

c) Determine the orbital radius of the satellite. State the answer in terms of the radius R of Earth.

Problem 3

Planets X, Y and Z orbit the same star. The orbital period of X is 100 days.

a) The orbital radius of Y is twice the orbital radius of X. Calculate the orbital period of Y.

b) The orbital period of Z is 50 days. Calculate $\dfrac{\text{orbital radius of Z}}{\text{orbital radius of X}}$.

Problem 4

A satellite orbiting a planet has an orbital period of 460 minutes and an orbital radius of 9.4 Mm.

a) The satellite orbits with uniform circular motion. Outline how this motion arises.

b) Show that the orbital speed of the satellite is about 2 km s^{-1}.

c) Deduce the mass of the planet.

d) Calculate the gravitational potential at the orbital radius of the satellite.

Problem 5

The graph shows how the gravitational potential V_g due to a planet varies with height h above the surface of the planet.

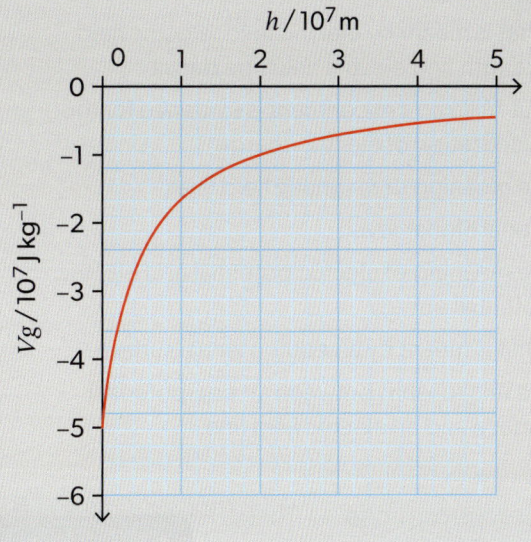

a) The radius of the planet is 5.0×10^6. Calculate the mass of the planet.

b) Calculate the escape speed from the surface of the planet.

A projectile of mass 1 kg is fired vertically upwards from the surface of the planet. The initial kinetic energy of the projectile is 3.0×10^7 J.

c) Estimate, using the graph, the maximum height from the surface of the planet that the projectile will reach.

Problem 6

The mass of Earth is 6.0×10^{24} kg and its radius is 6.4×10^6 m.

A satellite of mass 800 kg is initially at the surface of Earth. The satellite is to be placed in a circular orbit of radius 9.6×10^6 m.

a) Calculate the kinetic energy of the satellite in its final orbit.

b) Determine the minimum work that must be done on the satellite to place it in the orbit.

D.2 Electric and magnetic fields

You must know:

- the nature of electric charge
- that electric charge is conserved
- that electric charge can be transferred
  - by friction
  - by contact
  - by electrostatic induction
- Coulomb's law
- what is meant by electric field, electric field strength and electric field line density
- that Millikan's experiment is evidence for the quantization of electric charge
- the meaning of magnetic field lines

Additional higher level:

- the meaning of electric potential and electric potential energy
- what is meant by an equipotential surface
- the relationship between an equipotential surface and an electric field line.

You should be able to:

- solve problems involving the electric field between two parallel plates
- solve problems involving electric field strength and electric field line density

Additional higher level:

- solve problems involving the work done in moving an electric charge in an electric field
- solve problems involving electric potential, electric potential energy and electric potential gradient.

Approaches to learning

When learning about this topic, remember to be clear about the terms you use. Electric fields are derived from arrangements of electric charge; magnetic fields come from arrangements of magnetic poles. Magnetic poles are always in pairs. Single magnetic monopoles have never been observed.

D.2 Electric and magnetic fields

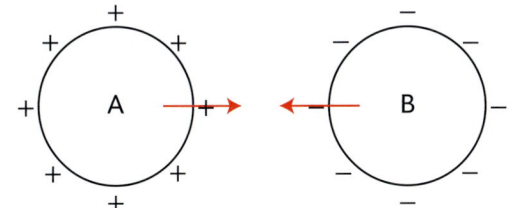

(a) Opposite charges attract.

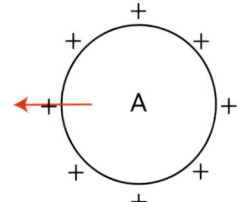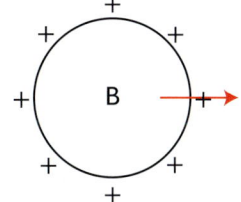

(b) Like charges repel.

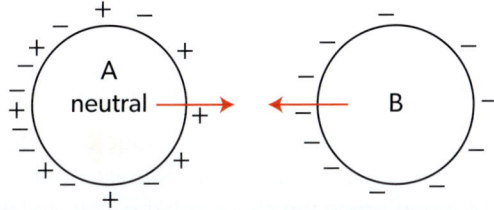

(c) Charge separation means that attraction occurs when one body is charged.

▲ Figure 1 Interactions between (a) unlike charged objects, (b) objects with the same sign of charge and (c) a charged object and an uncharged conductor

There are two types of charge: positive charge and negative charge. When the numbers of positive charges and negative charges are equal on an object, it is said to be uncharged.

- Positively charged objects attract negatively charged objects (Figure 1(a)).
- Two objects with the same sign of charge repel each other (Figure 1(b)).
- When an uncharged conductor is near a charged object, the charges inside the conductor rearrange their positions by repulsion and attraction. This enables the charged object to attract the uncharged object (Figure 1(c)).

Charge is conserved. When charges flow towards and away from a junction in an electric circuit, there can be no net storage of charge at the junction. The total amount of charge flowing towards the junction must equal the total amount of charge flowing away.

Charge can be transferred in several ways:

- By friction, where one object is rubbed with another (e.g. a plastic rod rubbed with a cloth).

  The two objects end up with opposite charges.

- By contact, where one charged object is touched against an uncharged object. Electrons are transferred to, or from, the uncharged object.

  Two objects charged by contact end up with the same sign of charge.

- By electrostatic induction, where a charged object induces a charge in a neutral object without touching it. The process is shown in Figure 2.

  When charged by induction, the objects end up with opposite charges.

 Charge is one of a number of quantities that are conserved. These include momentum (Topic A.2) and energy (when all types of energy and energy transfer are considered, Topic A.3).

Use the ideas of conservation to link your knowledge of different areas of physics together.

Assessment tip

When describing the movement of charge in conductors, remember that it is always the electrons that move. The positive charges in materials are generally attached to the atoms. Positive charge is an absence of electrons.

It might appear that there are exceptions to this because, in semiconductor theory, we talk about "holes" as being positive charges. But here too, a "hole" is really an absence of an electron. The "hole" has similar physical properties (such as mass and charge magnitude) to the electron that has been removed.

D Fields

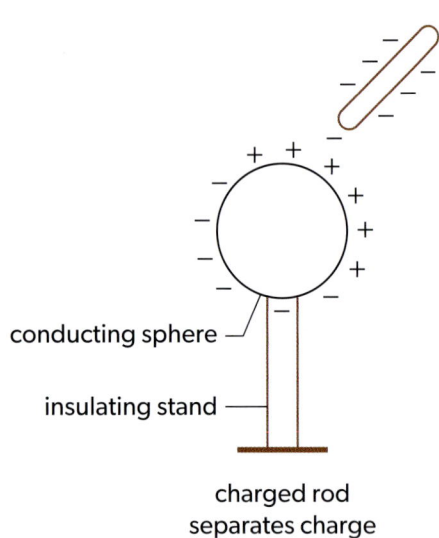
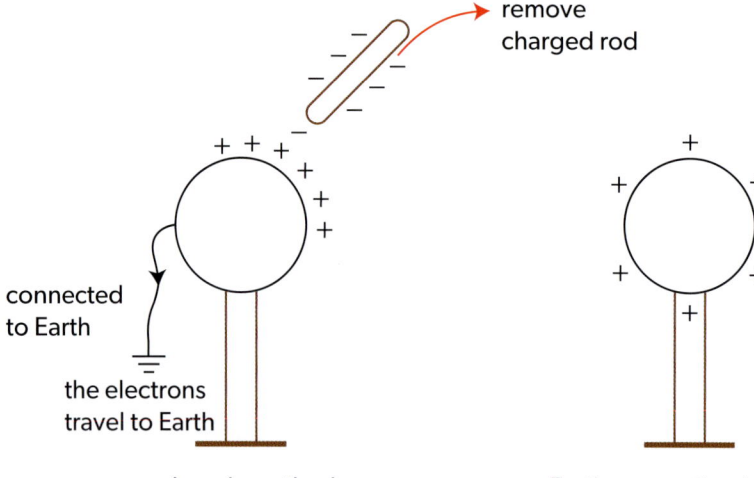

▲ Figure 2 Transferring charge to a conductor using electrostatic induction

Nature of science

In materials or states other than conductors, positive charges can also move. An example is that of the protons used in the particle beams at CERN. In earlier days of electrical physics, scientists believed that all electric current was due to the flow of positive charges. The work discussed at the start of the 20th century (Theme E) helped our understanding of the nature of the electron.

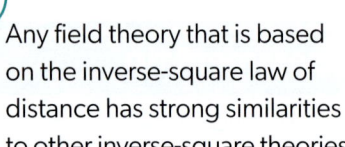

Any field theory that is based on the inverse-square law of distance has strong similarities to other inverse-square theories.

Approaches to learning

Electrostatic field theory has strong similarities to the gravitational field theory of Topic D.1. For that reason, the description of the theories follows a similar approach in this book. Use these similarities and the differences to strengthen your understanding of both topics.

In the process of electrostatic induction, the conductor is connected momentarily to Earth with a conducting wire. This allows electrons to flow to or from the ground (depending on the charge on the rod) until the system has no further ability to move charge. This process of connecting a conductor to Earth is called earthing or grounding. Earth is regarded as an infinite source (or sink) of electrons.

When an electric field is acting at a point in space, a charged object placed at that point will have a force acting on it due to the field. When the charge is free to move, it will be accelerated.

The law that describes the variation with distance of the force between two point charges is known as Coulomb's law.

Coulomb's law states that the force $F = k\dfrac{q_1 q_2}{r^2}$, where q_1 and q_2 are the magnitudes of the point charges that are separated by a distance r in a vacuum.

The constant of proportionality k in the law is known as the **Coulomb constant**, and it has the value 8.99×10^9 N m^2 C^{-2} (measuring charge in coulombs, force in newtons and distance in metres).

There is an alternative way to write the Coulomb law that replaces k with $\dfrac{1}{4\pi\varepsilon_0}$. This makes Coulomb's law $F = \dfrac{1}{4\pi\varepsilon_0} \times \dfrac{q_1 q_2}{r^2} = \dfrac{q_1 q_2}{4\pi\varepsilon_0 r^2}$.

The constant ε_0 is known as the **permittivity of free space** ("free space" means a vacuum)—it takes the value 8.85×10^{-12} C^2 N^{-1} m^{-2}. The reason for the replacement of k with $4\pi\varepsilon_0$ is to rationalize electric and magnetic units in SI.

The equation $F = \dfrac{q_1 q_2}{4\pi\varepsilon_0 r^2}$ applies when the charges are in a vacuum (or in air, which has a permittivity almost equal to that of a vacuum). When the charges are in a medium other than a vacuum or air, ε_0 must be replaced by $\varepsilon_0 \varepsilon_r$ where ε_r is the **relative permittivity** of the medium. This makes Coulomb's law $F = \dfrac{q_1 q_2}{4\pi\varepsilon_0 \varepsilon_r r^2}$.

Relative permittivity values (it has no units as it is a ratio) range from 2 for some plastics through about 80 for water to 10^6 for some crystals. Relative permittivity is a measure of how easily the molecules that make up the substance can be rotated in an electric field.

Coulomb's law can predict the direction of the electric force. When both q_1 and q_2 are positive, the force in Coulomb's law is also positive. The force must repel in this situation.

When one point charge is positive and the other is negative, F is negative, so the law predicts that the charges are attracted.

Figure 3 shows this. Two opposite charges are separated by a distance r. This is measured with displacement to the right considered positive. The product of q_1 and q_2 is negative, so the force must be in the opposite direction to the measurement of displacement: that is, to the left.

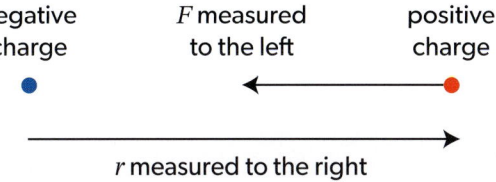

▲ Figure 3 Coulomb's law shows the direction in which the force on a charge acts

> **Approaches to learning**
>
> All the equations and constants that you require in this and other themes are contained in the *Physics data booklet*. Make sure that you are very familiar with the locations of all of these useful reference materials.

Approaches to learning

You have the choice in your learning. You may prefer to initially ignore the direction when calculating the magnitude of the force between charges and to then impose a force direction from your knowledge of the geometry and signs of the charges.

Example 1

Two identical spheres have charges of magnitude q and $2q$. When separated, the attractive force between them is F. The spheres touch to share charge and are then returned to their original separation. Calculate the force between the charges after the separation.

Solution

Call the original separation d.

$$F = \frac{kq \times 2q}{d^2} = \frac{2kq^2}{d^2}$$

This is an attractive force, so one of the charges must be positive and the other must be negative.

Therefore, when they combine the net remaining charge is q.

This is shared between the two spheres, so each has a charge of $\frac{q}{2}$.

The new force is $F' = \frac{k\left(\frac{q}{2}\right)^2}{d^2} = \frac{kq^2}{4d^2}$, which is repulsive and equal to $\frac{F}{8}$.

Field strength is defined in general as $\frac{\text{force acting on an object in a field}}{\text{magnitude of the property that responds to the field}}$, so that, for gravitation (Topic D.1), the gravitational field strength $g = \frac{F}{m}$, where m is the mass of a small test object in the gravitational field.

In electric field theory, the same general definition applies, so that electric field strength $E = \frac{F}{q}$, where F is the force acting on a small test charge of charge magnitude q.

The force F acting on a charge q is $F = qE$.

> There are strong links between this topic and Topic B.5.
>
> Potential difference V in current electricity is the work done W per unit charge moving between two points that have different potentials. When a total charge q is transferred, $V = \frac{W}{q}$ (and $1\,\text{V} \equiv 1\,\text{J}\,\text{C}^{-1}$). Electric field strength can be written as
>
> $$E = \frac{F}{q} = \frac{F \times \text{distance moved}}{q \times \text{distance moved}}$$
>
> $$= \frac{\text{energy}}{q} \times \frac{1}{\text{distance}}$$
>
> This is $\frac{\text{potential difference}}{\text{distance}}$.
>
> So $E = \frac{F}{q} = \frac{V}{d}$.
>
> This leads directly to an expression for the electric field strength E at a distance r from a charge of size Q.
>
> The test charge has a size q and the force between both charges is $F = k\frac{qQ}{r^2}$.
>
> Therefore $E = \frac{F}{q} = \frac{1}{q} \times k\frac{qQ}{r^2} = k\frac{Q}{r^2}$.

D Fields

> **Electric field strength** E is a vector. It acts in the direction of the force that would act on a positive point test charge, $E = \dfrac{F}{q}$. The unit of E is N C^{-1}. In fundamental units, this is kg m s^{-3} A^{-1}.
>
> The definition involves a "point test charge"—this is taken to be a charge so small that it does not disturb the field.
>
> E can also be expressed as $E = \dfrac{V}{d}$, where V is the potential difference between two points and d is the distance between them. This gives an alternative unit for E of V m^{-1}. V m^{-1} and N C^{-1} can be used interchangeably.

Example 2

Calculate the electric field strength at a distance of 0.15 m from a +12 nC charge.

Solution

$$E = k\dfrac{q}{r^2} = 8.99 \times 10^9 \times \dfrac{12 \times 10^{-9}}{0.15^2} = 4.8 \times 10^3 \text{ N C}^{-1}$$

Electric forces and electric field strengths both add vectorially. Figure 4 shows the vector addition of two electric forces.

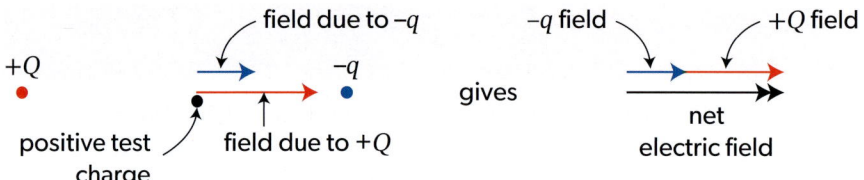

▲ **Figure 4** Electric forces and electric field strengths add vectorially because they are vector quantities

When charges move in current electricity (Topic B.5), energy must be transferred to them from the source of emf in the circuit. This energy is measured in terms of the energy transferred per unit charge, which is called the potential difference.

Of course, although the electric current is made up of the flow of the electrons, the energy transferred to an individual electron is very small. Because the work done on the electron is eV, the energy gained by an electron when it goes through a potential difference of 1 V is 1.6×10^{-19} J.

In atomic and nuclear physics, it is usual to replace the joule as the unit of energy by another unit known as the electronvolt.

One electron volt (abbreviated eV) is the energy transferred by one electron as it moves through a potential difference of one volt.

$1 \text{ eV} \equiv 1.60 \times 10^{-19}$ J

> Electric charge is a constant that is defined in the SI so that the charge on one electron is exactly $1.602176634 \times 10^{-19}$ C. This links to Topic B.5 and Theme E.

D.2 Electric and magnetic fields

Example 3

The kinetic energy of an electron is 300 eV.

Calculate the speed of the electron.

Solution

In order to calculate the speed, the energy must be expressed in J.

$300 \, eV = 300 \times 1.60 \times 10^{-19} = 4.8 \times 10^{-17}$ J

Since $E_K = \frac{1}{2}mv^2$ and the mass of the electron $m = 9.11 \times 10^{-31}$ kg (this value is given in the *Physics data booklet*):

Speed of the electron $v = \sqrt{\frac{2E_K}{m}}$

$= \sqrt{\frac{2 \times 4.8 \times 10^{-17}}{9.11 \times 10^{-31}}} = 1.0 \times 10^7 \, m\,s^{-1}$

Assessment tip

You should be able to use J and eV with equal ease.

Electric field lines

As with gravitational forces, electric field lines can be used to show the shape of electric fields.

Figure 5 shows the electric field lines for the radial fields of a positive charge and a negative charge. Because there are two types of charge, it is crucial with electric fields to show the direction of field lines. This is the direction in which a **positive** charge will move when free to do so.

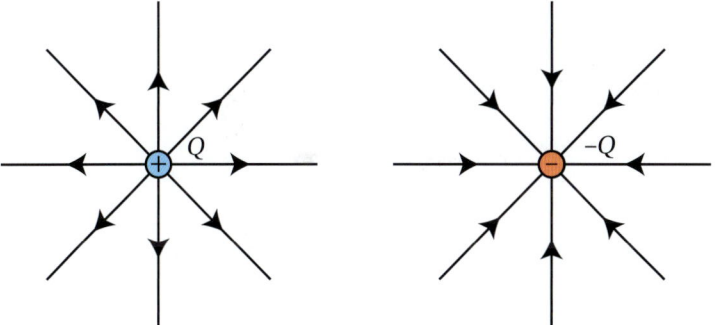

▲ Figure 5 The radial fields for an isolated positive charge and an isolated negative charge

Conventions for drawing electric field lines:
- start and finish only on charges of opposite sign
- an arrow shows the field direction
- if the lines are close together it means that the field is strong or the electric field line density is large
- the lines never cross
- the lines meet conducting surfaces at 90°.

Figure 6 shows the shape of the electric field due to other important charge configurations.

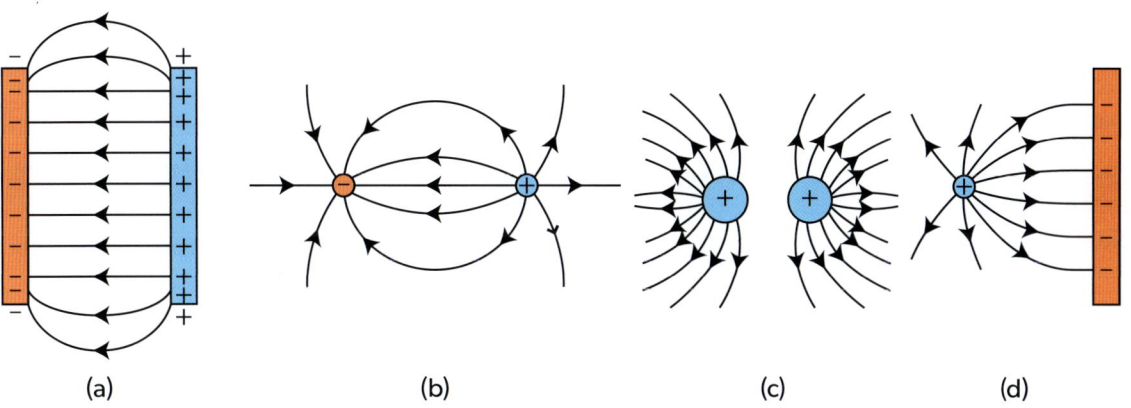

(a)　　　　(b)　　　　(c)　　　　(d)

▲ Figure 6 The electric field pattern due to (a) two parallel plates with opposite charge, (b) two isolated charges of opposite sign, (c) two isolated positive charges and (d) a single positive charge near to a negatively charged flat surface

The electric field between parallel charged plates, as shown in Figure 6(a), is of particular interest. The field between the plates is uniform, so you should draw the field lines parallel and equally spaced. However, at the edge of the plates the field changes from the uniform value inside to a zero value outside the plates (assuming that there are no other fields in the vicinity). This must happen gradually and is drawn showing a gradual increase in line separation as the field weakens. This weakening is known as an edge effect.

> **Assessment tip**
>
> Example 4, part (a) is a "show that" question. You must make every step clear, including the final calculation. To achieve this, quote your answer to at least one more significant figure than the value given in the question.

> **Example 4**
>
> A charged point sphere of mass 2.1×10^{-4} kg is suspended from an insulating thread between two vertical parallel plates that are 80 mm apart.
>
> A potential difference V of 5.6 kV is applied between the plates.
>
> The thread then makes an angle of 8.0° to the vertical.
>
> a) Show that the electrostatic force F on the sphere is given by 0.29 mN.
>
> b) Calculate:
>
>   (i) the electric field strength between the plates
>
>   (ii) the charge on the sphere.
>
> **Solution**
>
> a) Horizontally, $F = T \sin 8°$ and, vertically, $mg = T \cos 8°$, where T is the tension in the thread.
>
> $$F = \frac{mg}{\cos 8°} \sin 8° = 2.1 \times 10^{-4} \times 9.8 \times \tan 8° = 0.289 \text{ mN}$$
>
> b) (i) $E = \frac{V}{d}$ gives $\frac{5600}{80 \times 10^{-3}} = 70$ kV m^{-1}
>
>   (ii) $q = \frac{F}{E}$ gives $\frac{2.9 \times 10^{-4}}{7.0 \times 10^{4}} = 4.1 \times 10^{-9}$ C.

The field lines between two parallel charged plates meet the plate surfaces at 90°. This is because there can be no component of field (electric force) along the plate surfaces—this would accelerate charges on the plate surface. This right-angle between line and surface is true when sufficiently close to **any** surface.

The electric field E between the plates is given by $E = \frac{V}{d}$ as before. The field E is also given by $E = 4\pi k\sigma = \frac{\sigma}{\varepsilon_0}$, where σ is the charge per unit area on each plate (the total charge on a plate divided by the area of that plate). σ is also known as the **surface charge density** and has units C m^{-2}. The concept of surface charge density is not required for your examinations.

Each plate contributes half of the electric field, so the **electric field close to one plate** (Figure 6(d)) is $E = 2\pi k\sigma = \frac{\sigma}{2\varepsilon_0}$.

The physicist Robert A Millikan devised an elegant method to determine the charge on the electron using a pair of parallel plates and an oil drop (Figure 7). This is known as Millikan's experiment and provides evidence for the quantization of electric charge.

- A charged oil drop is selected from many others. The drop is charged by friction or by ionization (Figure 7(a)). The drop is allowed to fall at terminal speed with no electric field in the chamber (Figure 7(b)). The net resultant force on the drop is zero. This resultant force is made up of a drag force, a buoyancy force and the weight of the drop. This allows the weight of the drop to be estimated. (Topic A.2.)
- The electric field is used to hold the drop stationary in the chamber (Figure 7(c)). The electric force and the weight are now equal and opposite. This allows the electric force on the drop to be estimated and hence its excess charge calculated.
- The charge on the drop can be changed by ionization and the experiment repeated. Or another drop can be selected.
- When Millikan had measured many drops with many different charges, he found the charges (within experimental error) to be integer multiples of a lowest common denominator. Millikan took this value to be the charge of one electron.
- Today, this charge, known as the elementary charge, is defined to be $1.602176634 \times 10^{-19}$ C, based on many separate values obtained since Millikan's first experiment. The value given in the *Physics data booklet* is 1.60×10^{-19} C.

D.2 Electric and magnetic fields

The important outcome of Millikan's experiment is that electric charge is quantized. All electrons have exactly the same charge.

Example 5

A negatively charged oil drop of weight 3.0×10^{-14} N is held between two parallel horizontal plates that are oppositely charged.

a) The electric field between the plates is uniform.

   (i) Explain what this means.
   (ii) Sketch field lines to show the electric field due to the plates.

The plates are separated by 4.0 mm and the potential difference applied between them is 380 V.

b) Calculate the magnitude of the charge on the oil drop. State the answer in terms of the elementary charge e.

Solution

a) (i) Field strength is the force per coulomb acting on the oil drop.

   The field strength is the same everywhere between the plates.

   (ii) The field lines should be at 90° to the plates and parallel to each other and the field line separations should be constant. The direction of the field should be from the top to the bottom plate. Edge effects should be shown.

b) Field strength $E = \dfrac{V}{d} = 95\,000$ N C^{-1}

   Weight of drop = Eq. So $3.0 \times 10^{-14} = 9.5 \times 10^4 q$ and $q = 3.2 \times 10^{-19}$ C. This is equivalent to $2e$. The drop has an excess of two electrons.

Magnetic field lines

Magnetic fields can be visualized in a similar way to electric fields, with lines used to indicate the direction and strength of the field.

As with electric field lines, properties can be assigned to magnetic field lines.

Figures 8 to 11 show the magnetic field patterns for a bar magnet, a long straight (current-carrying) wire, a current-carrying circular coil and a solenoid with an air core.

Rules to help you to remember the field patterns and their directions are included.

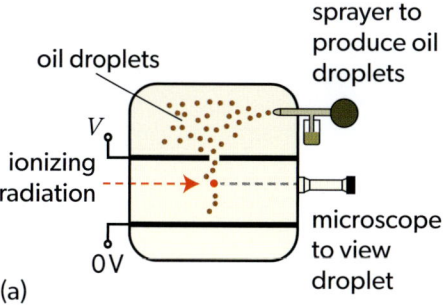
(a)

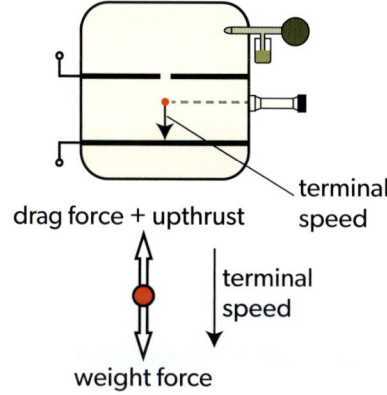
(b)

(c)

▲ Figure 7 The three stages of Millikan's experiment. (a) A charged drop is selected. (b) Its terminal speed is measured. (c) Its charge is estimated

Magnetic field lines:
- are in the direction in which a north pole would travel if free to do so
- cannot cross
- show the strength of the field—the closer the lines are, the stronger the field
- begin on north poles and end on south poles
- act as though they are elastic strings that repel each other but still try to be as short as possible.

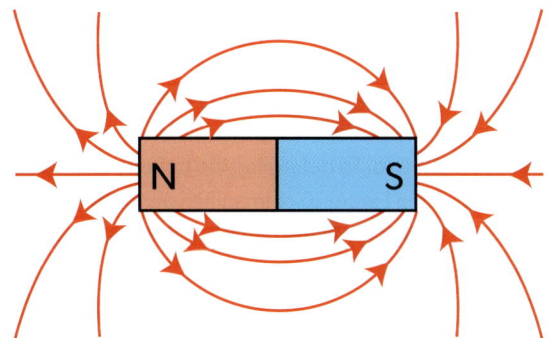

▲ Figure 8 The field pattern for a bar magnet. The field lines go from the north pole to the south pole, in the direction in which a free north pole would travel

155

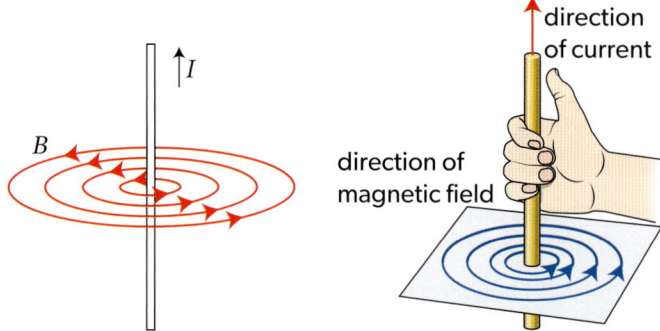

▲ Figure 9 The field pattern for a long straight wire carrying a current. The field lines are circles centred on the wire. A right-hand rule can be used to remember the pattern direction. Imagine holding the wire with your right-hand thumb pointing in the direction of the conventional current. Your fingers curl in the direction of the magnetic field

Approaches to learning

The term 'north pole' is odd. It seems to contradict theory because a 'magnetic north pole' is apparently attracted to the 'geographic North pole'. But we are always taught that 'like poles repel'. The dilemma is solved when you realise that 'north pole' is a shortened version of 'geographic north-seeking pole' and the N end of a magnet is the one that tries to point to the geographic north.

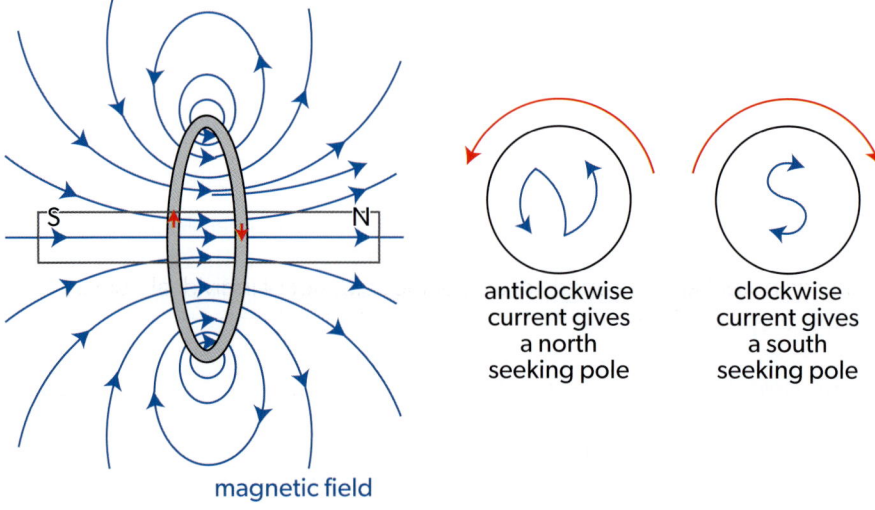

▲ Figure 10 The magnetic field pattern for a circular coil carrying a current. (a) The field lines closely resemble that of a bar magnet, which is superimposed on the coil to show this. (b) The view of the current direction when looking into the end of the coil from outside indicates the pole at that end of the coil

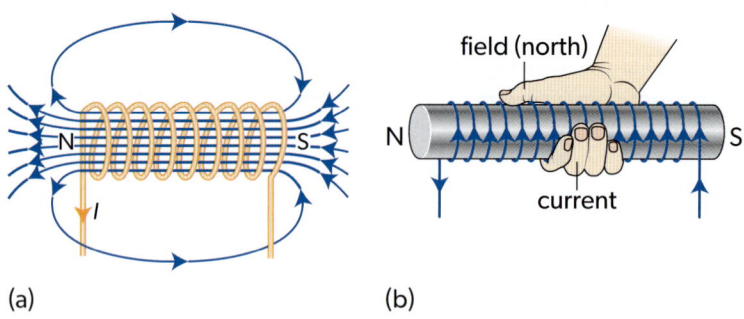

▲ Figure 11 The field pattern for a solenoid carrying a current. (a) A solenoid is a coil whose length is greater than the coil diameter—as with the coil, the field pattern resembles that of a bar magnet. (b) To determine the polarity at the end of the solenoid, use the same direction rule as in Figure 10 or imagine your right hand wrapped round with the wires. When your fingers are in the direction of the current, your thumb points to the solenoid's north pole

Example 6

Two parallel wires X and Y carry equal currents in opposite directions, X into the page of the paper and Y out of the page. Point P is at the same distance from the currents.

Explain the direction of the magnetic field at P.

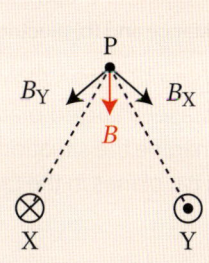

Solution

The total magnetic field B is the vector sum of the fields B_X and B_Y due to the individual currents. B_X is perpendicular to the line joining X and P, and B_Y is perpendicular to the line joining Y and P. The directions of these fields can be deduced from the right-hand rule and are shown in the diagram. B_X and B_Y have the same magnitude, so, from the symmetry of the arrangement, the overall field B is directed downwards.

Electric potential and electric potential energy

Electric potential is a property of an electric field, not of individual objects within the field. It is a quantity that is independent of the magnitude of the charge on the object on which the force acts.

- The electric potential is defined to be zero at infinity.
- In a radial field that obeys inverse-square law behaviour (that is, the force is proportional to $\frac{1}{r^2}$), electric potential depends on $\frac{1}{r}$, where r is the distance from the origin of the field (for example, a point charge). Therefore, there is a potential difference involved in moving a charge from one (non-infinity) point in a field to another.

> In Topic D.1, gravitational potential is also defined to be zero at infinity and is the work done in moving a unit mass from infinity to a particular point.
>
> Newton's law of gravitation, like the Coulomb law, is an inverse-square relationship.

The **electric potential** V_e in a radial field is given by $V_e = \frac{kq}{r}$, where q is the charge that produces the radial field. This follows from Coulomb's law and the relationship between field strength and potential.

The **electric potential energy** E_p for two point charges q_1 and q_2 brought from infinity to a separation of r is $E_p = k\frac{q_1 \times q_2}{r}$ (there is zero potential at infinity).

The work done W in moving a charge q in an electric field is $W = q \times \Delta V_e$, where ΔV_e is the **electric potential difference**.

The unit of electric potential is $JC^{-1} \equiv V$.

When there are many separate charges, each one has to be brought from infinity, so the total electric potential energy E_p is the sum of each separate energy transfer.

As with gravitation, a very important relationship between field strength and potential is

electric field strength $E = -\dfrac{\text{change in electric potential}}{\text{change in distance}} = -\dfrac{\Delta V_e}{\Delta r}$.

In a graph of electric potential V_e against distance r, the electric field strength at any point on the graph is the negative of the gradient of the graph at that point. The graphical relationships are shown in Figure 12.

D Fields

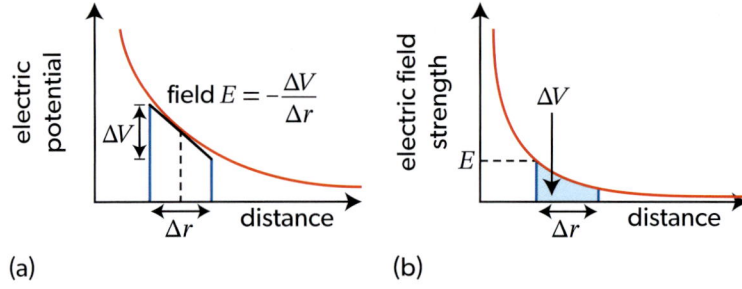

▲ Figure 12 Graph showing the relationships between (a) electric potential and distance and (b) electric field strength and distance

Figure 13 shows the uniform field and the equipotential surfaces between two parallel charged plates. The equipotential surfaces are in red dashed lines. (Imagine the situation in three dimensions to visualize the equipotential surfaces as flat planes in the centre of the space well away from the plate edges.)

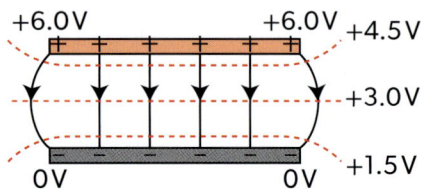

▲ Figure 13 The electric field pattern and equipotential surfaces for two parallel charged plates

Between the plates, the field lines are parallel and equally spaced—the field is uniform. Outside the plates, the curved edge effects show the system making the transition to zero field. The equipotential surfaces also curve away just outside the plates as shown, so that they are always at 90° to the field lines. No work is done when a charge moves on an equipotential surface.

> **Assessment tip**
>
> You can ignore the "effect" of edge effects in the IB Diploma Programme physics course but not their existence. Always draw them when you have to represent the field between two charged plates.

Figure 14 shows the equipotential shapes for a point charge. The equipotentials become closer together as their distance from the point charge decreases.

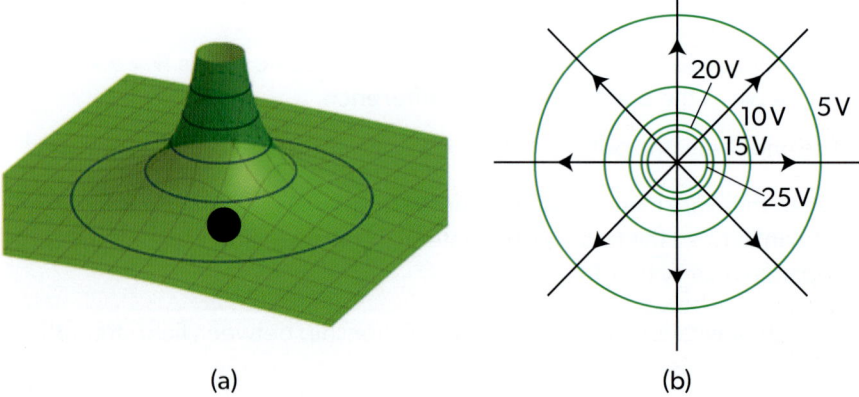

▲ Figure 14 (a) The shape of the equipotentials for a point charge shown in three dimensions. (b) The same diagram in cross-section. Each equipotential differs by 5 V from the adjacent one

> **Approaches to learning**
>
> There are many similarities between gravitational field theory and electric field theory. You should try to use these similarities (and the occasional differences) to improve your understanding of this theme. You can begin by linking the ideas in electric field theory here to the concept of gravitational equipotentials and gravitational field lines described in Topic D.1 on page 141.

> To help your understanding, link these new ways that describe energy change in a field to the meaning of potential difference in current electricity (Topic B.5).

> **Assessment tip**
>
> You can be asked to consider the equipotential surfaces for a situation that includes up to four point charges. This is where your knowledge of potential as a scalar quantity comes in. Remember that you can add potentials using normal (that is, non-vectorial) arithmetic to find, for example, the position where the electric potential of a system is zero.

Charged conducting spheres (both hollow and solid) provide an interesting example of electric field strength and electric potential in action.

The free electrons within a conducting sphere will always rearrange themselves to arrive at the smallest possible energy (within the constraints of conservation of charge). This is achieved by the charges maximizing their distances from each other.

For a hollow conducting sphere with a total charge Q:

- **outside the sphere** the electric field strength and electric potential behave identically to a point object placed at the sphere's centre with a charge Q
- **inside the sphere**:
  - the electric potential V everywhere is equal to the potential of the surface, using $V = \dfrac{kQ}{r}$, where r is the radius of the sphere
  - the electric field strength is zero because there is no potential gradient inside the sphere.

For a solid conducting sphere with a total charge Q, the situation is the same as for the hollow conducting sphere, both inside and outside. Here it is easier to recognize that the electric potential must be constant for the sphere because it is a conductor. Any conductor (treated as ideal with zero resistance) has no potential gradient across it and must therefore act as an equipotential volume.

For a hollow or solid conductor, no work is done in moving a small test charge inside the sphere. This is used in practice to shield electromagnetic signals in metal transmission cables from stray electric and magnetic fields outside the cables.

Example 7

Points A and B are separated by 200 mm.

Point A has a charge of +2.0 nC. Point B has a charge of –3.0 nC.

a) Explain why there is a point, X, on the line joining A and B at which the electric potential is zero.

b) Calculate the distance of the point X from A.

Solution

a) The potential due to A is always positive and the potential due to B is always negative. Therefore, there must be a point (X) at which the potentials sum to zero.

b) Taking x as the distance from A to the zero potential point:

$$\dfrac{k \times 2 \times 10^{-9}}{x} + \dfrac{k \times (-3 \times 10^{-9})}{0.20 - x} = 0$$

Solving this gives $x = 8.0$ cm.

Example 8

Three point charges of +1.0 nC each and a fourth charge Q are placed in the corners of a square. The 0 V equipotential line passes through the centre of the square, as shown.

a) Determine Q.

An electron is placed at point P on the 0 V equipotential and released.

b) Draw an arrow to show the initial acceleration of the electron.

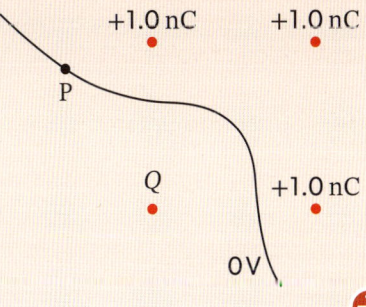

Solution

a) The centre of the square is a distance d from each charge. The potential at the centre is the sum of the potentials from the charges:

$$\frac{k}{d}(1.0 + 1.0 + 1.0 + Q) \times 10^{-9} = 0$$

Therefore, $Q = -3.0$ nC.

b) The acceleration of the electron has the same direction as the electric force acting on it at **P**, perpendicular to the 0 V equipotential. The electron will be accelerated in the overall direction of the positive charges.

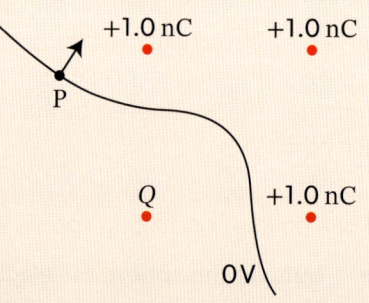

Example 9

A conducting sphere of radius 2.0 cm is initially uncharged. The sphere is touched with an electrode that has a constant potential of +5.8 kV.

a) Explain the direction of the flow of charge carriers between the sphere and the electrode.

b) Calculate the charge remaining on the sphere when the electrode is removed.

Solution

a) The potential on the surface of the sphere is initially zero; hence there is a potential difference between the sphere and the electrode. Owing to this potential difference, free electrons in the sphere are attracted to the electrode. When the electrode is brought into contact with the sphere, the electrons flow from the sphere to the electrode, leaving the sphere with net positive charge.

b) The surface potential of the sphere is now +5.8 kV. The potential is the same as if the net charge of the sphere was placed at its centre.

From $V_e = \frac{kq}{r}$,

$$\text{charge} = \frac{V_e r}{k} = \frac{5.8 \times 10^3 \times 2.0 \times 10^{-2}}{8.99 \times 10^9} = 13 \text{ nC}$$

Example 10

Two point charges, -8 nC and -5 nC, are 0.10 m from each other.

a) Calculate the electric potential energy of the system.

b) Comment on the sign of the answer in part (a).

c) Determine the work needed to reduce the separation of the charges to 2.0 cm.

Solution

a) $E_p = 8.99 \times 10^9 \times \dfrac{(-8.0 \times 10^{-9}) \times (-5.0 \times 10^{-9})}{0.10} = 3.6 \times 10^{-6}$ J

b) The force between the charges is repulsive, so positive work must be done against the electric force to bring the charges from infinite separation to their present position. The potential energy of the arrangement is equal to this work, so it is positive.

c) The work W is equal to the change of the electric potential energy between the initial and the final arrangement of the charges:

$$W = 8.99 \times 10^9 \times (-8.0 \times 10^{-9}) \times (-5.0 \times 10^{-9}) \times \left(\frac{1}{0.020} - \frac{1}{0.10}\right)$$

$$= 1.4 \times 10^{-5} \text{ J}$$

D.2 Electric and magnetic fields

Sample student answer

A cable consisting of many copper wires is used to transfer electrical energy from a generator to an electrical load. The copper wires are protected by an insulator.

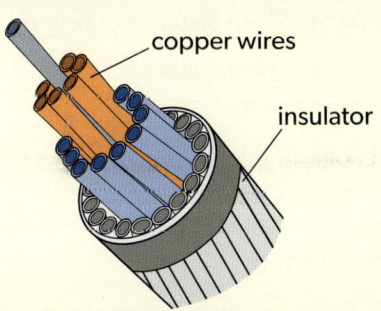

Both the copper wires and the insulator are exposed to an electric field.

Discuss, with reference to electric field, why there is a significant electric current only in the copper wires. [3]

This answer could have scored 2/3 marks:

> The copper wires contain charge carriers, which are negatively charged electrons. When there is an electric field in the wire, the electrons flow past the copper ions to carry the charge. However, the insulator has no charge carriers as it is made of a material with no delocalized electrons, so it cannot carry charge.

▼ The student does not quite get to the heart of the issue, which is that all the charges have a force exerted on them because of their presence in the electric field. When the charges are free to move, they are accelerated. This is the case for the electrons in the metal. As the student says, the charges in the insulator are not mobile, so although a force acts there is little or no current.

Sample student answer

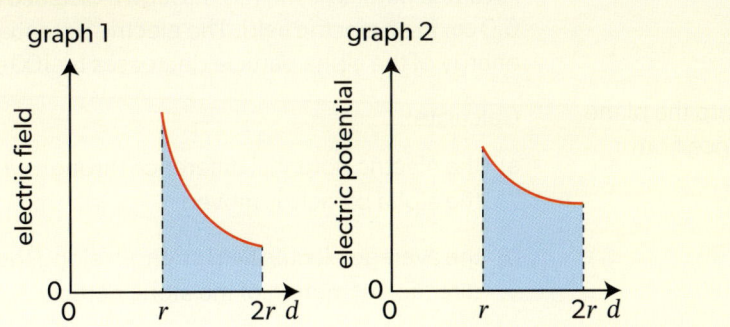

The work done by an external force in moving a test charge $+q$ from $d = 2r$ to $d = r$ is equal to:

A. $q \times$ shaded area under graph 1

B. $q \times$ shaded area under graph 2

C. $q \times$ average value of the electric field between $d = 2r$ and $d = r$ in graph 1

D. $q \times$ average value of the electric potential between $d = 2r$ and $d = r$ in graph 2

Select the answer you consider to be the best. [1]

This answer could have scored 1/1 marks:

> A

▲ One way to answer this is to remember the rule work done = force × distance. Although this rule only applies to a constant force, it helps you to remember that the area under a force–distance graph is also the **work done**. In this case, neither graph is force against distance. Ask yourself which area gives you the work done per unit charge, because each response multiplies a graphical quantity by q.

161

D Fields

Practice problems

Problem 1

Two parallel horizontal plates have a potential difference between them of 1.8 kV and are separated by 9.0 mm.

a) Calculate the magnitude of the electric field strength between the plates.

A negatively charged oil drop of mass 2.6×10^{-14} kg is suspended at rest between the plates.

b) (i) State the direction of the electric field strength.

   (ii) Calculate the charge on the drop. State the answer in terms of the elementary charge e.

Problem 2

Two point charges, $+q$ and $-q$, are separated by a distance d.

Point P is midway between the charges.

a) (i) Determine, in terms of q and d, the magnitude of the electric field strength at P.

   (ii) State the direction of the electric field at P.

Let $q = 8.0$ nC and $d = 5.0$ cm.

b) Calculate the magnitude of the electric force between the charges.

Problem 3

Two parallel wires carry equal currents into the plane of the paper. Point P is at the same distance from both wires.

P

Determine the direction of the magnetic field at P.

Problem 4

Two point charges, +5.0 nC and −5.0 nC, are separated by a distance of 8.0 cm.

a) Calculate the electric potential energy of the system.

b) Determine the work needed to double the distance between the charges.

Problem 5

a) Define electric potential at a point.

b) A metal sphere of radius 0.060 m is charged to a potential of 450 V.

   (i) Deduce the magnitude of the electric charge on the sphere.

   (ii) Determine the magnitude of the electric field strength at a distance of 0.12 m from the centre of the sphere. State an appropriate unit for your answer.

   (iii) Identify the magnitude of the gradient of electric potential at a distance of 0.12 m from the centre of the sphere.

Problem 6

An alpha particle is moved through a distance of 6.0 cm in an electric field. The electric potential energy of the alpha particle decreases by 100 eV.

Calculate:

a) the electric potential difference through which the alpha particle is moved

b) the average electric field strength along the direction of motion of the alpha particle.

D.3 Motion in electromagnetic fields

You must know:
- that there is a force on a current-carrying conductor when it is in a magnetic field
- that moving charged particles are deflected when in a magnetic field.

You should be able to:
- solve problems involving the force acting on a current-carrying conductor in a magnetic field
- solve problems involving the forces acting between two parallel current-carrying wires
- solve problems involving the motion of a charged particle in
  - a uniform electric field
  - a uniform magnetic field
  - perpendicularly orientated uniform electric and magnetic fields.

Moving charged particles are affected by the presence of electric and magnetic fields. The fields lead to forces that act on the particles.

Force on a charge moving in a magnetic field

The direction of the magnetic force is at right angles to the plane containing the magnetic field and to the velocity of a moving charge. The force itself is proportional to:
- the velocity of the charge v
- the magnitude of the charge q
- the magnetic field strength B.

The magnitude of the force F is given by $F = qvB \sin \theta$, where θ is the angle between the velocity of the charge and the magnetic field.

When the velocity of the charge and the magnetic field direction are at right angles (in other words, $\theta = 90°$ so $\sin \theta = 1$) the magnetic force acting on a moving charge is qvB.

Rules can be used to help you with force directions (see Figure 1).

An electric current in a conductor is essentially a flow of charged particles (electrons in this case—Topic B.5). The movement of charged particles gives rise to a magnetic field. It is not surprising that charges moving in a pre-existing magnetic field are also affected. The interaction of the moving charge and the magnetic field leads to a force acting on the charge.

The production of an induced emf and hence a current is explored in Topic D.4.

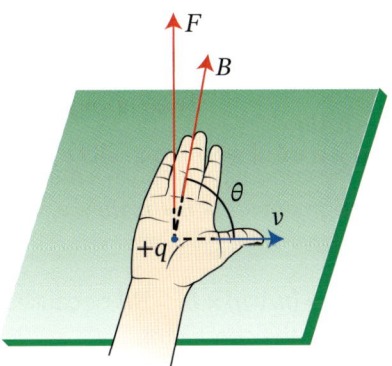

▲ Figure 1 A rule for relating F, B, the velocity v of the charge and the angle θ between B and v. This gives the force direction for a positive charge. The direction is reversed for a negative charge

163

D Fields

If you study the higher level of the IB Diploma Programme physics course, you will see another interpretation for the tesla in Topic D.4.

Magnetic field strength B is measured in tesla (T). The equation $B = \dfrac{F}{qv}$ shows that the fundamental units for tesla are kg s^{-2} A^{-1}.

The tesla is a large unit. The magnetic field strength of the Earth is about 50 μT.

Example 1

A beta particle moves at 15 times the speed of an alpha particle through the same magnetic field. Both particles are moving at right angles to the field.

Calculate $\dfrac{\text{magnetic force on the beta particle}}{\text{magnetic force on the alpha particle}}$.

Solution

From $F = qvB$:

$$\dfrac{F_\beta}{F_\alpha} = \dfrac{Bq_\beta v_\beta}{Bq_\alpha v_\alpha} = \dfrac{e}{2e} \times 15 = 7.5$$

Force on a current-carrying conductor in a magnetic field

There is another way to express the equation $F = qvB \sin \theta$.

Imagine a conductor that carries an electric current I of charge carriers, with n charge carriers per cubic metre (Figure 2).

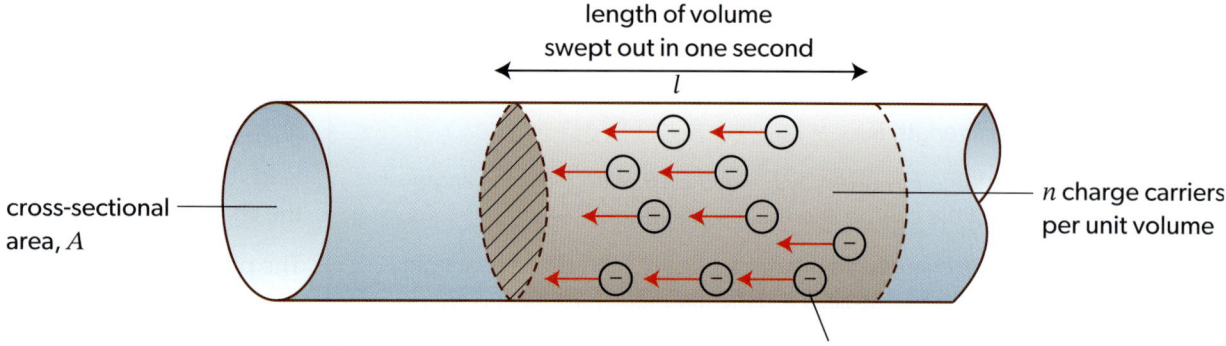

▲ Figure 2 Charge carriers moving through a conductor

The area of the conductor is A and the length of the segment is l. The charge on each carrier is q. There are nAl charge carriers in the segment. So the total charge in the segment is $nAlq$.

The total force acting on the conductor when in a magnetic field of strength B is $B \times$ total charge in the segment $\times v = B \times nAlq \times v$, where v is the average speed of each charge carrier.

Therefore, $F = B \times nAlq \times v = (nAvq) \times Bl = BIl$ because $I = nAvq$ relates the carrier speed to the current in the conductor.

The equation $F = qvB \sin \theta$ is equivalent to $F = BIl \sin \theta$, where θ is now the angle between the magnetic field and the direction of the current.

D.3 Motion in electromagnetic fields

Fleming's left-hand rule relates the directions of the conventional current (Topic B.5) and the field to the force.

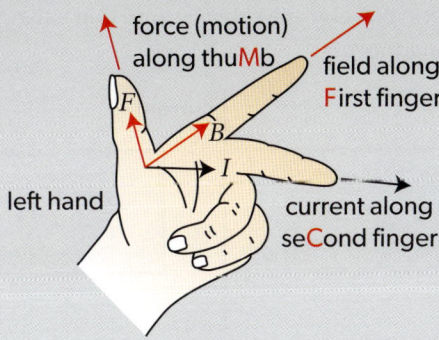

You may have been taught other rules. Make sure that you understand how to use them and **don't use the wrong hand!**

The effects of the forces between two parallel wires can be predicted using the magnetic field patterns from Topic D.2 and the imagined properties of the field lines. Figure 3 shows what happens: the wires run into and out of the page.

- In Figure 3(a), the currents are in the same direction and the net effect of the field line directions is that the lines link up. One "property" of field lines is that they try to be a short as possible. The wires will try to move together when free to do so.

- In Figure 3(b), where the wires have current in opposite directions, the field lines between the wires are in the same direction, which corresponds to a strong magnetic field. The system will try to resist this situation and the wires will have forces acting to move them apart.

The magnetic force F acting on one wire (call it wire 1) due to the magnetic field of the other wire (call it wire 2) must be:

F = (magnetic field due to wire 2 at the position of wire 1) $\times I_1 \times L_1$

where I_1 is the current in wire 1 and L_1 is the length of wire 1.

This means that the force per unit length acting on wire 1 is $\frac{F}{L_1} = B \times I_1$.

The magnetic field due to a current at a perpendicular distance r from a wire carrying a current I is given by $B = \mu_0 \frac{I}{2\pi r}$.

Therefore, for the case of wires 1 and 2 here, the force per unit length acting on wire 1 due to the magnetic field of wire 2 is $\frac{F}{L_1} = \left(\mu_0 \frac{I_1}{2\pi r}\right) \times I_2$ or $\frac{F}{L} = \mu_0 \frac{I_1 I_2}{2\pi r}$.

This expression is symmetrical, which shows that the magnitude of the force of wire 2 on wire 1 (at wire 1) is equal to that of wire 1 on wire 2 (at wire 2). But you might expect this from Newton's third law (Topic A.2).

The two expressions for the force acting on a moving charge and a current-carrying conductor are entirely equivalent. They do not express different ideas in physics but remind us of the link between moving charge and electric current.

The definition of the tesla also links to $B = \frac{F}{IL}$, which is in a similar form to the definitions of gravitational and electric field strength ($g = \frac{F}{m}$ and $E = \frac{F}{q}$), but it has two terms in the denominator (Topics D.1 and D.2).

Assessment tip

The symbol ⊙ is a common way to show you that a current or field direction is out of the page. For a current or field direction going into the page, the convention is to draw ⊗ or ⊕.

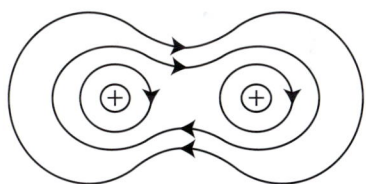

currents in same directions

(a)

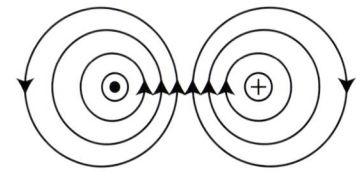

⊙ current out of page   ⊕ current into page

(b)

▲ Figure 3 The forces between two parallel wires when (a) the currents are in the same direction and (b) the currents are in opposite directions

Nature of science

Scientists change fundamental parts of their language as new ideas and innovations emerge. A good example of this type of change is the revision in the definition of the ampere. Until 2019, this was defined in terms of the magnetic force per metre acting between two parallel wires carrying current. The experimental requirements required to make such measurements were painstaking and lengthy. Since 2019, scientists now use the defined value of the electronic charge (known through measurement) and the relationship between current, charge and time (Topic B.5).

Example 2

A 1.5 cm long section of a current-carrying wire is in a magnetic field between two bar magnets.

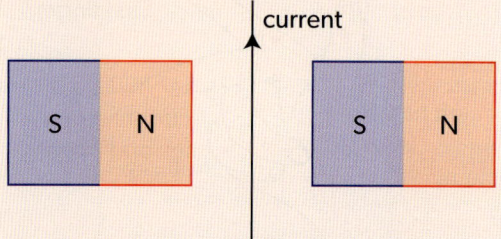

a) Deduce the direction of the magnetic force on the wire.

The current in the wire is 2.5 A. The average magnetic field strength between the magnets is 80 mT.

b) Calculate the magnitude of the magnetic force on the wire.

Solution

a) The magnetic field is from the north pole to the south pole, and hence to the right. Using Fleming's left-hand rule (or equivalent), the force on the wire is into the plane of the paper.

b) Assume that the field is practically uniform and the current is perpendicular to the field.

$F = BIL = 0.080 \times 2.5 \times 0.015 = 3.0$ mN

Example 3

A square-shaped conducting loop **ABCD** of side length 5.0 cm carries a current of 3.0 A in a clockwise direction. The loop is placed 2.0 cm from a long straight wire that carries a current of 6.0 A.

a) Explain the direction of the magnetic force acting on the loop due to the current in the wire.

b) Calculate the magnitude of the force.

Solution

a) The forces on segments **BC** and **DA** are equal and opposite (because these sections of the loop are at the same distance from the straight wire, but carry the current in opposite directions), so they do not contribute to the net force on the loop.

AB is attracted to the straight wire because its current is parallel to the current in the wire. Each of the two sections that make up **CD** is repelled from the wire. **CD** is further away from the straight wire than **AB**, so the repulsive force is less than the attractive force. The loop will therefore be attracted to the straight wire.

D.3 Motion in electromagnetic fields

b) Use $\frac{F}{L} = \mu_0 \frac{I_1 I_2}{2\pi r}$ to calculate the forces on AB and CD.

$$F_{AB} = \frac{4\pi \times 10^{-7} \times 3.0 \times 6.0}{2\pi \times 0.02} \times 0.05 = 9.0 \times 10^{-6} \text{ N}$$

$$F_{CD} = \frac{4\pi \times 10^{-7} \times 3.0 \times 6.0}{2\pi \times (0.02 + 05)} \times 0.05 = 2.6 \times 10^{-6} \text{ N}$$

Net attractive force on the loop = $F_{AB} - F_{CD}$ = 6.4×10^{-6} N

Sample student answer

The diagram shows a cross-sectional view of a wire.

The wire, which carries a current of 3.5 A into the page, is placed in a region of uniform magnetic field of field strength 0.25 T. The field is directed at right angles to the wire.

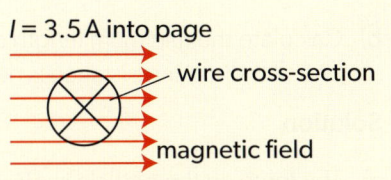

The speed of each charge carrier is 2.0×10^{-4} m s^{-1}.

Determine the magnitude and direction of the magnetic force on one of the charge carriers in the wire. [2]

This answer could have achieved 2/2 marks:

$v = 2.0 \times 10^{-4}$ m s^{-1} $I = 3.5$ A $B = 0.25$ T $q = 1.6 \times 10^{-19}$ C $\sin 90° = 1$

$F = qvB \sin\theta = 1.60 \times 10^{-19} \times 2.0 \times 10^{-4} \times 0.25$

$= 8.0 \times 10^{-24}$ N downwards

▲ The student has set out the calculation clearly and quoted a direction for the force.

Charged particles in an electric field

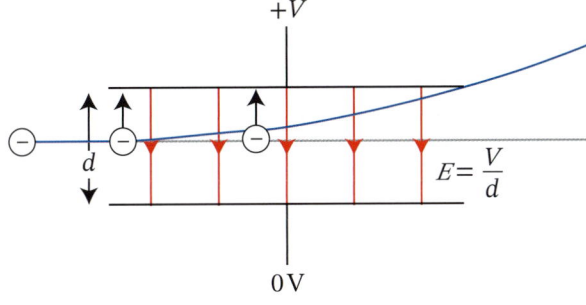

▲ **Figure 4** An electron is moving perpendicular to a uniform electric field. The force is constant in both magnitude and direction. This leads to a parabolic path for the electron while it is in the electric field

An electron enters a uniform electric field perpendicular to the field lines (Figure 4). A force acts in the opposite direction to these field lines.

The force on the charge is Ee and the acceleration $a = \frac{Ee}{m_e}$.

The electric field E is produced by the potential difference V between the plates, which are separated by a distance d, so $E = \frac{V}{d}$. Therefore, $a = \frac{eV}{m_e d}$.

In any practical situation, you can ignore the weight because the electric force is very much greater than the weight.

This idea links strongly to Theme A.

A constant force Ee acts on the charged particle so its acceleration a is also constant in both **magnitude** and **direction**. This is precisely the condition that you used in Topic A.1 to solve problems involving the constant acceleration of free fall. You can use the kinematic (*suvat*) equations here too.

The charged particles are usually in a vacuum and so are not subject to resistive drag. Like the trajectory of a projectile that has initial horizontal and vertical components, the path of the charged particle will be parabolic (Figure 4).

D Fields

Example 4

A precipitation system collects dust particles in a chimney. It consists of two large parallel vertical plates, separated by 4.0 m, maintained at potentials of +25 kV and −25 kV.

A small dust particle moves vertically up the centre of the chimney, midway between the plates. The charge on the dust particle is +5.5 nC.

a) Show that there is an electrostatic force on the particle of about 0.07 mN.

The mass of the dust particle is 1.2×10^{-4} kg and it moves up the centre of the chimney at a constant vertical speed of 0.80 m s^{-1}.

b) Calculate the minimum length of the plates so that the particle strikes one of them. Air resistance is negligible.

Solution

a) The force on the particle is $qE = \dfrac{qV}{d}$, where d is the distance between the plates. The potential difference V is 50 kV.

Force on the particle = $\dfrac{5.5 \times 10^{-9} \times 5.0 \times 10^{4}}{4.0} = 6.875 \times 10^{-5}$ N

b) Horizontal acceleration = $\dfrac{\text{force}}{\text{mass}} = \dfrac{6.875 \times 10^{-5}}{1.2 \times 10^{-4}} = 0.573$ m s^{-2}

The particle is in the centre of the plates, so it must move 2.0 m horizontally to reach a plate. Using $s = ut + \dfrac{1}{2}at^2$, and knowing that the particle has no initial horizontal component of velocity:

$2.0 = 0 \times t + \dfrac{1}{2}(0.573)t^2$

$t = \sqrt{\dfrac{2 \times 2.0}{0.573}} = 2.64$ s

Therefore, the length must be $2.64 \times 0.8 = 2.1$ m.

Approaches to learning

The motion of a charged particle in a magnetic field links to a number of other areas and can help you to consolidate your knowledge.

Circular motion appears in Topic A.2. The concept of electrons moving in a circle is a fundamental consideration for the beginning of an understanding of the Bohr model in Topic E.1.

Charged particles in a magnetic field

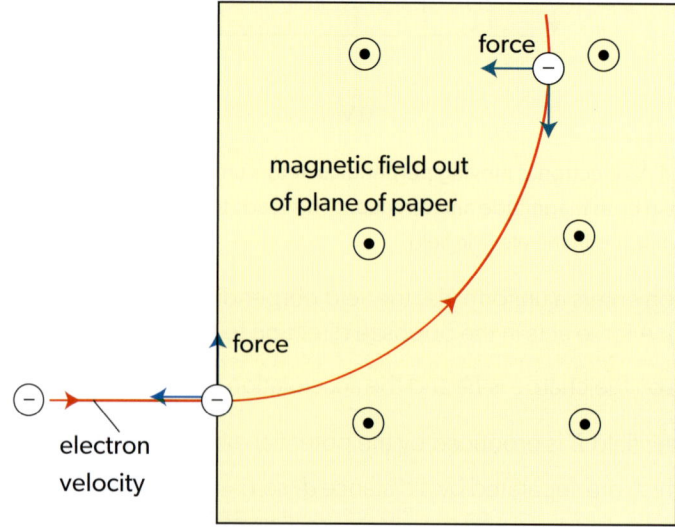

▲ Figure 5 The electron is moving perpendicular to a magnetic field. The force on the electron is at right angles to its motion so the magnetic force is providing a centripetal force that acts on the electron

D.3 Motion in electromagnetic fields

A charged particle of mass m enters a uniform magnetic field of strength B perpendicular to the field lines (Figure 5). The charge on the particle is q.

The magnetic force acts always at 90° to the velocity of the particle. This is the condition for a centripetal force, so the particle moves in a circular path. The speed of the particle does not change.

The force on the particle is Bqv, which must equal $\frac{mv^2}{r}$, where r is the radius of the orbit.

There are several useful ways to express the radius of the orbit:

$$r = \frac{mv}{qB} = \frac{p}{qB} = \frac{\sqrt{2mE_k}}{qB}$$

In these expressions, p is the momentum of the charged particle and E_k is its kinetic energy. The kinetic energy remains constant in the magnetic field, as does the magnitude of the momentum (the direction changes constantly because the motion is circular).

> **Assessment tip**
>
> A measurement of the radius can lead to the determination of the ratio $\frac{\text{charge}}{\text{mass}}$ for a charged particle (this is sometimes known as the **specific charge**).
>
> For a charged particle of mass m accelerated from rest through potential difference V before entering the magnetic field:
>
> $\frac{1}{2}mv^2 = eV$, so that $v = \sqrt{\frac{2eV}{m}}$. Because $r = \frac{mv}{eB}$, then $\frac{e}{m} = \frac{2V}{B^2 r^2}$.
>
> This is another useful derivation to remember for examinations.

When the direction of motion of the charged particle is not perpendicular to B, the relevant speed to use for the circular part of the motion is the component of velocity perpendicular to the field. The perpendicular component leads to circular motion as before (but with a larger radius as the momentum component is smaller). The charged particle will move at constant speed in the direction of B along a helical path.

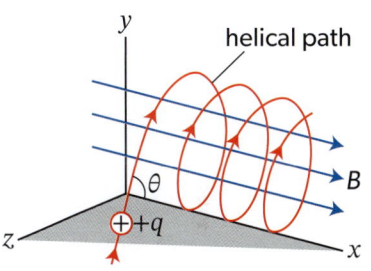

This links to ideas of vector components in Topics A.1 and T.3.

Example 5

A proton and a positive pion (a particle of a smaller mass than a proton, but carrying the same electric charge) travel along the same path at a speed of 1.5×10^7 m s^{-1} into a uniform magnetic field of magnitude 0.16 T. The magnetic field direction is at 90° to the initial direction of motion of the proton and the pion.

a) Calculate the radius of curvature of the path of the proton.

b) Comment on how the path of the pion will differ from that of the proton.

c) The magnetic field strength is decreased. Suggest how this affects the paths of the particles.

Solution

a) Rearranging and simplifying $\frac{m_p v^2}{r} = Bev$ gives $r = \frac{m_p v}{eB}$.

The DP Physics data booklet provides the mass of a proton m_p. Substituting gives:

$$r = \frac{1.673 \times 10^{-27} \times 1.5 \times 10^7}{1.6 \times 10^{-19} \times 0.16} = 0.98 \text{ m}$$

b) A pion has less mass than a proton. The particles have the same charge so only the mass is different. From the radius equation, r is smaller for the pion so its path is more curved.

c) Again, from the radius equation, when B is decreased, r increases, so the path of both particles is less curved.

Charged particles in perpendicular electric and magnetic fields

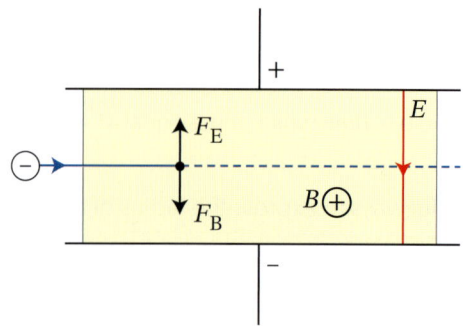

▲ **Figure 6** The electric force and magnetic force on the charged moving particle cancel out so that there is no deflection in the crossed fields

Figure 6 shows a negatively charged particle entering a region where two fields act. There is an electric field E vertically downwards and a horizontal magnetic field B into the page. These are called crossed fields.

The electric force on the particle is upwards and the magnetic force is downwards. When these two forces act on the charged particle with the same magnitude, there will be no resultant force on the particle (this assumes that the gravitational force is negligible).

For this condition, $F_E = F_B$ so $qE = qvB$, where q is the charge on the particle and v is its speed. This gives $v = \dfrac{E}{B}$ as the single speed at which the particle will travel without deflection. At all other speeds, the particle will be deflected upwards (at speeds less than v when $F_E > F_B$) or downwards (when $F_E < F_B$).

This arrangement can be used as a velocity filter (or selector) when particles of a certain speed are needed for an experiment.

Sample student answer

A beam of electrons e⁻ enters a uniform electric field between parallel conducting plates **R** and **S**. **R** and **S** are connected to a power supply. A uniform magnetic field B is directed into the plane of the page and is perpendicular to the direction of motion of the electrons.

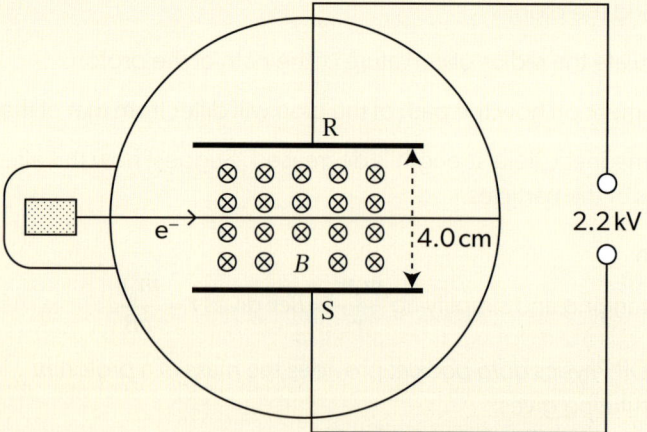

The magnetic field is adjusted until the electron beam is **undeflected** as shown.

a) Identify on the diagram the direction of the electric field between the plates. [1]

This answer could have achieved 1/1 marks:

D.3 Motion in electromagnetic fields

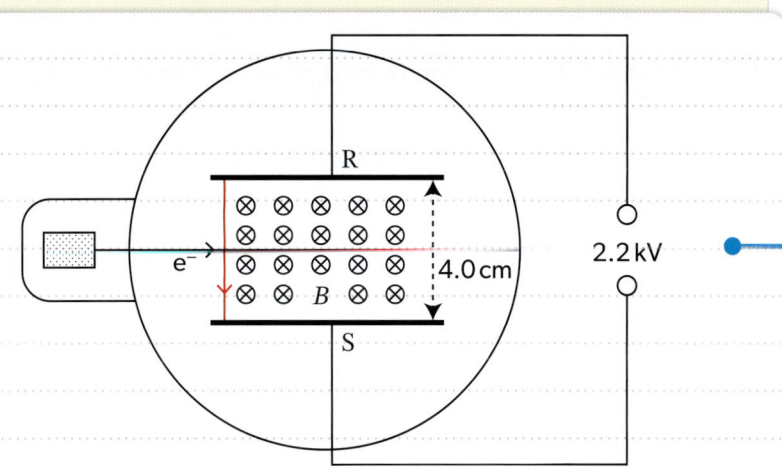

> ▲ The student has identified that the magnetic force acting on the electron is downwards in the diagram. The electron has a negative charge and the electric force on it must therefore be upwards. Consequently, the force acting on a positive charge in the same position must be downwards and this is also the direction of the electric field.

The following data are available:

Separation of the plates **R** and **S** = 4.0 cm

Potential difference between the plates = 2.2 kV

Velocity of the electrons = 5.0×10^5 m s^{-1}

b) Determine the strength of the magnetic field *B*. [2]

This answer could have achieved 2/2 marks:

$F = qE = Bqv \Rightarrow B = \dfrac{E}{v}$ $E = \dfrac{V}{d}$

$ = \dfrac{5.5 \times 10^4}{5.0 \times 10^5}$ $= (2.2 \times 10^4)/(4.0 \times 10^{-2})$

$ = 0.11$ T $= 5.5 \times 10^4$ V m^{-1}

> ▲ The solution is correct and gains full marks. However, the layout is not perfect. The value for *E* could be taken to be the final answer, whereas it is an intermediate value that is fed into the $\dfrac{E}{v}$ equation. Try to avoid this type of layout if you can.

c) The velocity of the electrons is now increased. Explain the effect that this will have on the path of the electron beam. [2]

This answer could have achieved 1/2 marks:

$B = \dfrac{E}{v}$

∴ When velocity of the electrons increases, B field becomes stronger and the electrons will travel in a path acting towards S.

> ▼ The student claims that B is stronger and this is not true. The magnetic force increases. The conclusion is correct.

Practice problems

Problem 1

A straight wire carries a current of 0.90 A.

a) The wire is in a uniform magnetic field of magnitude 0.25 T directed at 30° to the wire. Calculate the magnetic force on a 5.0 cm long section of the wire.

b) The wire is now placed parallel to another straight wire that carries a current of 4.0 A in the same direction. The distance between the wires is 20 cm.

   (i) Outline why there is a force between the wires.

   (ii) Calculate the force per unit length of the wires.

Problem 2

Two parallel metal plates, A and B, are fixed vertically 20 mm apart and have a potential difference of 1.5 kV between them.

a) Sketch a graph showing the potential at different points in the space between the plates.

A plastic ball of mass 0.5 g is suspended midway between the plates by a long insulating thread. The ball has a conducting surface and carries a charge of 40 nC.

b) The ball is released from rest. Deduce the subsequent motion of the ball.

Problem 3

A proton from the Sun passes above the north magnetic pole moving parallel to the Earth's surface at a speed of 1.2×10^6 m s^{-1}. At this point, the magnetic field of the Earth is vertically downwards with a magnitude of 5.8×10^{-5} T.

a) Calculate the radius of the path of the proton when it is above the pole.

A helium nucleus moving with the same initial velocity as the proton is also above the pole.

b) Compare, without calculation, the paths of the proton and the helium nucleus.

Problem 4

An electron moves undeflected in a region of crossed electric and magnetic fields that are perpendicular to each other and to the velocity of the electron.

The electric field strength is 1.4 kV m^{-1}. The speed of the electron is 2.0×10^5 m s^{-1}.

a) Calculate the magnetic field strength.

b) The magnetic field is switched off. Describe the subsequent motion of the electron.

D.4 Induction

You must know:

- that a magnetic flux exists in a region where there is a magnetic field
- that an electromotive force (emf) is induced in a conductor
  - when it moves relative to a magnetic field
  - when there are changes in the magnetic flux that links the conductor
- that Lenz's law is a consequence of energy conservation
- Faraday's law of induction
- what is meant by self-induction.

You should be able to:

- solve problems including magnetic flux and magnetic flux density
- solve problems involving Faraday's law of induction
- explain how rotating a coil in a uniform magnetic field leads to the induction of a sinusoidally varying emf.

D.4 Induction

Scientists in the 1820s realized that relative motion of a conductor in a magnetic field leads to the production of an induced emf and to an induced current in the conductor. This introduced new concepts of magnetic flux, magnetic flux density and magnetic flux linkage.

Just as in Topics D.1, D.2 and D.3, we use the idea of field lines to visualize a field. The closer the magnetic field lines, the stronger the magnetic field. This leads immediately to the idea of defining a field strength in terms of the density (closeness) of field lines. Early workers called the magnetic field lines a magnetic flux and they used the term magnetic flux density for the field line density.

> **Magnetic flux** Φ is proportional to the number of field lines passing through a surface. (This can be the number of lines enclosed by a one-turn coil placed in the magnetic field.)
>
> The unit of magnetic flux is the weber (Wb).
>
> Magnetic flux is the product of the magnetic flux density B (magnetic field strength) and area A, so $\Phi = B \times A$, where A is the area through which the conductor moves and the normal to the area is parallel to the direction of the magnetic field.
>
> The units of **magnetic flux density** can therefore be expressed as $Wb\,m^{-2}$ as well as in tesla (T). $1\,T \equiv Wb\,m^{-2}$.
>
> When there is an angle θ between the magnetic field and the normal to the area A, the equation for magnetic flux becomes $\Phi = BA \cos \theta$.

The density of field lines indicates the strength of the field, so the magnetic field strength that you met in Topics D.2 and D.3, measured in tesla (T), can also be called the magnetic flux density (in symbols, $\Phi = BA$). "Density" in this case means the number of field lines per square metre.

When a coil has more than one turn (N turns), then each turn in the coil links one set of lines and the total magnetic flux linkage is (flux for one turn) × (number of turns)—in other words, NBA or $N\Phi$.

Faraday imagined a conductor moving across (cutting) magnetic field lines. When this happened, he observed an emf being generated in the conductor. Faraday could not explain why this happened, but we can. In the 200 years since Faraday, we have developed more knowledge of electrical conduction and the important role of the charge carriers (electrons in a metal).

Nature of science

The word "flux" is an old English word that means a flow of something. Although it is not in common usage now, the word has stuck for magnetic phenomena. Scientists such as Faraday had the idea that the field lines were moving (flowing) from pole to pole. This idea has disappeared and nowadays we think of the field lines as static.

It is the nature of science that concepts change either dramatically in a paradigm shift or more gradually as ideas evolve.

Approaches to learning

In Topics D.2 and D.3, the symbol B is used for the quantity magnetic field strength. In this topic, it is used for the magnetic flux density too. You can use either term in this course. Similarly, the two units, $Wb\,m^{-2}$ and T, can be considered to be identical.

There are subtle differences in the meaning of field strength and flux density, but these are beyond the IB Diploma Programme physics course. You can treat them as identical, especially when dealing with magnetic fields in air or in a vacuum.

D Fields

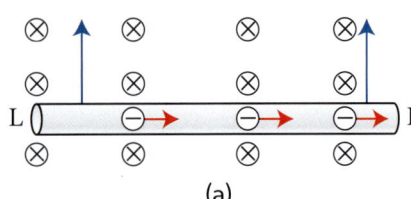 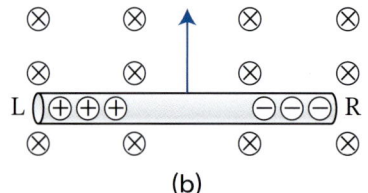 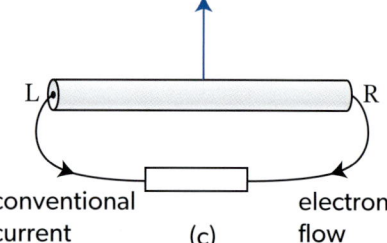

▲ Figure 1 An emf is induced in a rod moving in a magnetic field: (a) immediately after the rod begins to move, (b) when the rod is moving at constant speed and (c) when an external conducting link is connected between L and R

- Figure 1(a) shows the free electrons in a conductor (a metal rod). The conductor is moving through a uniform magnetic field. The magnetic field direction is into the page and the rod is moving upwards in the plane of the page.

- The conductor carries the free electrons with it so that each electron can be regarded as a current moving up the page, each individual current parallel to every other. A force (according to Topic D.2) acts on each electron (because it is being treated as a current moving in a magnetic field). To work out the direction of this force, remember that a free electron moving up the page is equivalent to a positive charge moving down. Fleming's left-hand rule with conventional current shows that the force on the electron is to the right—perpendicular to the field and the direction of motion of the rod.

- This drift of the electrons to the right produces a negative charge at R and an electron deficit (positive charge) at L (Figure 1(b)). Charges accumulate at the ends until the electric field along the rod is so large that further charge movement along it stops. An induced electromotive force (emf) has been generated across the rod between R and L.

- When the circuit is completed (Figure 1(c)), a flow of electrons occurs. Inside the rod, the conventional current flows from R to L. The electrons flow out of end R of the rod, and this is a conventional current in the external circuit from L to R. These electrons now move in a magnetic field. As a result, an additional force acts on them (Topic D.3). However, this force acts downwards to prevent the rod from moving upwards.

There is a conventional current acting from R to L inside the rod—with the magnetic field into the paper, the force is acting downwards and opposes the original force upwards on the rod. This leads to Lenz's law which is a simple statement of the direction of the induced emf.

Faraday's law goes on to make a quantitative statement about the magnitude of the electromagnetic effect. Essentially, the faster the rod moves upwards, the greater the emf that will be generated in the rod and hence the greater the rate of charge flow (electric current) through the rod.

> **Lenz's law** can be expressed in a number of ways. The most common way is probably the statement that "the emf induced in the system is always such as to oppose the change creating it".
>
> Notice that the emphasis is on the generation (induction) of the emf. An induced current appears when there is an induced emf and a complete circuit for the flow of charge.
>
> Faraday's law is a quantitative statement that the emf ε induced is equals the rate of change of magnetic flux. This can be written as $\varepsilon = \dfrac{\Delta \Phi}{\Delta t}$.
>
> When Lenz's law (that the induced emf opposes the change) is combined with Faraday's law, then we can write $\varepsilon = -\dfrac{\Delta \Phi}{\Delta t}$. Although this formulation, including the negative sign, was first proposed by Neumann it is still usually known as **Faraday's law**.
>
> When the field and the normal to the plane in which the rod moves are not perpendicular (with an angle θ between them), then the law becomes $\varepsilon = -N\dfrac{\Delta \Phi}{\Delta t}$ $= -N\dfrac{\Delta(BA\cos\theta)}{\Delta t}$, where B and A are the magnetic flux density and the area swept out, respectively.

> **Lenz's law** states that work must be done to overcome this opposing force and this is where the transferred energy in the system comes from.
>
> Lenz's law is equivalent to conservation of energy. If both forces acted upwards, then energy would be created in the system. This is not possible.
>
> **Faraday's law** states that the induced emf in a circuit is equal to the rate of change of magnetic flux linkage through the circuit.

You can apply Faraday's law to the situation in Figure 1.

When the conductor of length l moves at speed v in a uniform field of magnetic flux density B, the change in the flux every second is Bvl because the rod sweeps out an area of vl every second—this is the change in area per second, so the change of magnetic flux per second is $B \times vl$.

Faraday and others in the 19th century developed the idea that conductors moving in magnetic fields "cut" field lines—this is a simple idea to grasp.

Imagine a single-turn coil that encloses an area A. Suppose the magnetic flux density changes by ΔB in a time Δt.

Faraday's law predicts that the emf generated will be $\varepsilon = -NA\dfrac{\Delta B}{\Delta t}$, where N is the number of turns, A is the fixed area of the coil and $\dfrac{\Delta B}{\Delta t}$ is the rate at which the magnetic flux density changes with time.

Example 1

A pair of parallel conducting rails form a ramp that makes an angle of 20° with the horizontal. A conducting rod of length $L = 0.40$ m slides down the ramp without friction at a constant speed $v = 0.85$ m s⁻¹. The rails are electrically connected at the bottom of the ramp and form a closed circuit with the rod. The system is in a uniform vertical magnetic field of flux density $B = 0.12$ T.

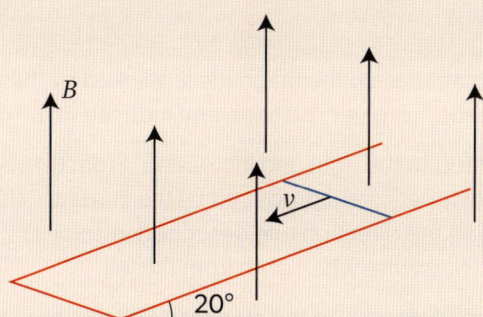

a) Calculate the magnitude of the emf induced in the rod.

b) Discuss the forces acting on the rod in the direction of motion.

Solution

a) Component of magnetic field normal to ramp $= B \cos 20°$.

emf $\varepsilon = (B \cos 20°)vL = 0.12 \times \cos 20° \times 0.85 \times 0.40 = 38$ mV.

b) There are two forces acting in the direction of motion.

- Component of weight down the ramp $F_g = mg \sin 20°$, where m is the mass of the rod.

- A magnetic force due to the current induced in the rod.

Current $= \dfrac{\varepsilon}{R}$

where R is the resistance of the rod (assuming that the rest of the circuit has zero resistance).

Magnitude of magnetic force $F_m = BIL = \dfrac{B\varepsilon L}{R}$

From Lenz's law, it acts opposite to the motion of the rod, so up the ramp.

Because the velocity is constant, the net force on the rod is zero. Therefore, the forces have the same magnitude, $F_g = F_m$.

Assessment tip

There are many ways to change the flux linked by a wire or a coil.

- The magnetic field can move while the conductor remains fixed in space.

- The magnetic flux density can change in magnitude while the coil remains fixed. Because there is an overall change in magnetic flux density from Φ_1 to Φ_2, the emf induced is $N\dfrac{(\Phi_2 - \Phi_1)}{\Delta t}$.

- A coil can be rotated in a constant magnetic field. The angle of rotation will dictate the magnitude of the emf induced. Imagine a coil of area A that is initially perpendicular to the direction of the magnetic field B. When the coil is flipped over through 180° in a time t, then an emf of magnitude $2\dfrac{BA}{t}$ will be induced in the coil. This is because the magnetic flux direction has reversed inside the coil.

You must take care though. For a coil moving entirely in a uniform magnetic field and not changing its orientation in the field, no emf is induced. Or rather, two equal and opposite emfs are induced in the coil as it cuts the field lines in opposite senses and these two emfs cancel out.

D Fields

Example 2

A coil with five turns has an area of 0.25 m². The magnetic flux density in the coil changes from 60 mT to 30 mT in a time of 0.50 s.

Calculate the magnitude of the emf induced in the coil.

Solution

$\varepsilon = -NA \dfrac{\Delta B}{\Delta t}$

Magnitude of $\varepsilon = 5 \times 0.25 \times \dfrac{(60 - 30) \times 10^{-3}}{0.50} = 75$ mV

Example 3

A bar magnet is moved towards a stationary circular coil. The coil is perpendicular to the direction of motion of the magnet.

a) Explain why an emf is induced in the coil.

b) Determine:

  (i) the direction of the current induced in the coil

  (ii) the nature of the magnetic force between the magnet and the coil.

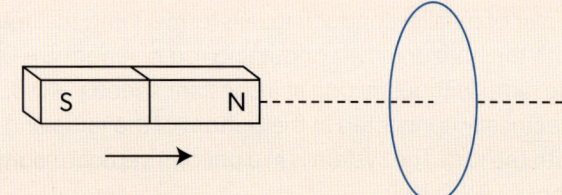

Solution

a) The magnetic field strength at the location of the coil increases, so the magnetic flux linked by the coil also increases. From Faraday's law, the change in the magnetic flux gives rise to an emf induced in the coil.

b) (i) The original magnetic field through the coil is directed to the right and increasing. From Lenz's law, the induced current must oppose this change, so the magnetic field produced by the current in the coil must have its north-seeking pole pointing to the left. Using the direction rule described in Topic D.2 (see Figure 10 on page 156), the current induced in the coil is anticlockwise when seen from the left.

  (ii) From Lenz's law again, the magnetic force does negative work on the magnet (so that energy is transferred from the moving magnet to the current in the coil). The force is therefore repulsive.

Sample student answer

A vertical metal rod of length 0.25 m moves in a horizontal circle about a vertical axis in a uniform horizontal magnetic field.

The metal rod completes one circle of radius 0.060 m in 0.020 s in the magnetic field of strength 61 mT.

Determine the maximum emf induced between the ends of the metal rod. [3]

This answer could have achieved 2/3 marks:

- Speed of rotation = $2\pi \times 0.06 \div 0.02 = 0.6\pi$ m s⁻¹
- Maximum emf is $BvI = 61 \times 10^{-3} \times 0.25 \times 0.6\pi = 9.15$ mV

▼ This attempt has one error. The calculation of the speed of rotation is incorrect (it should be 6.0π m s⁻¹). The final answer would have been better to two significant figures.

▲ The solution is well explained and the steps are clear.

Self-induction

An induced emf can arise inside a single conductor that links its own magnetic flux. When this occurs, it is known as **self-induction**.

The single-turn coil in Figure 2 has a changing current in it—suppose that the magnitude of the current is increasing. As a result, the magnetic field strength will be getting larger too. This can be visualized as the field lines in the centre of the coil moving inwards so that they become closer to each other as time goes on—the density of the lines (magnetic flux density) is increasing. By Lenz's law, the system will try to prevent this increase in the magnetic field. It can only do so by providing an emf that opposes the change—in other words, by providing an additional emf that is in opposition to whatever was causing the increase in the original charge flow.

Refer to Topic D.2 (Figure 10) to see how the magnetic field pattern for the coil in Figure 2 arises.

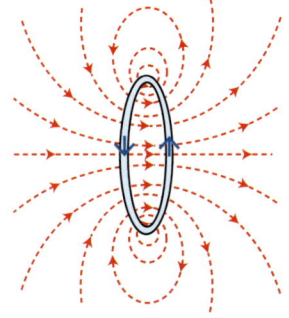

▲ Figure 2 A coil links its own magnetic flux and there is a self-induction effect

Example 4

A wire of negligible resistance forms a coil. A current is established between the ends **A** and **B** of the coil, in the direction shown.

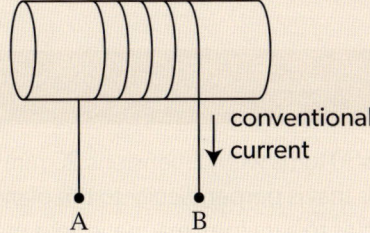

The current is decreasing.

a) Explain why there is a potential difference across the coil.

b) Deduce which end of the coil is at a higher potential.

Solution

a) The current in the coil gives rise to a magnetic field. The coil links its own magnetic flux, and, since the flux is changing, there is a self-induced emf in the coil. This emf causes the ends of the coil to be at different potentials.

b) From Lenz's law, the direction of the self-induced emf is such that it opposes the decrease of the current. To achieve this effect, the self-induced current must have the same direction as the original current, from **A** to **B** in the coil but from **B** to **A** in the part of the circuit that is *external* to the coil. Therefore, **B** is at a higher electrical potential than **A**.

Coil rotating a magnetic field

Figure 3 shows a single-turn coil of wire rotating in a uniform magnetic field. In Figure 3(a), zero flux links the coil because the coil is parallel to the field lines. When the coil has turned through 90° it is perpendicular to the field lines and the magnetic flux linkage has changed to its maximum magnitude, inducing an emf while the coil is turning. When the coil turns through another 90° it is in its original orientation, except that the sides of the coil have reversed. The magnetic flux linkage has changed again. The coil rotates through a final 180° to return to its original orientation.

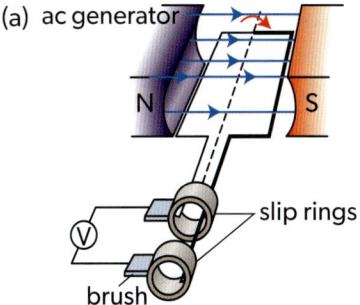

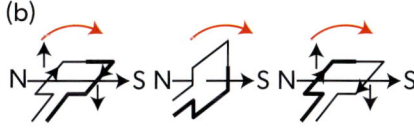

▲ Figure 3 A simple ac generator. The two sides of the coil are drawn with different thicknesses to show the orientation of the coil in various positions

The emf in the coil varies continuously between a positive peak and a negative trough. With suitable electrical connections, there will be an alternating current in the load connected to the coil (in this case, a voltmeter). Slip rings transfer energy from the coil to the load. The rings are attached to the coil ends and rotate with it. Stationary brushes, often made of conductive carbon, press onto the rings and connect to the external load.

Figure 4 shows the output of such a generator when it is rotating at a constant speed—the waveform is sinusoidal. Quantities associated with the varying output are its frequency, the peak emf and the peak current.

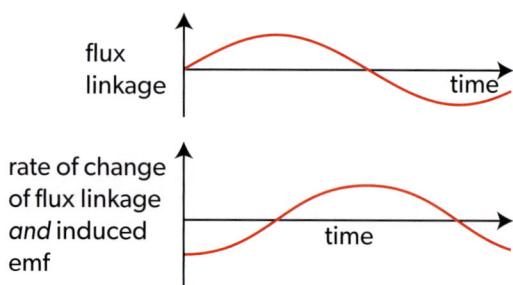

▲ **Figure 4** The variation of emf with time for one cycle of the coil in Figure 3

Assessment tip

When the rotation speed doubles:
- the frequency doubles because the time for one rotation halves
- the rate of change of magnetic flux linkage doubles and so the peak emf doubles too.

Approaches to learning

The term 'electromotive force' is properly used for the induced pd that arises when a conductor cuts flux. Emf is an old term that has its roots in the early discoveries of electromagnetism. It is now also used for those electrical devices that transfer energy to an electrical sink. In this case, the use of the word is more difficult to understand.

For electromagnetic induction, however, a force clearly acts on the electrons that are being moved through the magnetic field.

When learning about emf—whether in an electromagnetic induction or a current electricity context—try to focus on the dimensions of emf and potential difference, JC^{-1}.

Example 5

A rectangular coil of 650 turns with dimensions 20 mm × 35 mm rotates about a horizontal axis that is perpendicular to the plane of the diagram. The axis is at right angles to a uniform magnetic field of flux density 2.5 mT. At one instant, the plane of the coil makes an angle θ with the vertical.

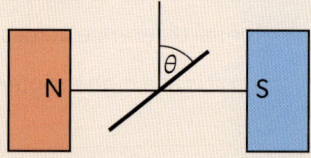

a) Identify the value of θ for the magnitude of the magnetic flux through the coil to be a minimum.

b) Calculate the magnetic flux passing through the coil when θ is 30°.

c) Determine the maximum flux linkage through the coil as it rotates.

Solution

a) The coil must be parallel to the field for the flux to be zero (this is the minimum magnitude). θ is either 90° or 270°.

b) The angle between the **normal** to the coil and the magnetic field is θ.

Magnetic flux $\Phi = BA \cos \theta$
$= 2.5 \times 10^{-3} \times 20 \times 10^{-3} \times 35 \times 10^{-3} \times \cos 30° = 1.5$ μWb.

c) The maximum flux linkage is when the coil is perpendicular to the field ($\theta = 0°$), which corresponds to a maximum flux of 1.75 μWb.

Maxiumum flux linkage = $650 \times 175 \times 10^{-6}$ = 1.1 mWb turns

Sample student answer

A square coil **ABCD** rotates in a magnetic field. The coil has one turn.

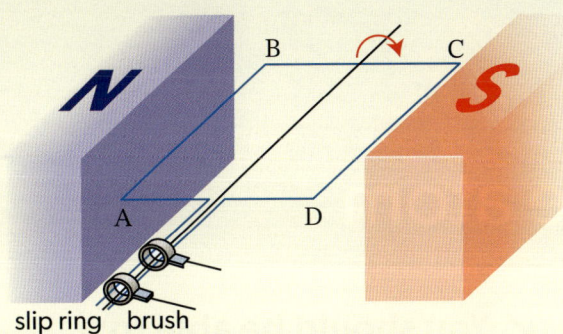

The ends of the coil are connected to slip rings and brushes. The plane of the coil is shown at the instant when it is parallel to the magnetic field.

Explain, with reference to the diagram, how the rotation of the generator produces an electromotive force (emf) between the brushes. [3]

This answer could have achieved 1/3 marks:

> The rotation of the square coil changes its flux over time and according to Faraday's law, an emf will be produced as emf is the rate of change of flux. In the diagram, as AB rotates upwards, experiencing a change in flux in the magnetic field between two ends of the magnet, the force is upwards, the field direction is from N → S (rightwards). Using Fleming's left-hand rule, the current will flow from A to B.

▼ A good answer to this question requires references to:
- the concept of flux *linkage*
- Faraday's law and the rotation of the coil.

The emphasis in this answer on the direction of forces and currents is not required by the question.

Practice problems

Problem 1

The graph shows the variation of magnetic flux Φ through a coil with time t.

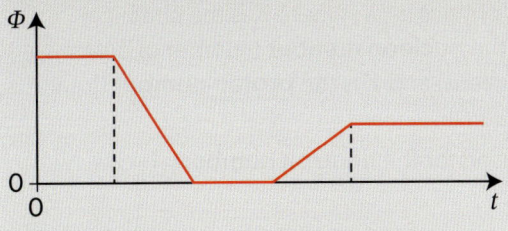

a) Sketch the variation of the magnitude of the emf in the coil with time over the same time period.

b) Explain your answer to part (a).

Problem 2

A square loop with sides 8.0 cm is made from copper wire of radius 1.5 mm. A magnetic field perpendicular to the loop changes at a rate of 5.0 mT s^{-1}. The resistivity of copper is 1.7 10^{-4} Ω m.

Determine the current in the loop.

Problem 3

A vehicle with a radio antenna of length 0.85 m is travelling horizontally with a speed of 95 km h^{-1}. The horizontal magnetic field due to the Earth where the vehicle is located is 0.055 mT.

a) Deduce the maximum possible emf that can be induced in the antenna due to the Earth's magnetic field.

b) Identify the relationship between the velocity of the vehicle and the magnetic field direction for this maximum to be attained.

Problem 4

Two identical circular coils, X and Y, are arranged parallel to each other and wound in the same sense. When direct current is switched on in X, the current direction is clockwise in X.

Predict the direction of any induced current in Y when:

a) the current in X is switched on

b) the current in X is switched off.

E Nuclear and quantum physics

E.1 Structure of the atom

You must know:
- details of the Geiger–Marsden–Rutherford experiment
- how emission and absorption spectra arise
- that emission spectra can be used to determine chemical composition

Additional higher level:
- that high-energy deviations from Rutherford scattering give information about nuclear forces
- details of the Bohr model for the hydrogen atom including:
  - the existence of discrete energy levels for the atom
  - the quantization of angular momentum for the hydrogen atom
  - the existence of quantized electron orbits
- what is meant by nuclear density.

You should be able to:
- use the $^A_Z X$ nuclear notation
- solve problems involving the frequency of emitted and absorbed photons in atomic and nuclear transitions

Additional higher level:
- solve problems involving the closest approach of a charged particle to a nucleus
- solve problems involving the relationship between the radius of a nucleus and its nucleon number
- solve problems involving the Bohr model for the hydrogen atom.

A neutral atom consists of a positively charged **nucleus** with **electrons** outside the nucleus. The nucleus itself contains **protons** that are positively charged and **neutrons** that have no charge. In a neutral (non-ionized) atom, the number of protons equals the number of atomic electrons.

An ion forms when a neutral atom gains or loses one, or more, electrons.

The notation used in nuclear physics is $^A_Z X$, where X is the chemical symbol for the element, A is the **nucleon number** (number of protons + number of neutrons in the nucleus) and Z is the **proton number** (total number of protons in the nucleus).

Sometimes the symbol N is used for the **neutron number**: $A = Z + N$.

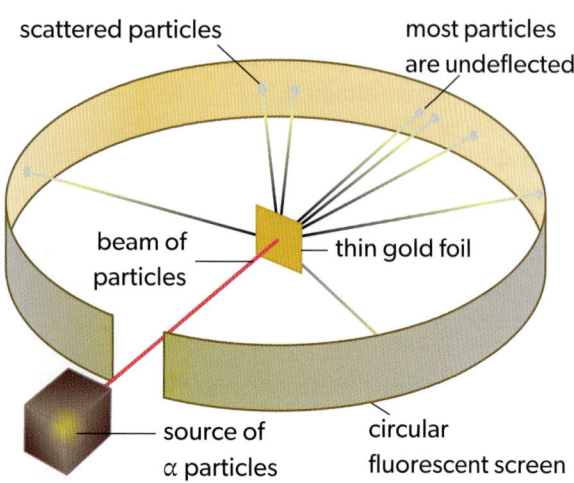

▲ Figure 1 The Geiger–Marsden–Rutherford experiment

In the early 20th century, Rutherford suggested that Geiger and Marsden should investigate alpha particle scattering. They found that a small number of the alpha particles were scattered through very large angles by a thin gold foil (Figure 1). This result allowed Rutherford to propose that the nucleus is small, dense and positively charged, and that atomic electrons exist outside the nucleus.

The conclusions from the experiment are explored in Example 1.

Later in the century, it became clear that there were both positive (protons) and neutral (neutrons) particles in the nucleus. Scattering experiments then began to reveal an increasing complexity in the interactions (forces) that exist in the nucleus.

E.1 Structure of the atom

Example 1

In the Geiger and Marsden's experiment, most alpha particles pass through gold foil with no change in direction or energy loss. A small number of particles deviate from their original direction by angles more than 90°.

Explain, with reference to Rutherford's atomic model and the forces acting in the nucleus:

a) why some alpha particles are deflected through large angles

b) why most of the alpha particles are not deviated.

Solution

a) A few alpha particles approach very close to a gold nucleus. Both the nucleus and the alpha particle are positively charged and so there is a repulsive force between them.

   The force varies as $\frac{1}{r^2}$, where r is the distance between the centres of the alpha particle and the gold nucleus. When r is very small, the force is very large and the alpha particle undergoes a significant deviation.

b) According to the Rutherford model, the spaces between the gold nuclei are very large compared with their diameter. The majority of the alpha particles do not approach a gold nucleus closely enough to experience a significant force.

A few alpha particles are deflected so that they return along the same path by which they arrived. The angle of deflection in this case is 180°. This is called head-on scattering.

The point at which the alpha particle reverses its direction of motion is where all its initial kinetic energy E_α is transferred to electric potential energy E_p stored in the nucleus–alpha system. This can be used to estimate the radius of the scattering nucleus. At the instant when the alpha particle is stationary, a comparison of E_α with E_p gives $\frac{1}{2}m_\alpha v_\alpha^2 = k\frac{(+2e) \times (+Ze)}{r}$, where r is the distance of closest approach, m_α is the mass of the alpha particle, v_α is the initial speed of the alpha particle, Z is the proton number of the nucleus and k is the Coulomb constant. This leads to an equation for the closest approach distance: $r = k\frac{2 \times (+2e) \times (+Ze)}{m_\alpha v_\alpha^2}$, as used in Example 2.

This equation assumes that the nucleus does not recoil (i.e. that it remains stationary). The larger the value of v_α, the closer the alpha particle will be to the centre of the nucleus before deflection, so this way of estimating a nuclear radius is inexact.

As the initial energy of the alpha particle increases, the distance of closest approach will decrease. In principle, a very large E_α should lead to the prediction of a very small nuclear diameter. In practice, for large initial kinetic energies the alpha particle scattering does not obey the assumptions that Rutherford made in his theoretical explanation of the Geiger and Marsden results (Figure 2).

Rutherford had to assume that Coulomb's law applied to the interaction, so the principal assumption he made was that there was a repulsive force between the nucleus and the alpha particle. We now know that there is also an attractive strong nuclear interaction that acts between the alpha particle and the nucleus. This interaction only operates over short nuclear distances and modifies the Rutherford prediction when the alpha particle is close to the nucleus. An alpha particle requires an initial energy of about 30 MeV in order to be close enough to "feel" this attractive force.

The kinetic energy of the alpha particle links this physics to Topic A.3. The electric potential energy links to Topic D.2.

The Coulomb constant k in the equation $F = k\frac{q_1 \times q_2}{r^2}$ also links to Topic D.2.

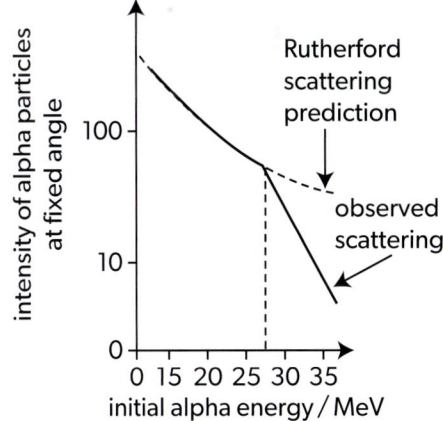

▲ Figure 2 When the initial E_α is greater than about 28 MeV, the Rutherford assumptions break down because the alpha particles are now influenced by the strong nuclear interaction

The strong nuclear attraction is discussed in Topic E.3.

E Nuclear and quantum physics

Example 2

a) Calculate the kinetic energy of an alpha particle travelling at 2.00×10^7 m s^{-1}. Ignore relativistic effects. Mass of alpha particle = 6.64×10^{-27} kg.

b) Determine the closest distance of approach for a head-on collision between the alpha particle and a gold nucleus, $^{197}_{79}$Au. Assume that the gold nucleus does not recoil.

Solution

a) Kinetic energy of the alpha particle $= \frac{1}{2}mv^2$
$= \frac{6.64 \times 10^{-27} \times (2.00 \times 10^7)^2}{2} = 1.33 \times 10^{-12}$ J

b) The gain in electric potential energy when the nucleus and alpha particle are at their closest separation r must equal the loss in kinetic energy. The alpha particle charge is $+2e$ and the gold nucleus charge is $+Ze$.

Gain in potential energy $= 1.33 \times 10^{-12} = \frac{kq_1q_2}{r} = \frac{k \times 2e \times Ze}{r}$

Substituting $Z = 79$ gives $1.33 \times 10^{-12} = \frac{k \times 2 \times 79 \times (1.6 \times 10^{-19})^2}{r}$

so $r = 2.7 \times 10^{-14}$ m.

> **Assessment tip**
>
> The values of k and e can be found together with many other constant in the *Physics data booklet*. You will be provided with an unmarked copy of this booklet for your examinations.

> The evolution of a neutron star is described in Topic E.5, page 222.

Nuclear radius measurements show that nuclear density is roughly constant whatever the nucleon number A of the nuclide. This density is very high, approximately 2×10^{17} kg m^{-3}. This value is approached only by that of a neutron star—a matchbox full of which has a mass of three billion tonnes.

> The nuclear radius R varies with nucleon number A as $R = R_0 A^{\frac{1}{3}}$, where $R_0 = 1.20 \times 10^{-15}$ m ($\equiv 1.20$ fm) is a constant called the Fermi radius. The value is given in the *Physics data booklet*.

The expression for the radius of a nucleus is derived from the assumption that nuclear density ρ is constant for all nuclei. When M is the mass of a nucleon, then $\rho = \frac{\text{mass}}{\text{volume}} = \frac{A \times M}{\frac{4}{3}\pi R^3}$ so that $R^3 = \frac{3M}{4\pi\rho}A$ and $R = \sqrt[3]{\frac{3M}{4\pi\rho}} \times A^{\frac{1}{3}}$.

Example 3

The radius of the $^{12}_{6}$C nucleus is 3.0×10^{-15} m. Calculate the radius of the $^{20}_{10}$Ne nucleus.

Solution

$R = R_0 A^{\frac{1}{3}} \Rightarrow R_{Ne} = R_C \times \left(\frac{A_{Ne}}{A_C}\right)^{\frac{1}{3}} = 3.0 \times 10^{-15} \left(\frac{20}{12}\right)^{\frac{1}{3}} = 3.6 \times 10^{-15}$ m

Spectra can be emitted from hot objects (Figure 3). emission spectra from a low-pressure gas and continuous spectra from hot solids and high-pressure gases. Absorption spectra are seen when continuous radiation emerges from a cold gas.

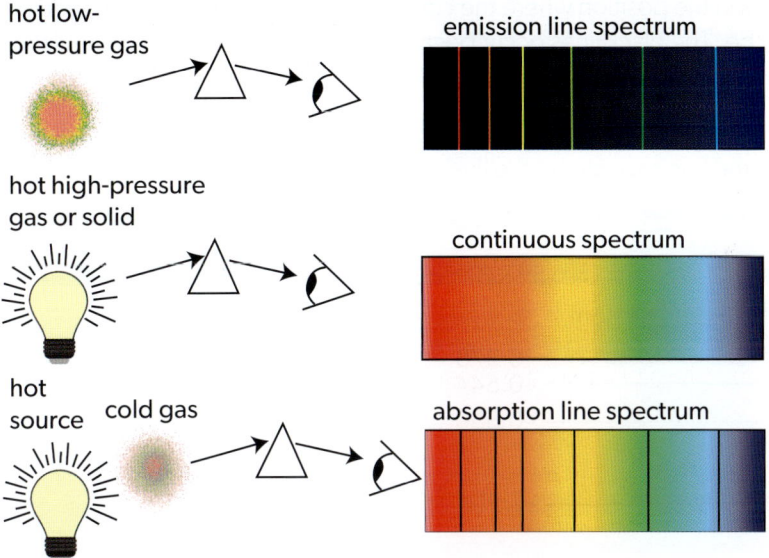

▲ Figure 3 The formation of continuous, emission and absorption spectra

An emission spectrum, or emission line spectrum (Figure 3), consists of a series of discrete lines at very precise frequencies that are produced when hot, low-pressure gas atoms de-excite from an excited energy state.

The individual atoms initially absorb energy to go into the higher energy state. However, this state is unstable and, within a short time, the atom drops back to a lower-energy state. As it does so, it transfers the energy from itself as a quantum of energy.

The emission spectrum is produced from the atoms of a gas at low pressure. As the pressure of the gas increases (still keeping the gas at a high temperature) the gas atoms interact so that the spectrum becomes continuous. This is now close to the black-body radiation behaviour that was discussed in Topic B.1.

Assessment tip

Calculations involving photon energy and wavelength are simpler if you substitute a pre-calculated value of the product hc into the equation $E = \dfrac{hc}{\lambda}$. The *Physics data booklet* lists hc expressed in two different units: $hc = 1.99 \times 10^{-25}$ J m or 1.24×10^{-6} eV m. These are in the section on unit conversions (page 4 of the booklet). Choose one or the other depending on what energy units are appropriate for the problem.

Joseph von Fraunhofer used his discovery of the diffraction grating in 1814 to investigate some mysterious dark lines that cross the continuous spectrum of the Sun. These dark lines are an absorption spectrum. The absorption spectrum is explained by the presence of a cold, low-pressure gas between the hot source of the continuous spectrum and the observer. Some of the radiation incident on the cold gas from the continuous spectrum is at exactly the correct frequency for the cold gas to absorb it. This stimulates the atoms in the cold gas into a higher energy state. As the atoms relax back to the ground

The **quantum** is one "packet" of energy that constitutes the whole of the energy difference between the two states. The energy from one transition is never transferred in more than one quantum.

The energy is transferred to and from the emitting atom in the form of a single **photon**.

The energy E of the photon is related to its frequency f by $E = hf$.

This can also be written as $E = \dfrac{hc}{\lambda}$, where λ is the wavelength of the radiation and c is speed of electromagnetic radiation in a vacuum.

The constant h is known as the Planck constant and it takes the value 6.63×10^{-34} J s.

The value of h was suggested by Plank when he explained some anomalous results in the nature of black-body radiation at the end of the 19th century (see Topic E.2).

Approaches to learning

Diffraction gratings are described in Topic C.3. This is an opportunity for you to use two separate parts of the course to enhance your learning of both.

E Nuclear and quantum physics

state, they re-radiate the energy, but not necessarily in the original direction from which the incident radiation came. Energy has therefore been removed from the original direction and the continuous spectrum appears to have black lines in the position where the cold gas emission spectrum lines would normally be. The black lines occur because there is less energy in the radiation at this wavelength than at adjacent non-absorbed wavelengths.

The sequence of lines is called an absorption spectrum and when the sequence relates to the Sun's spectrum, the lines are known as Fraunhofer lines.

Example 4

The diagram shows some of the energy levels of a hydrogen atom.

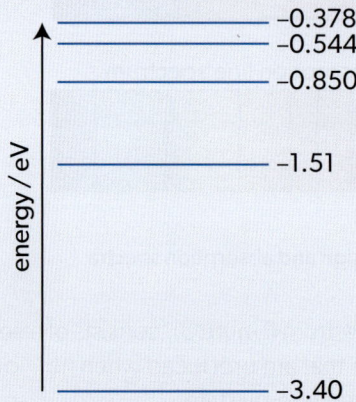

a) Determine, in eV, the energy of a photon of wavelength 656 nm.
b) Identify the transition that gives rise to a photon of this wavelength.
c) Explain why the lines in an emission spectrum involving the atom returning to the same final state become closer together as the wavelength of the emitted photons decreases.

Solution

a) The most convenient way to carry out the calculation is by using the value of hc expressed in eV m.

$$E = \frac{hc}{\lambda} = \frac{1.24 \times 10^{-6}}{656 \times 10^{-9}} = 1.89 \text{ eV}$$

b) The 1.89 eV must be the energy difference between the two levels concerned. The only possibility on the diagram is the change between −1.51 eV and −3.40 eV.

c) The higher energy levels are closer together than the lower energy levels. The energy differences between the first excited level and the second excited level and above become more equal and the difference in the wavelength of the emitted photons decreases too.

Niels Bohr proposed a model of the hydrogen atom to account for:
- the known emission spectrum of hydrogen
- the observations of Geiger and Marsden
- the problem that the concept of an electron orbiting a nucleus contradicted classical physics principles.

Earlier, Johannes Rydberg had shown that the wavelengths λ of the spectral lines of the hydrogen emission spectrum obeyed $\frac{1}{\lambda} = R_H \left(\frac{1}{n_1^2} - \frac{1}{n_2^2} \right)$, where n_1 and n_2 are integers, and R_H is a constant that has the empirical value 1.097×10^7 m^{-1}. This agreed with earlier work by Balmer after whom one set of the hydrogen lines (with $n_2 = 2$) is named.

Bohr recognized that the Rydberg formula describes energy transitions and he proposed an atomic model in which the hydrogen nucleus contained a single proton with an electron orbiting the nucleus. In order to do this, he had to make four assumptions:

1. There are certain allowed discrete orbits that he called stationary states. This use of the word "stationary" is close to the old terminology of "stationary waves" for standing waves (Topic C.4). He thought of the electron as being in a standing wave orbit around the proton.

2. An atom cannot transfer energy while it remains in a stationary state.

3. An atom only transfers energy when it changes between one stationary state and another.

4. The angular momentum of an electron in a stationary state is quantized in integer values of $\frac{h}{2\pi}$.

Bohr was able to explain the whole of Rydberg's work using a single energy-level diagram (Figure 4).

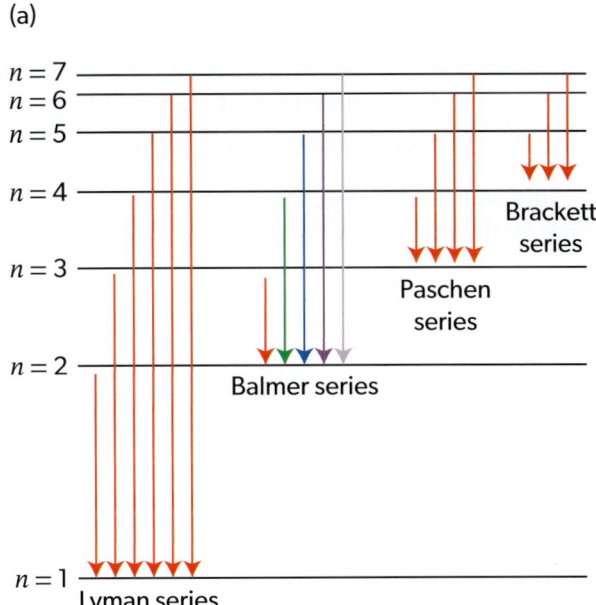

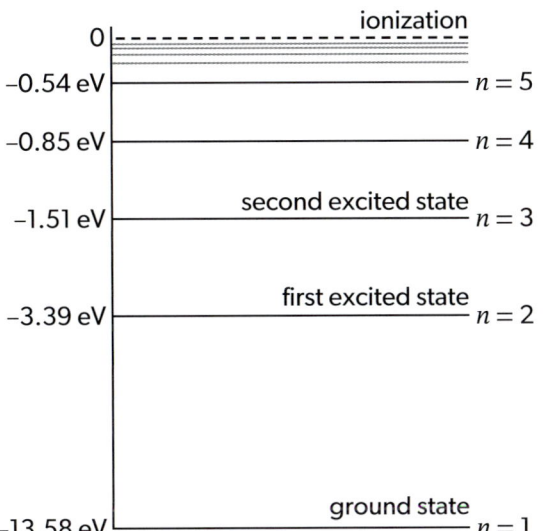

> The lowest energy level (–13.6 eV) is the **ground state**.
>
> All states below the ionized condition (electron removed from the atom) are **bound states**.
>
> The symbol n that labels the state in Figure 4 is the **principal quantum number**.
>
> Bohr showed that the total energy E of an electron in the nth state is $E = -\frac{13.6}{n^2}$ eV.

▲ Figure 4 (a) The energy transitions that lead to four spectral line series for hydrogen.
(b) The energy values that correspond to the atomic energy states for hydrogen

E Nuclear and quantum physics

Example 5

a) State what is meant by an atom in an excited state.

b) A hydrogen atom is in the second excited state ($n = 3$). Determine the shortest wavelength of radiation that this atom can emit.

Solution

a) An atom that has absorbed energy, causing its electrons to occupy higher energy states than the ground state.

b) The shortest wavelength corresponds to the greatest energy of the emitted photon. This means that the atom undergoes transition to the ground state ($n = 1$).

The energy of the photon is equal to the difference between atomic energy levels:

$$E = 13.6 \times \left(\frac{1}{1^2} - \frac{1}{3^2}\right) = 12.1 \text{ eV}$$

The corresponding wavelength is $\lambda = \frac{hc}{E} = \frac{1.24 \times 10^{-6}}{12.1} = 103$ nm.

This radiation is in the UV region of the electromagnetic spectrum.

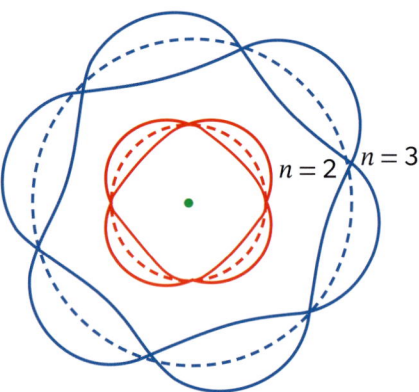

▲ Figure 5 The standing wave orbitals for $n = 2$ and $n = 3$ in the hydrogen atom

The constant $a_0 = \frac{h^2}{4\pi^2 m_e k e^2}$ has the value 5.29×10^{-11} m and corresponds to the **Bohr radius of the electron orbiting in the hydrogen ground state**. It can be regarded as the radius of the (ground state) hydrogen atom.

When the fourth assumption (quantized angular momentum) is introduced and the de Broglie wavelength is used (Topic E.2, page 193), it is possible to show that $m_e v r = \frac{nh}{2\pi}$, where r is the orbital radius of the electron and m_e is the electron mass. This assumes that the orbitals resemble standing waves with n loops in the orbit for the nth energy state (Figure 5).

It is possible to show, after equating the centripetal force to the electric attractive force between proton and electron $\left(\frac{m_e v^2}{r} = k\frac{e^2}{r^2}\right)$ that $r = \left(\frac{h^2}{4\pi^2 m_e k e^2}\right) \times n^2$ so that $r \propto n^2$.

Sample student answer

a) Rutherford constructed a model of the atom based on the results of the alpha particle scattering experiment. Describe this model. [2]

This answer could have achieved 2/2 marks:

> Rutherford proposed that an atom is mostly empty space with a tiny positively charged nucleus at its centre. The nuclei are orbited by electrons.

▲ This answer gives the core information that there is a small positive nucleus surrounded by empty space. In fact, the Geiger–Marsden–Rutherford experiment can say little directly about the electrons except by inference.

b) Bohr modified the Rutherford model by introducing the condition $mvr = n\frac{h}{2\pi}$. Outline the reason for this modification. [3]

This answer could have achieved 1/3 marks:

> Bohr noticed that electrons can only exist at certain energy levels because atoms absorb and emit light of only certain fixed wavelength. In order to explain why electrons only exist in certain energy levels, Bohr said that electrons behave like standing waves where n is an integer representing the energy level. He got this by quantizing angular momentum of an electron.

▼ The student has misunderstood the question. What is required is the reasons that provoked Bohr to construct the model, not the assumptions of the model. Bohr knew that atoms radiate energy and he was clear that this was to do with electron energy changes. Classical physics predicted that electrons must spiral into the nucleus when they radiate energy and that the atom should be unstable. These points, together with the point about discrete energy levels (which the student does mention), are the essentials of a good answer.

E.1 Structure of the atom

c) (i) Show that the speed v of an electron in the hydrogen atom is related to the radius r of the orbit by the expression $v = \sqrt{\dfrac{ke^2}{m_e r}}$, where k is the Coulomb constant. [1]

This answer could have achieved 1/1 marks:

$$F = \dfrac{ke \times e}{r^2} = \dfrac{mv^2}{r} \qquad \dfrac{ke^2}{r^2} = \dfrac{mv^2}{r} \qquad v^2 = \dfrac{ke^2}{m_e r}$$

$$v = \sqrt{\dfrac{ke^2}{m_e r}}$$

▲ This solution is correct and easy to follow.

(ii) Using the answers in parts (b) and (c)(i), deduce that the radius r of the electron's orbit in the ground state of hydrogen is given by the expression

$$r = \dfrac{h^2}{4\pi^2 k m_e e^2}$$ [2]

This answer could have achieved 2/2 marks:

$$m_e v r = \dfrac{nh}{2\pi} \Rightarrow m_e \times \sqrt{\dfrac{ke^2}{m_e r}}\, r = \dfrac{h}{2\pi} \qquad [n = 1 \text{ for ground state}]$$

$$m_e^2 \dfrac{ke^2}{m_e r} r^2 = \dfrac{h^2}{4\pi^2} \Rightarrow m_e k e^2 r = \dfrac{h^2}{4\pi^2} \Rightarrow r = \dfrac{h^2}{4\pi^2 m_e k e^2}$$

▲ This solution is correct and easy to follow. The student has taken care to show the direction of the answer using the arrow to link line 1 and line 2.

(iii) Calculate the electron's orbital radius in part (c)(ii). [1]

This answer could have achieved 1/1 marks:

$$r = \dfrac{(6.63 \times 10^{-34})^2}{4\pi^2 \times 9.11 \times 10^{-31} \times 8.99 \times 10^9 \times (1.6 \times 10^{-19})^2} = 0.0053 \times 10^{-8}$$

$$= 5.3 \times 10^{-11} \text{ m}$$

▲ Again, the student sets the work out well and gains the straightforward mark for applying the equation from the previous part.

Practice problems

Problem 1
In the Geiger–Marsden–Rutherford experiment it was observed that most alpha particles passed through the gold foil undeflected or with very small deflections, and a very few alpha particles were deflected through large angles.

a) State what force causes alpha particles to be deflected from their original direction.

b) Explain how the results of the experiment support the nuclear model of the atom.

Problem 2
The diagram shows the energy levels of an atom.

$$0 \qquad \underline{\hspace{5cm}}$$

$$-2.42 \times 10^{-19} \text{ J} \qquad \underline{\hspace{5cm}} \text{ level 2}$$

$$-5.48 \times 10^{-19} \text{ J} \qquad \underline{\hspace{5cm}} \text{ level 1}$$

$$-2.18 \times 10^{-18} \text{ J} \qquad \underline{\hspace{5cm}} \text{ ground state}$$

E Nuclear and quantum physics

An electron with a kinetic energy of 2.0×10^{-18} J makes an inelastic collision with an atom in the ground state.

a) Calculate the speed of the electron just before the collision.

b) (i) Deduce whether the electron can excite the atom to level 2.

(ii) Calculate the wavelength of the radiation that will result when an atom in level 2 falls to level 1.

(iii) State the region of the spectrum in which the radiation in part (b)(ii) belongs.

Problem 3
The radius of a carbon-12 ($^{12}_{6}C$) nucleus is 3.1×10^{-15} m.

a) Determine the radius of a magnesium-24 ($^{24}_{12}Mg$) nucleus.

b) Sketch a graph to show the variation of nuclear radius with nucleon number.

Annotate your graph with both the C-12 and Mg-24 nuclei.

Problem 4
Alpha particles are scattered by silver-107 ($^{107}_{47}Ag$) nuclei.

a) Determine the distance of closest approach for a head-on collision of an alpha particle with initial energy 18 MeV and a silver nucleus. Assume that the electric force is the only force acting.

The number of alpha particles scattered at very large angles deviates from the Rutherford model at initial energies greater than 18 MeV.

b) Explain how this observation allows the nuclear radius of silver-107 to be estimated.

Problem 5
a) Outline how an emission line spectrum of a gas is formed.

b) Calculate the wavelength of radiation emitted by hydrogen atoms that undergo a transition from energy level $n = 3$ to energy level $n = 2$.

c) A beam of electrons passes through a gas sample that contains hydrogen atoms in the ground state. Determine, in eV, the minimum energy of the electrons so that the wavelength calculated in part (b) is present in the emission spectrum of the sample.

E.2 Quantum physics

You must know:
- that the photoelectric effect and the Compton effect are evidence for the particle nature of light
- that the diffraction of particles is evidence for the wave nature of matter
- that matter shows wave–particle duality
- Einstein's explanation of the photoelectric effect
- what is meant by the Compton effect.

You should be able to:
- solve problems involving the de Broglie wavelength
- solve problems involving the photoelectric effect using Einstein's photoelectric equation
- solve problems involving the Compton effect
- demonstrate which aspects of photoelectricity and the Compton effect cannot be explained by wave theory.

The Einstein explanation of the photoelectric effect

When ultraviolet radiation is incident on a metal sheet that is negatively charged, the sheet loses negative charge. This does not occur with radiation of longer wavelengths (smaller frequencies) or when the sheet is positively charged. The sheet is losing charge by losing electrons.

Einstein's explanation of photoelectricity included the following.

- Light of frequency f consists of **photons**, each with energy $E = hf$, where h is the Planck constant, which has the value 6.63×10^{-34} J s.
- This light must be incident on the metal surface.
- Only one photon incident on the metal can interact with each electron.
- No electrons are emitted when the photon energy is below a minimum value—this corresponds to the **threshold frequency** f_0.
- The minimum photon energy for emission, known as the **work function** $\Phi\,(= hf_0)$, is associated with the energy required to overcome the attractive forces that oppose the removal of the electron from the metal.
- Excess energy $(hf - hf_0 = hf - \Phi)$ remaining when the electron has left the surface is transferred to the kinetic energy of the electron. This represents the maximum kinetic energy E_{max} that an emitted electron can have.
- An increase in the incident light intensity increases the number of incident photons but does *not* change the energy of the incident photons. Therefore, an increase in intensity increases the number of electrons emitted per second, but does not change their maximum kinetic energy.

These led to the Einstein photoelectric equation $E_{max} = hf - \Phi$, which can also be written $\frac{1}{2} m_e v_{max}^2 = hf - \Phi$, where v_{max} is the maximum speed of the emitted electrons.

Figure 1 shows the experiment Millikan devised to test Einstein's result.

Monochromatic light is incident on the cell. When the photon frequency f is sufficiently great, the cathode emits electrons. However, increasing the reverse potential difference between cathode and anode (the anode is negative in this experiment) gives a potential difference at which electrons can no longer reach the anode. This is shown by the current in the circuit becoming zero. The excess energy of the emitted electrons is not sufficient for them to overcome the reverse potential difference between anode and cathode.

Figure 2 shows a graph of the variation of V_s with $\frac{1}{\text{wavelength}}$ for two different metal surfaces. The lines for the two different metals have the same gradient (which depends only on h, c and e) but are offset by a value that depends on their individual work functions.

> The Bohr model of the hydrogen atom relied on the idea that energy is quantized (Topic E.1, page 184). This was a concept originally developed by Planck to explain the breakdown of the classical theory of electromagnetic radiation at small wavelengths (the "ultraviolet catastrophe"). The theory was used by Einstein to explain the photoelectric effect. It was then used by other scientists to develop 20th-century physics. Part of this new science was the recognition that matter has a wave–particle duality (page 192).

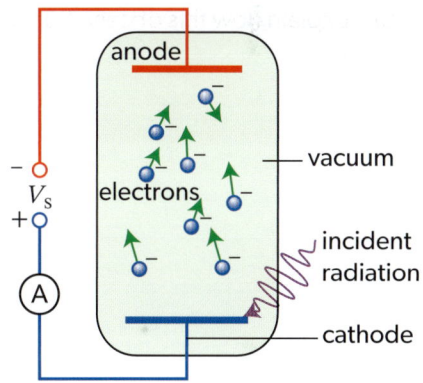

▲ Figure 1 A cell used to measure h. Electromagnetic radiation falls on the cathode, which emits photoelectrons. Cathode here simply means "electron emitter". This sign of the potential is unusual because chemists normally define the cathode to be negatively charged

> The potential difference V_s at this point is known as the **stopping potential**, and $eV_s = hf - \Phi$. This can be written as $V_s = \frac{hc}{e\lambda} - \frac{\Phi}{e}$, where λ is the wavelength of the incident radiation. Compare this with $y = mx + c$ for a straight line.

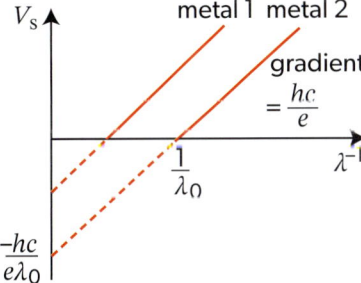

◀ Figure 2 The graph obtained in the Millikan experiment to verify Einstein's photoelectric equation

E Nuclear and quantum physics

Assessment tip

The stopping voltage for the photoelectric effect, in volts, is numerically equal to the maximum kinetic energy of the photoelectrons, expressed in eV.

For example, when a voltage of 1.0 V prevents the photoelectric current from flowing in a cell, then the electrons leave the cathode with a maximum kinetic energy of 1.0 eV.

Calculations involving the photoelectric effect are, therefore, simpler when photon energy and work function are also expressed in eV.

Example 1

Electromagnetic radiation of frequency f is incident on a metal. The maximum kinetic energy E_{max} of photoelectrons from the surface is measured. The graph shows the variation of E_{max} with f.

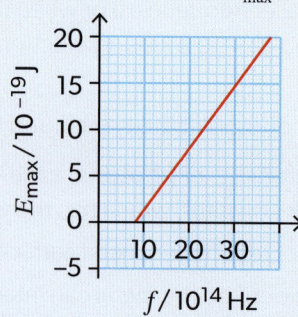

a) Calculate, using the graph, the Planck constant h.

b) Determine the minimum energy required to remove an electron from the metal surface.

Solution

a) The equation for the graph is $E_{max} = hf - \Phi$.

Comparing this with $y = mx + c$ shows that the gradient (m) is the Planck constant h.

Using the coordinates $(38 \times 10^{14}, 20 \times 10^{-19})$ and $(8 \times 10^{14}, 0)$ from the graph:

gradient $h = \dfrac{20 \times 10^{-19} - 0}{38 \times 10^{14} - 8 \times 10^{14}} = 6.7 \times 10^{-34}$ J s

b) The minimum energy (Φ) is the energy found by extrapolating the graph to the y-axis.

This value is about 5×10^{-19} J.

Alternatively, $\Phi = hf_0$, where $f_0 = 7.5 \times 10^{14}$ Hz is the intercept on the x-axis. $\Phi = 6.7 \times 10^{-34} \times 7.5 \times 10^{14} = 5.0 \times 10^{-19}$ J.

Example 2

A photoelectric cell with a magnesium cathode is illuminated by monochromatic UV radiation of wavelength 280 nm. The work function of magnesium is 3.7 eV.

a) Calculate, in eV, the energy of the photons incident on the cathode.

A potential difference V is applied across the cell. The magnesium cathode is at a positive potential relative to the anode.

b) Determine the minimum value of V so that the photoelectric current in the cell is zero.

The photoelectric effect does not occur in this cell when visible light is used for the radiation source.

c) (i) Explain this observation.

(ii) Explain why this observation is inconsistent with the classical wave theory of light.

Solution

a) $E = \dfrac{hc}{\lambda} = \dfrac{1.24 \times 10^{-6}}{280 \times 10^{-9}} = 4.4$ eV

b) The photoelectrons emitted from the magnesium cathode have the maximum kinetic energy $4.4 - 3.7 = 0.7$ eV. The stopping potential difference for this UV source is therefore 0.7 V.

c) (i) The shortest wavelength of visible light is about 400 nm, which corresponds to a photon energy of $\frac{1.24 \times 10^{-6}}{400 \times 10^{-9}}$ = 3.1 eV. This is less than the work function for magnesium, so photons of visible light have insufficient energy to remove electrons from magnesium.

(ii) Wave theory predicts that light energy should be continuously transferred to the electrons regardless of the frequency. Therefore, the photoelectric emission should occur at all frequencies, with a possible time delay for low-intensity radiation.

Compton scattering

In 1923, Compton observed that, when X-radiation was scattered by carbon, the wavelengths of the X-rays increased after they were scattered (Figure 3).

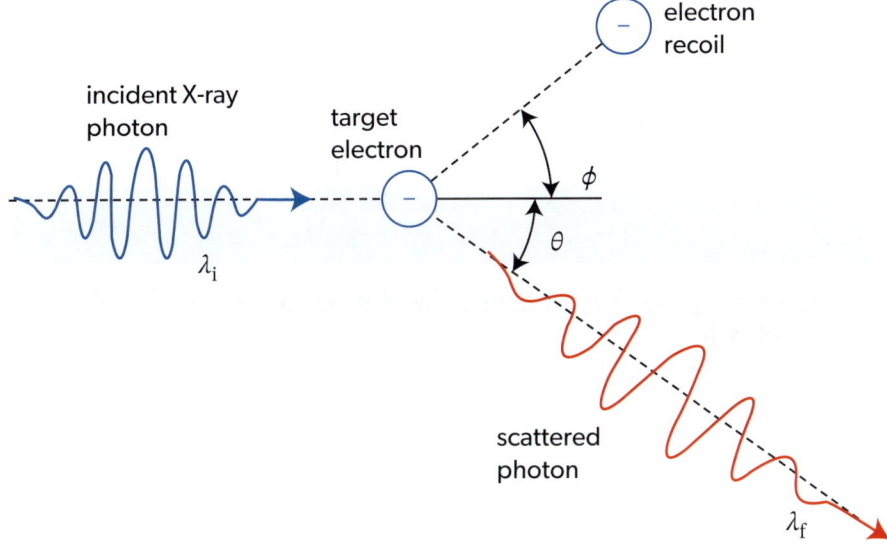

▲ Figure 3 The incident X-ray photon interacts with a free electron. The electron gains energy and momentum, and recoils. The X-ray photon must have lost energy, so it has a lower frequency (greater wavelength)

The incident X-radiation is monochromatic. Compton found that, after scattering, most radiation continues straight-on with the original wavelength. However, any X-rays scattered out of the beam have a wavelength that depends on the angle θ at which they are detected (Figure 3). As the scattering angle θ increases, so does the wavelength. (In fact, there are two wavelengths observed at each value of θ: one wavelength corresponds to scattering from a free electron; the other is due to scattering by one of the target carbon atoms.)

The free electrons are produced by ionization of the carbon atoms by an X-ray photon (this in itself will change the photon frequency slightly as it loses a small amount of energy). The remaining energy of the X-ray photon is involved in the interaction with the now free electron. The electron recoils, carrying away momentum and energy—this also changes the energy of the photon.

The situation can be analysed using the rules for the conservation of momentum in two dimensions that were used in Topic A.2. The analysis leads to an equation for the wavelength shift $\Delta\lambda = \lambda_f - \lambda_i$, where λ_f and λ_i represent the final and initial wavelengths, respectively:

$$\Delta\lambda = \lambda_f - \lambda_i = \frac{h}{m_e c}(1 - \cos\theta)$$

This equation is provided in the *Physics data booklet*.

When the electron is scattered through an angle φ and the photon through angle θ, then $\frac{1}{\tan\varphi} = \left(1 + \frac{hf}{m_e c^2}\right)\tan\left(\frac{\theta}{2}\right)$, where f is the initial photon frequency.

> While photoelectricity gives clear evidence for wave–particle duality, even stronger evidence is provided by **Compton scattering**, when electromagnetic radiation interacts with an electron. Both classical mechanics (Topic A.2) and the Planck–Einstein–de Broglie view of waves and matter are required to explain the Compton interaction.

Assessment tip

Only the wavelength shift for the photon is required for the IB Diploma Programme physics course.

This analysis relies on:
- the energy of the photon being given by $E = hf = \dfrac{hc}{\lambda}$
- the momentum of the photon being given by $p = \dfrac{hf}{c} = \dfrac{h}{\lambda}$.

Wave theory cannot account for the following.
- Photoelectricity: This is because the photoelectric effect is observed as soon as the photon is incident on the surface. Wave theory suggests that no matter how small the intensity of the light, given enough time, energy will accumulate and release an electron for all wavelengths. On the other hand, Einstein's explanation of the photoelectric effect predicts that when the photon energy is below a threshold value, no electron will be released—this is what is observed in practice.
- Compton scattering: This is because the classical wave theory for scattering of light by a charged particle does not predict wavelength shifts for small radiation intensities. When the radiation intensity is large, classical wave theory predicts other effects that are different from the Compton effect.

Assessment tip

The factor $\dfrac{h}{m_e c}$ in the Compton effect equation has a constant value that can be calculated conveniently using the electron mass in MeV c^{-2} (see Topic E.3 for units of mass in atomic physics):

$\dfrac{h}{m_e c} = \dfrac{hc}{m_e c^2} = \dfrac{1.24 \times 10^{-6}}{0.511 \times 10^6}$
$= 2.43 \times 10^{-12}$ m

The value of hc is given as one of the conversions in the *Physics data booklet*.

Example 3

A photon of energy 18 keV is scattered at an angle of 120° off a stationary electron.

a) Explain why the emerging photon has a different energy than the incident photon.

b) Calculate the change of the wavelength of the photon.

c) Determine, in eV, the energy of the recoil electron.

Solution

a) The electron recoils and gains kinetic energy and momentum as a result of interaction with the photon. Energy conservation implies that the energy of the photon decreases by the amount transferred to the electron.

b) Wavelength shift $= \dfrac{h}{m_e c}(1 - \cos\theta) = \dfrac{1.24 \times 10^{-6}}{0.511 \times 10^6}(1 - \cos 120°)$
$= 3.6 \times 10^{-12}$ m

c) Wavelength of initial photon $\lambda_i = \dfrac{hc}{E} = \dfrac{1.24 \times 10^{-6}}{18 \times 10^3} = 6.89 \times 10^{-11}$ m

Wavelength of emerging photon $\lambda_f = \lambda_i + \Delta\lambda$
$= 6.89 \times 10^{-11} + 3.6 \times 10^{-12} = 7.25 \times 10^{-11}$ m

The energy E of the electron is equal to the energy difference between the incident and the emerging photon:

$E = 18 \times 10^3 - \dfrac{1.24 \times 10^{-6}}{7.25 \times 10^{-11}} = 900$ eV

Wave–particle duality

Under appropriate conditions, electromagnetic radiation can demonstrate both wave-like properties (Theme C) and particle-like properties (Topics E.1 and E.2). This leads to the idea of wave–particle duality. In 1924, Louis de Broglie suggested that this duality applied also to matter. Under certain circumstances, matter could be observed to demonstrate properties that showed it was behaving as matter waves.

He suggested that particles have a de Broglie wavelength. This hypothesis was verified by the American physicists, Davisson and Germer. They fired a beam of electrons at a crystal of nickel (Figure 4).

> The **de Broglie wavelength** λ is given by $\lambda = \dfrac{h}{p}$, where p is the momentum of the particle.
>
> The wavelength of an electron that has been accelerated from rest through a potential difference V, travelling at non-relativistic speed, can be calculated by equating the kinetic energy to the electrical energy: $\dfrac{1}{2}m_e v^2 = eV$.
>
> This can be used to write an expression for the electron's momentum: $p = m_e v = \sqrt{2m_e eV}$. Because $\lambda = \dfrac{h}{p}$, this leads to $\lambda = \dfrac{h}{\sqrt{2m_e eV}}$.

Under these conditions, the electrons demonstrate a wave behaviour, with the nickel atoms acting as a diffracting slit or grating—in this case, the diffracted beam is scattered backwards rather than forwards as in the optical case. The intensity of the electron beam varies with scattering angle θ in a way that is very similar to the pattern obtained when a wave (light or sound, for example) interacts with an aperture. The position of the minimum between $\theta = 0$ and the intensity maximum gives information about the separation of the nickel atoms in the crystal.

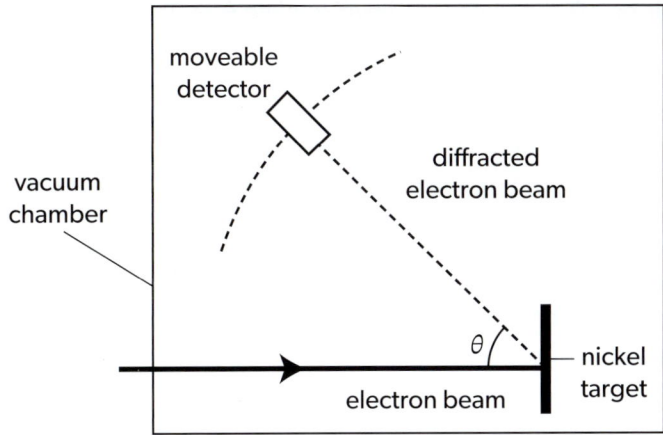
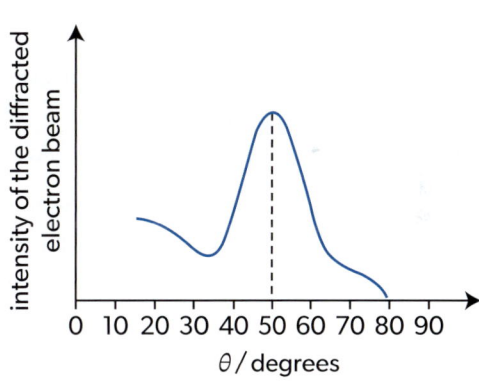

▲ **Figure 4** The Davisson and Germer experiment together with a typical graph of the variation of the beam intensity with angle

Example 4

a) Calculate the Broglie wavelength of an electron moving at a speed of 9.0×10^6 m s^{-1}.

b) Explain why electrons of this speed can be used to investigate crystal structures using electron diffraction.

Solution

a) $\lambda = \dfrac{h}{m_e v} = \dfrac{6.63 \times 10^{-34}}{9.11 \times 10^{-31} \times 9.0 \times 10^6} = 8.1 \times 10^{-11}$ m

b) The separation of atoms is of the order of 10^{-10} m, so is roughly the same as the wavelength of the electron found in part (a).

E Nuclear and quantum physics

Sample student answer

An apparatus is used to investigate the photoelectric effect. A caesium cathode C is illuminated by a variable light source. A variable power supply is connected between C and the collecting anode A. A current is observed on the ammeter when violet light illuminates C. With V held constant the current becomes zero when the violet light is replaced by red light of the same intensity. Explain this observation. [3]

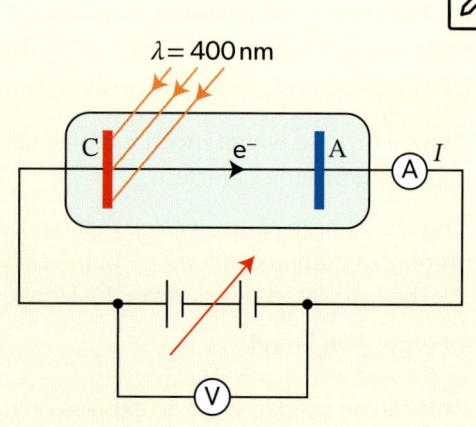

This answer could have achieved 2/3 marks:

According to the quantum model of light where light is made up of photons ($E = hf$), the energy provided by the light is dependent on the frequency of light. When red light, which has lower frequency, shines on the plate, the energy is not enough to remove the electrons. Therefore, electrons don't flow from C to A → no current → 0 A. Violet light has higher frequency, which is enough to liberate the electrons off the plate C to travel to plate A → There's a current.

▲ There is a statement that indicates that the light arrives as photons and that each photon has an energy $E = hf$.

▼ This question is all about comparison. The answer needs to be clearer about the differences between red and violet light and what these differences lead to. It can be inferred that the violet energy is greater than the red energy and this mark was awarded—but it is not the job of the examiner to do work for a candidate. There is no link to the work function—the term is not mentioned. There is no link between the energy of a red light photon and the fact that this is less than the work function.

This answer could have achieved 0/3 marks:

When violet light, which has lower wavelength than red light, is used, the electrons have kinetic energy given by $KE = hf - \Phi$, where Φ is work function of caesium. Since violet light has a lower wavelength than red light, the electrons emitted when violet light is shone on the cathode have greater kinetic energy than the electron emitted when red light is shone on the cathode. Hence, while the electrons have sufficient kinetic energy to overcome electrostatic forces of repulsion from the anode when violet light is shone on the cathode, electrons that are emitted when red light is shone on the cathode do not have sufficient kinetic energy to do so. This is why electrons cannot reach the anode when violet light is replaced with red light, and subsequently why the current is zero when violet light is replaced by red light.

▼ This answer is entirely in terms of the kinetic energy of the electrons. An attempt is made to explain Einstein's photoelectric equation, but again this is wrong as it is in terms of electrons.

Practice problems

Problem 1
Electromagnetic radiation incident on a metal causes a photoelectron to be emitted from the surface.

a) State and explain **one** aspect of the photoelectric effect that suggests the existence of photons.

The work function of sodium is 2.3 eV.

b) (i) Outline what is meant by work function.

Electromagnetic radiation of wavelength 320 nm is incident on sodium.

(ii) Determine the maximum kinetic energy of the electrons emitted from the sodium.

Problem 2
Electrons are emitted instantaneously from a metal surface when monochromatic light of wavelength 420 nm is incident on the surface.

a) Explain why the energy of the emitted electrons does not depend on the intensity of the incident light.

b) Suggest why the electron emission is instantaneous.

The work function of the metal is 2.3 eV and one electron is emitted for every 4500 photons incident on the surface. The surface area of the metal is 4.5×10^{-6} m^2.

c) (i) Determine, in joules, the maximum kinetic energy of an emitted electron.

(ii) Determine the initial electric current from the surface when the intensity of the incident light is 3.8×10^{-3} W m^{-2}.

Problem 3
Electrons are emitted from a heated cathode and accelerated in a vacuum through a potential difference as a narrow beam. This beam is fired at a polycrystalline graphite target in a chamber. The inside surface of the chamber is coated with fluorescent material that emits light when the electrons release their energy to it.

The electrons reach the inside surface travelling at a speed of 4.0×10^7 m s^{-1}.

a) (i) Calculate the de Broglie wavelength of the electrons.

(ii) Describe the pattern of light you would expect to see emitted by the fluorescent material.

(iii) Explain why the pattern suggests that electrons have wave-like properties.

b) Explain **one** aspect of the experiment that suggests that electrons have particle-like properties.

Problem 4
An X-ray photon of wavelength 8.27×10^{-11} m interacts with a free, stationary electron. The photon is scattered at an angle of 75° relative to the original direction.

a) Calculate the wavelength of the scattered photon.

b) Determine the energy of the recoil electron.

E.3 Radioactive decay

You must know:

- what is meant by an isotope
- what is meant by nuclear binding energy, mass defect and mass–energy equivalence
- that there is a strong nuclear interaction in the nucleus
- the variation of binding energy per nucleon with nucleon number
  - including the approximately constant value for $A > 60$ (AHL only)
- the evidence for the strong nuclear interaction (AHL only)
- the nature of radioactive decay
- the nature of the emission from the nucleus for
  - alpha (α) decay
  - beta-minus (β^-) decay
  - beta-plus (β^+) decay
  - gamma-photon (γ) emission
- the properties of the emitted alpha and beta particles and the emitted gamma photons
- of the existence of neutrinos and antineutrinos
- that the continuous energy spectrum in beta emission is evidence for the neutrino
- the effect of background radiation on radioactive decay experiments
- what is meant by activity, count rate and half-life in radioactive decay

Additional higher level:

- the link between activity and rate of decay
- what is meant by the decay constant
- that presence of discrete nuclear energy levels can be inferred from the alpha spectra and gamma photon spectra observed during radioactive decay
- the relationship between decay constant and probability of decay per unit time.

You should be able to:

- solve problems involving isotopes, nuclear notation and decay equations
- describe, interpret and solve problems involving the variation of binding energy per nucleon with nucleon number
- interpret the ratio neutron : proton in terms of nuclear stability and probable decay type (AHL only)
- solve problems for radioactive decay that involve half-life, decay constant, and changes in activity, number of nuclei and observed count rates
  - for integral values of half-life
  - for any decay period (AHL only).

Using the nuclear notation introduced in Topic E.1, page 180, two of the known isotopes of carbon are $^{12}_{6}C$ and $^{14}_{6}C$. $^{12}_{6}C$ has 6 neutrons and 6 protons and is stable. $^{14}_{6}C$ has 8 neutrons and 6 protons and decays with the emission of a beta-minus particle to form the stable nuclide of nitrogen $^{14}_{7}N$.

A further 14 carbon isotopes are believed to exist with nucleon numbers ranging from $A = 8$ to $A = 23$. Some of these isotopes are very short lived and transform to other elements through the process of natural radioactive decay. During radioactive decay, ionizing radiation is emitted and the nature of the nucleus changes.

E.3 Radioactive decay

Radioactive decay is random and spontaneous.

> **Isotopes** of an element have the same number of protons but a different number of neutrons.
>
> Isotopes have the same chemistry (i.e. the same chemical reactions) but different physics (i.e. different masses or different radioactive decay schemes).
>
> **Nuclide** is the term used for a particular species of nucleus.
>
> **Random** means that we cannot predict when a nucleus will decay.
>
> **Spontaneous** means that radioactive decay cannot be affected by changing the temperature, pressure, or any other physical or chemical conditions.

> **Assessment tip**
>
> Only use the term "isotope" when dealing with two or more nuclei of the *same* chemical element.
>
> Do not confuse radioactive decay and nuclear fission.
>
> - Radioactive decay occurs naturally and most fission events are induced by the arrival of a neutron (Topic E.4).
> - Although some nuclides fission spontaneously, this is comparatively rare.

Some of the emissions in radioactive decay, together with some of their properties, are listed in Table 1.

	Decay name			
	Alpha (α)	Beta-minus (β⁻)	Beta-plus (β⁺)	Gamma (γ)
Nature of emission	helium ($^4_2$He^{2+}) nucleus 2n + 2p	electron emitted from the nucleus e⁻	positron (anti-electron) emitted from the nucleus e⁺	electromagnetic radiation (photon) of large energy
Origin	overall change in neutron : proton ratio	neutron changes to proton	proton changes to neutron	removal of excess energy from nucleus nucleus moves to de-excited state
Ionizing power of radiation	strong	medium	quickly annihilated by nearby electrons	weak
Penetration	few cm of gas or few mm of paper	few cm of aluminium		many metres of lead or concrete
Notes	limited number of final energy states for a particular decay	broad energy spectrum of emitted electrons electron antineutrino is also emitted and carries off some energy	broad energy spectrum of emitted positrons electron neutrino is also emitted and carries off some energy	discrete energies linked to nuclear energy levels
Decay equation	$^A_Z X \rightarrow\,^{A-4}_{Z-2} Y +\,^4_2 \alpha^{2+}$	$^A_Z X \rightarrow\,^A_{Z+1} Y +\,^0_{-1}\beta^- + \bar{\nu}_e$	$^A_Z X \rightarrow\,^A_{Z-1} Y +\,^0_{+1}\beta^+ + \nu_e$	$^A_Z X^* \rightarrow\,^A_Z X + \gamma$

▲ **Table 1** Emissions in radioactive decay, together with their properties

> **Assessment tip**
>
> For **decay equations**, the nuclide that is decaying is on the left-hand side of the equation. All the products are shown on the right. An arrow is used to separate the two sides.
>
> Equations for decay processes must balance in terms of both A and Z.
>
> For gamma decay you can put a γ symbol on the right or write "energy" or "energy in form of gamma photon". In most radioactive decays, energy is also transferred in kinetic or other forms, but it is not usual to include this in the equation.
>
> Use the correct notation for the neutrino (ν) and the antineutrino ($\bar{\nu}$). When it is an electron (anti)neutrino, it is usual to add a subscript "e" to the symbol: ($\bar{\nu}_e$). The symbol used here (ν) is the Greek lower case letter pronounced "nu".

E Nuclear and quantum physics

Approaches to learning

When you are learning about topics in Theme E it is important to distinguish between fission and fusion. Fission is when a nucleus splits into two or more smaller nuclei—this can be spontaneous or initiated by the arrival of a neutron. Fusion is where two nuclei join together to make a new larger nuclide.

In both cases, energy must be released in some form for the reaction to proceed.

You can read more about this in Topics E.4 and E.5.

Example 1

Write down nuclear equations for:

a) alpha decay of radium-223 ($^{223}_{88}$Ra) into an isotope of radon (Rn)

b) negative beta decay of silicon-32 ($^{32}_{14}$Si) into an isotope of phosphorus (P)

c) positive beta decay of fluorine-18 ($^{18}_{9}$F) into an isotope of oxygen (O).

Solution

a) Decay particles are a radon nucleus and an alpha particle.

$$^{223}_{88}\text{Ra} \rightarrow {}^{219}_{86}\text{Rn} + {}^{4}_{2}\alpha$$

b) Decay particles are a phosphorus nucleus, an electron and an electron antineutrino.

$$^{32}_{14}\text{Si} \rightarrow {}^{32}_{15}\text{P} + {}^{0}_{-1}\beta^- + \bar{\nu}_e$$

c) Decay particles are an oxygen nucleus, a positron and an electron neutrino.

$$^{18}_{9}\text{F} \rightarrow {}^{18}_{8}\text{O} + {}^{0}_{1}\beta^+ + \nu_e$$

There are now known to be several species of neutrino. The type associated with beta decay is the **electron neutrino,** with the particle emitted in β⁻ decay being the **electron antineutrino**—this particle is given the symbol $\bar{\nu}_e$ (the bar over the symbol signifies "anti"). When a decay involves the emission of a β⁺ particle, the accompanying neutrino is the electron neutrino ν_e.

Neutrinos are difficult to observe. Large numbers emitted in fusion reactions in the Sun pass through the Earth every second. The particles can only be observed when a few interact indirectly with nuclei under special circumstances. The properties of the neutrino include neutral charge and effectively zero mass.

Alpha decay and beta (β⁻) decay are very different in energy terms. The alpha particles emitted from one particular decay (for example, when radium ($^{223}_{88}$Ra) decays to radon) only show one or two distinct energy values. However, for beta decay, there is a continuous spectrum of beta-particle energies from zero to a maximum value.

In alpha emission, one active nucleus decays into the daughter nucleus and the alpha particle. When the initial momentum before decay is zero, the final momentum must also be zero. The alpha particle (mass m_α) and the daughter nucleus (mass m_d) move in opposite directions with speeds in the ratio $\frac{v_\alpha}{v_d} = \frac{m_d}{m_\alpha}$.

A two-particle emission with a single fixed (discrete) nuclear energy change can lead to only a single solution for the speeds of the two emitted particles.

Review the link between the speed ratio and the conservation of momentum ideas from Topic A.2.

When a nucleus has been emitted in alpha or beta decay it is often left in an excited state. The nucleus will undergo one or more energy transitions to arrive at a more stable condition. An example of this occurs when a metastable form of cobalt-60 (known as Co-60^m) decays (Figure 1).

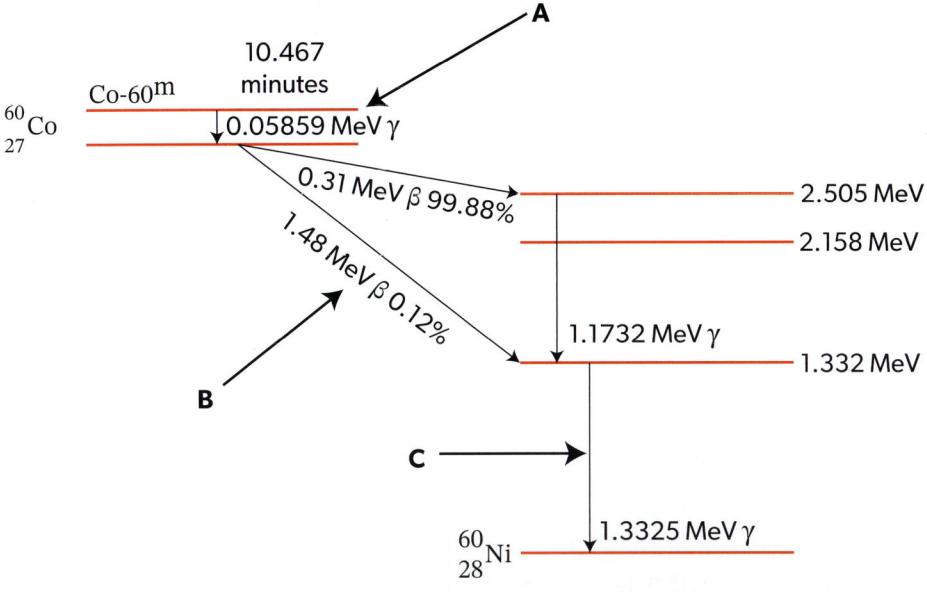

▲ Figure 1 The emitted gamma spectrum for the decay of Co-60^m

- The first gamma emission is to a more stable form of $^{60}_{27}\text{Co}$, which then decays (step **A**) ...

- ... to one of two excited forms of nickel-60, which themselves subsequently emit gamma photons (step **B**) to reach...

- ... the ground state of the nickel nuclide (step **C**).

The spectrum of the gamma photons is discrete. Like the alpha particle energy spectrum, this is strong evidence for the existence of discrete energy levels in the nucleus.

> The gamma photon spectrum links to the ideas of **discrete atomic energy levels** that you met in Topic E.1.

However, in β$^-$ decay, emitted electrons are observed to have a complete range of energies from zero to a maximum that is slightly less than the maximum energy believed to be available (Figure 2).

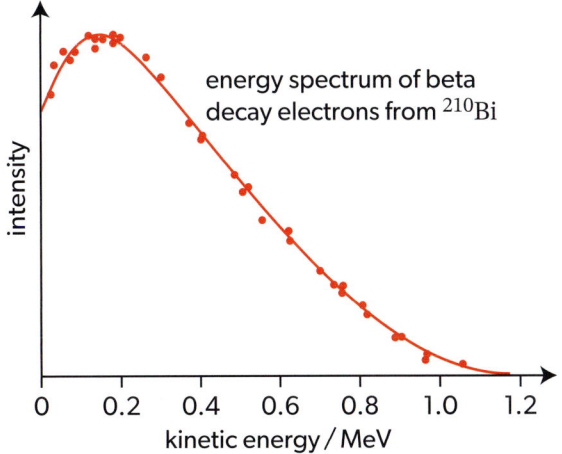

▲ Figure 2 The energy spectrum of the beta particle emitted in the decay of bismuth-210

In 1930, Pauli suggested that the continuous nature of the spectrum of β$^-$ particles meant that there must be three emitted particles. This led Fermi, in 1933, to explain beta decay by using the existence of the neutrino (the third particle), a small non-interactive particle, which was finally observed in the 1950s.

E Nuclear and quantum physics

Example 2

a) Calculate, in joules, the maximum beta-minus particle energy from Figure 2.

b) All the beta particles emitted by bismuth-210 arise from identical energy changes in the bismuth nucleus.

Explain how Figure 2 suggests that an electron antineutrino must also be emitted.

Solution

a) The maximum energy is 1.2 MeV.

This corresponds to $1.2 \times 10^6 \times 1.60 \times 10^{-19} = 1.9 \times 10^{-13}$ J.

b) The total energy available from the decay is constant. The existence of a beta energy spectrum must mean that there is no unique way to distribute the energy between the beta particle and the daughter product.

There are an infinite number of ways to distribute the energy three ways. This is an indication that three particles are involved. The energy is shared between the daughter nucleus, the electron and the antineutrino.

Sample student answer

Rhodium-106 ($^{106}_{45}$Rh) decays into palladium-106 ($^{106}_{46}$Pd) by beta-minus (β^-) decay.

The diagram shows some of the nuclear energy levels of rhodium-106 and palladium-106.

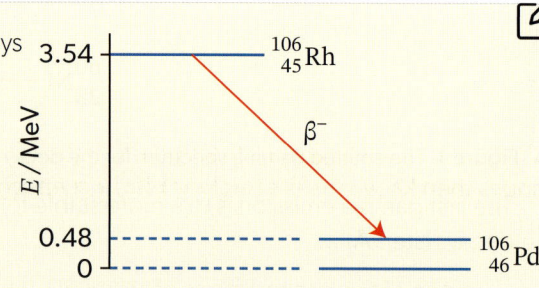

a) Calculate, in joules, the maximum energy of the emitted β^- particle. [2]

This answer could have achieved 2/2 marks:

▲ The answer shows a correct subtraction that leads to the final answer in joules. Quoting this answer to two significant figures would have been better than three here.

The energy of the beta particle is $3.54 - 0.48 = 3.06$ MeV. This is $3.06 \times 10^6 \times 1.60 \times 10^{-19} = 4.90 \times 10^{-13}$ J.

b) Suggest why the energy emitted by the electron in part (a) is a maximum value. [3]

This answer could have achieved 2/3 marks:

▲ The student correctly suggests that there are three particles, and that the energy has to be shared between them, with the beta-minus particle only gaining the full energy when the other two energies are zero.

There are three products altogether. There is the palladium, the beta particle and a neutrino.

The energy released is split between all three and we can't say how much goes to each one except that the total must be correct. The beta particle can only have the maximum energy when the other two are zero.

▼ However, the student does not give enough detail about the third particle, which is an electron antineutrino.

c) Suggest why the decay is followed by the emission of a gamma photon. [1]

This answer could have achieved 1/1 marks:

▲ The answer recognizes that the nucleus is left in an excited state and loses the excess energy via the emission of a gamma photon.

The palladium nucleus is not in its ground state and it releases the gamma ray to go to its ground state.

d) Calculate the wavelength of the gamma ray photon in part (c). [1]

This answer could have achieved 1/1 marks:

▲ The calculation is clear and accurate.

$\lambda = \dfrac{hc}{E} = \dfrac{1.24 \times 10^{-6} \text{ eV m}}{0.48 \text{ MeV}} = 2.583 \times 10^{-12} \text{ m} \approx 2.6 \times 10^{-12} \text{ m}$

E.3 Radioactive decay

The unified atomic mass unit is used in nuclear physics. Mass can be expressed in energy terms.

The magnitude of the **unified atomic mass unit** (abbreviated **u**) is very close to the mass of a proton or neutron (these differ slightly from each other).

1 u is defined to be the mass of $\frac{1}{12}$ of the mass of a stationary carbon-12 atom. This means that $1\,u \equiv 1.661 \times 10^{-27}$ kg.

Energy and mass are considered interchangeable using $\Delta E = c^2 \Delta m$ (also written more familiarly as $E = mc^2$).

Mass can be expressed in energy units, $1\,\text{kg} \equiv 1 \times (3 \times 10^8)^2 = 9 \times 10^{16}$ J.

$1\,\text{eV} \equiv 1.6 \times 10^{-19}$ J, so 1 eV is equivalent to 1.8×10^{-36} kg.

When expressed using eV in this way, the unit of mass is $eV\,c^{-2}$.

So $2\,eV\,c^{-2} \equiv 3.6 \times 10^{-36}$ kg.

Assessment tip

Make sure that you are familiar with the various units that can be used for mass. You should be confident about converting mass values between these units.

The numerical values for the unit conversions are given in the *Physics data booklet*: $1\,u \equiv 1.661 \times 10^{-27}$ kg $\equiv 931.5\,\text{MeV}\,c^{-2}$.

Precise determinations of atomic mass show that the mass of a nucleus is less than the total mass of its constituent parts (proton mass plus neutron mass). This difference is known as the **mass defect**. This can also be described in energy terms through the equivalence of mass and energy, and is then known as the **nuclear binding energy**.

Binding energy has its origins in the forces inside the nucleus. Protons are positively charged and repel each other through Coulomb's law. There must be an additional attractive interaction that opposes this repulsion to hold the nucleus together.

An attractive **strong nuclear interaction** (strong nuclear force) acts between the protons and neutrons. At very small distances, the strong interaction is effectively repulsive but, at larger distances within the nucleus (up to 3 fm), it is attractive and binds protons and neutrons together. The strong interaction is short range and cannot be observed at atomic distances (around 0.1 nm). At distances of about 1 fm the strong attraction is roughly 100 times stronger than the electromagnetic repulsion.

Energy must be added to a nucleus to separate it into its component parts (the protons and neutrons). Similarly, to form a nucleus by bringing its individual nucleons together from infinity, energy must be transferred out of the nucleus. This is the source of the **binding energy**, otherwise known as the **mass defect**.

Example 3

a) Explain why the mass of any nucleus is less than the combined mass of its individual nucleons.

The mass of an atom of nitrogen-14 ($^{14}_{7}\text{N}$) is 14.003074 u.

b) Calculate, for a nucleus of nitrogen-14, its:

   (i) mass defect

   (ii) nuclear binding energy.

Solution

a) Energy is needed to separate the individual nucleons, which are bound by the attractive strong nuclear force. Conversely, energy is released when a nucleus is formed from its component nucleons. From mass–energy equivalence, the release of energy is equivalent to a loss of mass of the nucleus compared with its separate nucleons.

b) (i) An atom of nitrogen-14 has 7 protons, 7 neutrons and 7 electrons. The mass of the nucleus is equal to the atomic mass minus the combined mass of the electrons:

$$m_{nucleus} = 14.003074 - 7 \times 0.000549 = 13.999231 \, u$$

The mass defect is the difference between the combined mass of the nucleons and the mass of the nucleus:

$$\Delta m = 7 \times 1.007276 + 7 \times 1.008665 - 13.999231 = 0.112356 \, u$$

(ii) The binding energy is the energy equivalent of the mass defect. In energy units, $1 \, u = 931.5 \, \text{MeV} \, c^{-2}$, so:

binding energy $= 0.112356 \times 931.5 = 104.7 \, \text{MeV}$

$$\text{Binding energy per nucleon} = \frac{\text{total binding energy for a nucleus}}{\text{number of protons + number of neutrons in nucleus}}$$

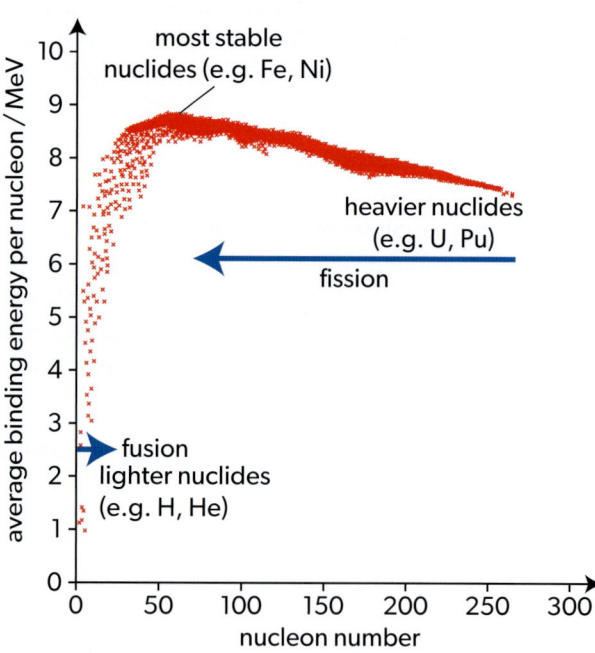

▲ **Figure 3** The graph of variation of binding energy per nucleon with nucleon number for the stable nuclides

You may wonder how the four initial protons become two protons and two neutrons in the fusion example. This is explained in more detail in Topic E.5.

The change in binding energy per nucleon for each reaction is greater for fusion than fission. On the face of it, fusion will be better than fission for power generation. Unfortunately, many of the engineering problems raised by nuclear fusion have yet to be solved (Topics E.4 and E.5).

Plotting a graph of binding energy per nucleon against nucleon number (Figure 3) demonstrates nuclear stability. Some heavy nuclides undergo induced fission (the nucleus becomes unstable when a moving neutron interacts with it—see Topic E.4). Other light nuclides can fuse to form larger nuclei (see Topic E.5). The graph illustrates why this is possible.

Remember that the binding energy per nucleon is a measure of the energy that is released when a nucleus forms from all its constituents. For fusion, four hydrogen nuclei (zero binding energy per nucleon as there is only one proton) form a helium nucleus (larger binding energy per nucleon and more stable). The overall change is that energy is transferred away from the system to achieve this stability.

Example 4

Radon-222 ($^{222}_{86}$Rn) decays into polonium-218 ($^{218}_{84}$Po) with the emission of an alpha particle.

Binding energies per nucleon are:
$^{222}_{86}$Rn: 7.6945 MeV, $^{218}_{84}$Po: 7.7315 MeV, α: 7.0739 MeV.

a) Determine the energy released in this decay.

b) The radon nucleus is stationary immediately before the decay. Explain which of the decay products has a greater kinetic energy.

Solution

a) Total binding energy of radon-222 = 222 × 7.6945 = 1708.2 MeV

Total binding energy of the decay product
= 218 × 7.7315 + 4 × 7.0739 = 1713.8 MeV

The energy released is equal to the difference between the binding energies:

1713.8 − 1708.2 = 5.6 MeV

b) The combined momentum of the alpha particle and the polonium nucleus is zero, so the momentum of the alpha particle is equal but opposite to the momentum of the polonium nucleus. From $E_k = \dfrac{p^2}{2m}$, the particle of smaller mass (the alpha particle) has a greater kinetic energy.

> The equation linking kinetic energy to momentum is discussed in Topic A.2.

The shape of the graph in Figure 3 has implications for both fusion and fission. The most stable (tightly bound) nucleus is nickel-62, with the greatest value of binding energy per nucleon. This is closely followed by two iron isotopes, Fe-58 and Fe-56. Above a nucleon number of about 60, the shape is virtually flat with only a small change (about 1 MeV) in the binding energy per nucleon, indicating that the nuclides with largest A are less stable.

This is due to the different distance dependencies of the electromagnetic and strong nuclear interactions (electromagnetic has an inverse-square dependence (Topic C.2) and strong nuclear has an exponential dependence). This means that, above a certain nuclear radius, the electromagnetic forces are more effective at repulsion than the strong nuclear forces are at attraction when a further proton is added to the nucleus.

Another way to view nuclear stability is by using a plot of neutron number N against proton number Z (Figure 4). This graph shows all the known nuclides, stable and unstable. The most stable nuclides are confined to a narrow zone in the centre of the plot. This zone lies along the line $N = Z$ for small Z. However, for nuclides with $Z > \sim 20$ the stability line deviates and the stable nuclides now have more neutrons than protons. The extra neutrons are supplying additional attractive strong nuclear interaction to overcome the increased range of the electromagnetic repulsion mentioned above.

There are unstable zones on each side of the line of stability but the type of decay that occurs in these zones depends on the region of the graph.

- Neutron-rich nuclides tend to decay by beta-minus decay, in which a neutron is converted to a proton with the emission of an electron and an electron antineutrino.

- Proton-rich nuclides tend to decay by beta-plus decay, with the emission of a positron and an electron neutrino.

E Nuclear and quantum physics

- For high Z, alpha decays (removing neutrons and protons in equal number) and spontaneous fission (leading to unpredictable, but more stable, nuclides; see Topic E.4) can occur.
- There are some highly unstable neutron or proton emitters well away from the zone of stability, which emit their excess particles directly rather than through the beta-decay conversion processes.

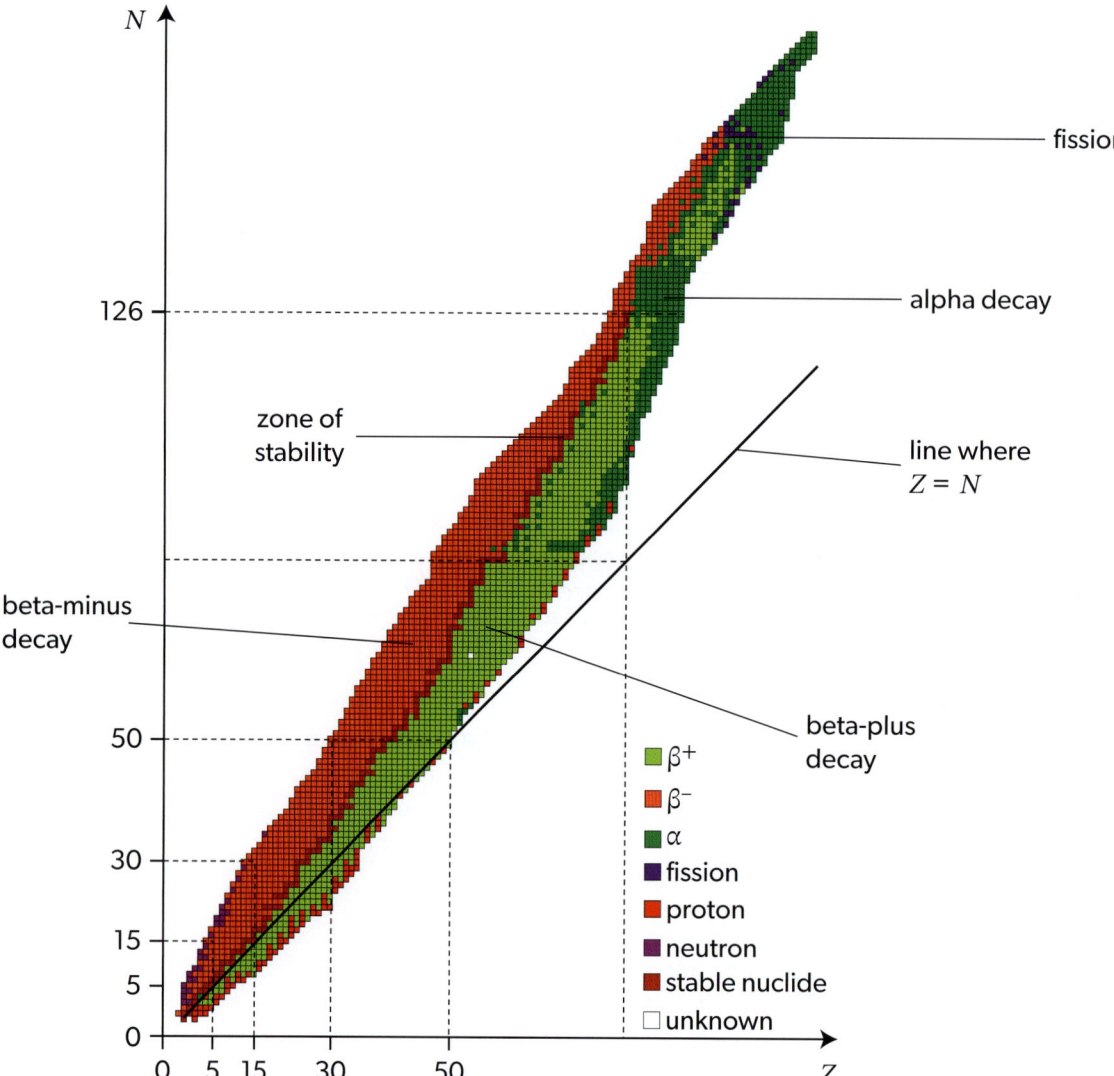

▲ Figure 4 The zone of stability for the plot of N against Z for the known nuclides (stable and unstable)

Example 5

Cobalt-59 ($^{59}_{27}$Co) is the only stable isotope of cobalt.

a) State the neutron number of cobalt-59.

b) Outline why stable isotopes of elements heavier than calcium ($Z = 20$) always have more neutrons than protons.

c) Predict the likely decay mode of the radioactive cobalt-60.

Solution

a) $N = 59 - 27 = 32$

b) The strong nuclear force that holds nucleons together has a shorter range than the repulsive electrostatic force between the protons so, for heavier (and larger) nuclei, additional neutrons are needed to provide an attractive force without contributing to electrostatic repulsion.

c) Cobalt-60 is "neutron-rich" (it has one neutron more than the stable cobalt-59) and will therefore decay by negative beta emission, in which one of its neutrons will be converted to a proton.

E.3 Radioactive decay

Sample student answer

A nucleus of phosphorus-32 ($^{32}_{15}P$) decays by beta minus (β^-) decay into a nucleus of sulphur-32 ($^{32}_{16}S$). The binding energy per nucleon of $^{32}_{15}P$ is 8.398 MeV and for $^{32}_{16}S$ it is 8.450 MeV.

Determine the energy released in this decay. [2]

This answer could have achieved 1/2 marks:

$32 \times 8.398 = 268.74$

$32 \times 8.45 = 270.4$

$M_e = 0.511$ MeV c^{-2}

$= 2.171$ MeV

▼ The mistake here is to include the mass of the beta-minus particle in the calculation. The particle is not bound to anything else so has no binding energy. In the same way, hydrogen-1 has zero binding energy as it consists of one proton in the nucleus.

This answer could have achieved 2/2 marks:

$(8.450 \times 32) - (8.398) \times (32) = 32(8.450 - 8.398)$

$= 1.664$ MeV

▲ Although not at all well explained, this does arrive at the correct answer and scores full marks even though this was a "determine" question, where a high quality of explanation is expected.

Natural sources of radiation give rise to background radiation.

Each nucleus of a naturally decaying element has an identical chance of decay per second. The total number of decays in one second, the decay rate, is proportional to the number of nuclei of the element in a sample. This leads to behaviour where the time to halve the number of original nuclei is constant, irrespective of how many were originally in the sample. This time is known as the **half-life**.

Half-life $T_{\frac{1}{2}}$ is the time taken for half of the nuclei initially present in a pure sample of a radioactive nuclide to decay.

Alternatively, the half-life is the time taken for the initial activity of a pure sample of the radioactive nuclide to halve.

Half-life can be determined from a graph of corrected count rate against time.

A nuclide with a short half-life has a high activity. Nuclides with very long half-lives (e.g. some uranium nuclides found in rocks) are not particularly radioactive because they take so long to decay and their initial concentration is small.

All radioactive measurements are affected by background radiation. This is the radiation produced by natural and man-made sources that are not part of the experiment. The background radiation originates:

- in rocks on Earth
- from cosmic rays that are incident at the top of the atmosphere
- from man-made sources such as nuclear energy and nuclear weapons testing.

To allow for the background radiation in an experiment, remove the source you are using from the laboratory and measure the count for a long period of time—at least several minutes. This can be converted to a count rate and subtracted from every count measurement you make using the source. It is a good idea to measure the background count rate before and after the actual experiment.

Activity is the total rate at which a sample is decaying.

The unit of activity is the becquerel (abbreviated Bq). 1 Bq is an activity of one disintegration every second.

Count rate is the measured number of counts being detected in one second.

Activity and count rate are different—it is difficult to count every emission from a decaying sample. Some emissions will be absorbed by the sample itself, by its container, by any gas between the container and detector, and by the detector itself.

Assessment tip

You may not be given corrected count rates in an examination question either in graphical or numerical form. You may need to deduce the background correction from other data.

E Nuclear and quantum physics

Nature of science

The diagram is a pie chart that shows some of the contributions to the background radiation. Some contributions arise naturally, such as that from the rocks that make up the Earth. Other contributions arise from man-made and artificial sources.

It is important that scientists communicate the nature and the risks that arise from these radiations with their different origins and intensities.

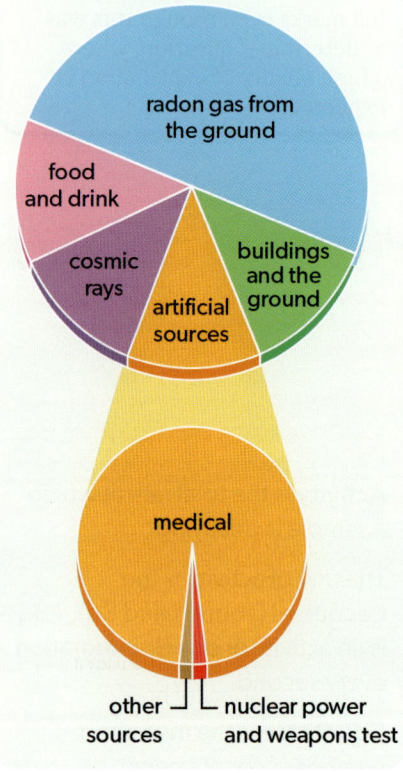

Example 6

The decay activity of a pure radioactive sample was measured every 30 s. The background activity is 0.4 Bq.

Time t / s	0	30	60	90	120
Count rate / Bq	14.0	9.6	6.6	4.6	3.2

Determine the half-life of the sample.

Solution

Subtract the background activity (0.4 Bq) from each count rate value.

Time t / s	0	30	60	90	120
Count rate / Bq	14.0	9.6	6.6	4.6	3.2
Count rate minus background activity / Bq	13.6	9.2	6.2	4.2	2.8

Plot these values on a graph:

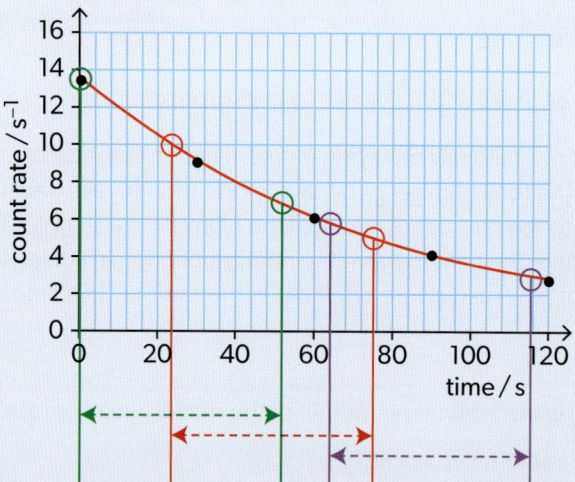

Determine the half-life for at least three different time intervals on the graph (this graph shows the half-lives from 14 to 7, 10 to 5 and 6 to 3 counts per second).

The average of these three values is 52 s.

Radioactive decay has many uses in industry, medicine and elsewhere.

- Living organisms absorb carbon while they are alive. The Earth's naturally occurring carbon contains both stable carbon-12 and unstable carbon-14, which has a half-life of 5700 years. This decays to nitrogen-14 via β^- emission. When the organism dies, there is no longer any absorption of carbon and the unstable carbon-14 decays. The ratio of $\frac{\text{carbon-14}}{\text{carbon-12}}$ previously in equilibrium with the environment (and constant) becomes smaller. Assuming that the value of $\frac{\text{carbon-14}}{\text{carbon-12}}$ in living organisms does not change with time, a measurement of the ratio will indicate the length of time for which the carbon-14 has been decaying and hence the time since the death of the organism.

- Geologists can determine the age of rocks using nuclides such as uranium-238 and uranium-235, which have half-lives of around 4.5 billion years and 700 million years, respectively.

- Technetium-99m can diagnose disease in the bones, the heart and other organs. The technetium is a metastable product of the decay of molybdenum and has a half-life of about six hours. It is usually extracted on site at a hospital. This radioactive tracer is often combined with biologically active materials that take it to areas in the body quickly. The low-energy gamma photons, that are emitted as the technetium decays to the stable form, can be detected outside the patient and are used for diagnosis. They pass through tissue with little or no damage and there are no damaging alpha or beta particles emitted during the decay. Radioactive iodine is used in a similar way to image, and then perhaps treat, the thyroid gland.

- Beta+ emitters (such as fluorine-18) are used to produce positrons in the body that almost immediately combine with electrons to produce two gamma photons. These move off in opposite directions (to conserve momentum). The time difference between their detection in a gamma camera allows a precise determination of the position of the positron when it was annihilated. This allows the detection of infection sites in the body. Oxygen-15 is another beta+ emitter that can be combined with hydrogen to give water that can be injected into the body. These techniques are called PET scans (positron emission tomography).

- The gamma-knife technique focuses intense streams of gamma photons emitted by decaying cobalt-60 nuclei onto small regions in the body to destroy tumours.

- Intense gamma-photon sources are used to sterilise materials in industry, medicine and food production.

- Gamma and beta sources are used for thickness measurements in manufacturing.

- River flow rates and water-purity measurements can be enhanced using radioactive tracers. These "label" fluids or solids and their movement in a river or ocean can then be tracked.

> **Assessment tip**
>
> Half-life is a value that relates to a large sample of decaying atoms.
>
> An individual nucleus does not have a half-life. In the AHL part of this topic there is the concept of a decay constant, which is related to the probability that an individual nucleus will decay in the next second.
>
> When determining a half-life from a graph, always carry out the determination more than once (three times is ideal) and then take the average of your three values. It is best to use large activity values for all three determinations.
>
> It would be very tempting in the sample student answer below to choose 480 Bq to 240 Bq and then 240 Bq to 120 Bq and 120 Bq to 60 Bq. However, by the time the activity is at 60 Bq the fractional uncertainty (Topic T.3) in the read-off is much greater than at 480 Bq. A better plan is to use (480 − 240) Bq, (400 − 200) Bq and (300 − 150) Bq for the activity ranges.
>
> If you are studying the IB Diploma Programme physics course at standard level, you will only be asked to solve problems that involve integer numbers of half-lives.

Sample student answer

The graph shows the variation with time t of the activity A of a sample containing phosphorus-32 ($^{32}_{15}$P).

Determine the half-life of $^{32}_{15}$P. [3]

This answer could have achieved 1/3 marks:

> 12 days

▲ The answer is correct.

▼ This is a "determine" question. The command verb implies that full details are required for the answer. A close look at the graph shows that there are markings at a count rate of 480 Bq and at 240 Bq. So this answer probably comes from only one determination. This view is supported by the lack of any quoted averages.

The 3 marks are likely to be awarded for making more than two determinations of half-life (preferably in the left-hand half of the curve, and try not to go 480 – 240 – 120 – 60). Then a correct determination of the mean is required and finally a correct answer is needed within an appropriate range and to an appropriate number of significant figures.

E Nuclear and quantum physics

Each identical nucleus of a particular nuclide has the same probability of decay per unit time.

$$\text{Probability of decay} = \frac{\text{number of nuclei that decay in the next time interval}}{\text{initial number of nuclei in the sample at the beginning of the time interval}}$$

The **decay constant** λ of a radioactive nuclide is the probability of decay per unit time when the time interval tends to zero. This is so that the *change* in the number of nuclei in the time interval is small when compared with the initial number present.

The SI unit of the decay constant is s^{-1} but you may find this specified in hour^{-1}, day^{-1} or year^{-1}. In such cases, you need to keep the decay constant unit consistent with the time units in the question.

> **Assessment tip**
>
> The negative sign in $\frac{dN}{dt} = -\lambda N$ is important to show that N *decreases* as time *increases*.

The decay constant λ provides the link between the number of nuclei N and their rate of decay per unit time A, where $A = -\lambda \times N$. As $A = \frac{dN}{dt}$, this can also be written as $\frac{dN}{dt} = -\lambda N$.

> Radioactive decay is an example of exponential change with its characteristic half-life behaviour. Exponential growth or change is very common in many branches of science.
>
> - In lightly damped oscillations (Topic C.4), the amplitude of the oscillator typically changes exponentially because the same fraction of the energy at the start of a cycle is removed during the cycle.
> - When an electrical system is discharged through a resistor, the behaviour is exponential because the same fraction of charge leaves the charged system every second.
> - Some sea creatures with shells grow exponentially (up to a certain limit) because their rate of growth is proportional to the amount of food they can ingest through their mouth aperture, which is proportional to their size. This leads to a logarithmic shape in the shell of the creature.

The equation $\frac{dN}{dt} = -\lambda N$ leads to $\frac{dN}{N} = -\lambda dt$, which can be solved to give the important equation $N = N_0 e^{-\lambda t}$, where t is the time and N_0 is the initial number of undecayed nuclei.

The sample activity at time t is $A = \lambda N_0 e^{-\lambda t}$.

There is an important relationship between $T_{\frac{1}{2}}$ and λ. After one half-life, the number of atoms has halved. Therefore, $\frac{N_0}{2} = N_0 e^{-\lambda T_{\frac{1}{2}}}$ and taking logarithms of both sides gives $\ln(0.5) = -\lambda T_{\frac{1}{2}}$, or $T_{\frac{1}{2}} = \frac{\ln 2}{\lambda}$.

As the number of unstable nuclei (mother nuclei) in a sample decreases, the number of nuclei formed (daughter nuclei) must increase. Figure 5 shows the variation of mother and daughter nuclei with time when the daughter product is stable.

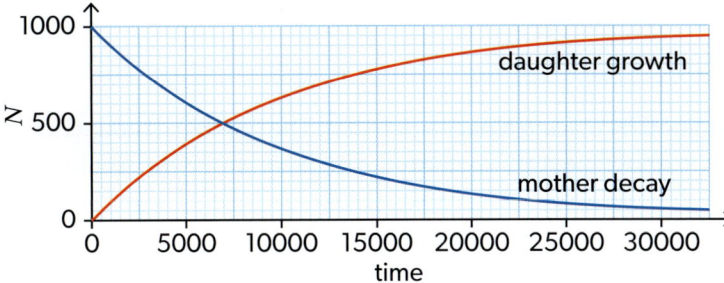

▲ **Figure 5** Decay and growth curve for the unstable and product nuclei in radioactive decay. The product is stable

The determination of λ for nuclides with long half-lives involves obtaining the nuclide in a pure sample and then measuring the mass of the sample. This leads to the number of atoms N in the sample. Then the total activity A of the sample can be measured (this involves determining the counts collected by a detector of finite size and using this to estimate the counts that would be obtained over a complete sphere surrounding the sample). The equation $A = -\lambda N$ is used to calculate the decay constant.

Reasonably short half-lives can be measured by:

- measuring the background count rate in the laboratory
- taking readings of count rate against time until the value is close to that of the background
- subtracting the background count rate from each reading
- assuming that the corrected count rate is proportional to activity
- plotting a graph of $\ln A$ against time
- finding the gradient of this graph, which gives $-\lambda$.

Nature of science

Measurements and observations are part of the nature of science. When half-lives are long—that is, they cannot be measured by standard laboratory techniques within reasonable time spans—a different technique is required, as explained here.

For very short half-lives, of the order of fractions of a second, advanced types of measurements are required involving, for example, irradiation of nuclides in a reactor.

Example 7

Radioactive iodine (I-131) has a half-life of 8.04 days.

a) Calculate, in seconds^{-1}, the decay constant of I-131.

b) Calculate the number of atoms of I-131 required to produce a sample with an activity of 60 kBq.

c) Deduce the time taken for the activity of the sample in part (b) to decrease to 15 kBq.

Solution

a) To calculate the decay constant in s^{-1}, the half-life must be expressed in s.

$$\lambda = \frac{\ln 2}{8.04 \times 24 \times 60 \times 60} = 1.0 \times 10^{-6} \text{ s}^{-1}$$

b) Number of atoms $N = \dfrac{A}{\lambda} = \dfrac{6.0 \times 10^4}{1.0 \times 10^{-6}} = 6.0 \times 10^{10}$

c) A decrease from 60 to 15 kBq is $\dfrac{1}{4}$: in other words, two half-lives, which is 16 days.

E Nuclear and quantum physics

Practice problems

Problem 1
The half-life of a radioactive nuclide was determined by placing it near a detector. The number of counts in 30 seconds every 10 minutes from the start of the experiment was recorded. The table shows the results.

t / minute	0	10	20	30	40	50	60
Number of counts in 30 s	70	52	45	33	28	24	20

a) Explain what is meant by half-life.

b) The background count rate at the position of the detector is 10 counts in 30 s.

  (i) Determine the half-life for this nuclide.

  (ii) Predict the observed count rate four half-lives from the start of the experiment.

Problem 2
Plutonium-240 ($^{240}_{94}$Pu) decays to form uranium (U) and an alpha particle. The following data are available.

Nucleus	Binding energy per nucleon / MeV
plutonium-240	7.5560
uranium	7.5865
alpha particle	7.0739

a) State the equation for this decay.

b) Determine, in joules, the energy released when one nucleus decays.

Problem 3
Aluminium-26 ($^{26}_{13}$Al) decays by the emission of a positron to form a nucleus of magnesium (Mg). The atomic mass of $^{26}_{13}$Al is 25.986892 u and that of Mg is 25.982593 u.

a) State the equation for this decay.

b) Determine, in MeV, the energy released in this decay.

Problem 4
Radioactive sodium ($^{22}_{11}$Na) has a half-life of 2.6 years. A sample of this nuclide has an initial activity of 5.5×10^5 Bq.

a) Explain what is meant by the **random nature** of radioactive decay.

b) Sketch a graph of the activity of the sodium sample for a time period of 6 years.

c) Calculate:

  (i) the decay constant, in s^{-1}, of $^{22}_{11}$Na

  (ii) the number of atoms of $^{22}_{11}$Na in the sample initially

  (iii) the time taken, in seconds, for the activity of the sample to fall from 100 kBq to 75 kBq.

Problem 5
Caesium-137 ($^{137}_{55}$Cs) decays by negative beta decay to form a nuclide of barium (Ba).

a) Write down the nuclear reaction for this decay.

b) The half-life of caesium-137 is 30 years. Determine the fraction of the original caesium that remains after 200 years.

Problem 6
Tin-112 ($^{112}_{50}$Sn) is the lightest stable isotope of tin.

a) Predict the likely decay mode of the radioactive isotope tin-111 ($^{111}_{50}$Sn).

b) The half-life of tin-111 is 35.3 minutes. Calculate the probability that a nucleus of tin-111 will decay in the next second.

E.4 Fission

You must know:
- the processes of spontaneous fission and neutron-induced fission
- what is meant by a chain reaction
- the roles of control rods, moderators, heat exchangers and shielding in nuclear power plants
- about the properties and management of nuclear fission products.

You should be able to:
- solve problems involving the energy transfer in fission reactions
- discuss the issues surrounding nuclear power and the management of its waste products.

Spontaneous nuclear fission is rare. Only a few elements undergo this process, in which a nuclide (typically, thorium-232, uranium-235 or uranium-238) splits into two or more smaller daughter products with the additional emission of neutrons. As the daughter nuclides must be smaller than the original nucleus that fissions, neutrons must be released or converted into protons via beta-minus decay to reduce the neutron : proton ratio (Topic E.3, page 203). This is unlikely to happen in a single disintegration and the daughter products are themselves likely to be radioactive with short half-lives. As with neutron-induced fission, it is not possible to predict the daughter elements.

More commonly, neutron-induced fission occurs when a moving neutron interacts with a nucleus. The fission occurs for nuclides such as U-235 (Figure 1). A neutron collides with a nucleus of uranium-235 and is absorbed.

This produces a nucleus of the highly unstable uranium-236 (U-236). The U-236 splits into two (or more) nuclear fragments, with the release of several fast-moving neutrons.

This occurs in about 80% of the neutron collisions that produce U-236. In the remainder, the U-236 de-excites with the emission of a gamma photon.

It is not possible to predict the exact outcome in terms of daughter nuclides or numbers of emitted neutrons. The two nuclei are unlikely to be of equal size. Figure 2 shows the observed distribution of nuclides produced in U-235 fission.

Energy is released in each fission because the daughter products are at higher positions on the binding energy per nucleon chart (Figure 3, in Topic E.3 on page 202) than the U-235. The energy transfer in each fission event can be evaluated (see Example 1).

> This topic links strongly to much of the physics described in Topic E.3. To get the most out of this topic, you should be familiar with:
> - decay equations
> - ideas of nuclear stability
> - binding energy.

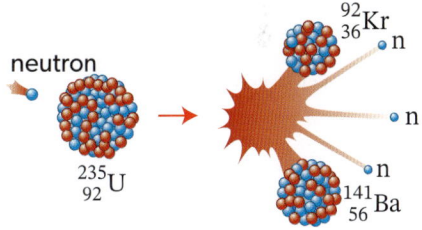

▲ **Figure 1** An incoming neutron causes a U-235 nucleus to fission into two smaller nuclear fragments with the additional emission of high-speed neutrons. The intermediate U-236 nuclide is not shown

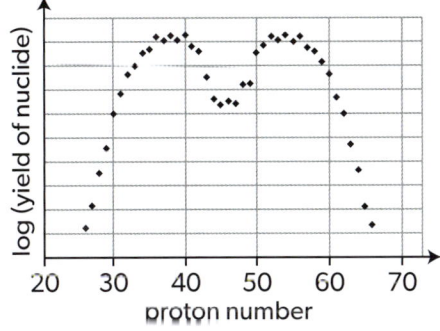

▲ **Figure 2** The two peaks in the distribution show that the production of two fission products with equal proton number is a rare event. Note that the y-axis is a logarithmic scale

E Nuclear and quantum physics

Example 1

Consider the following neutron-induced fission reaction:

$${}^{235}_{92}\text{U} + {}^{1}_{0}\text{n} \rightarrow {}^{132}_{50}\text{Sn} + {}^{101}_{42}\text{Mo} + 3{}^{1}_{0}\text{n}$$

The atomic masses are given in the table.

Nuclide	Atomic mass
${}^{235}_{92}\text{U}$	235.0439 u
${}^{132}_{50}\text{Sn}$	131.9178 u
${}^{101}_{42}\text{Mo}$	100.9103 u

a) Calculate, in MeV, the energy released in this reaction.

b) Estimate, using the result of part (a), the energy transferred when 1.0 kg of pure U-235 fissions. State the answer in joules.

Solution

a) Two new neutrons are released and so the mass difference for the reaction is:

$$\Delta m = m_{\text{U-235}} - (m_{\text{Sn-132}} + m_{\text{Mo-101}} + 2m_n)$$

$$= 235.0439 - (131.9178 + 100.9103 + 2 \times 1.0087) = 0.1984 \text{ u}$$

Energy released $\Delta mc^2 = 0.1984 \times 931.5 = 184.8$ MeV

b) The molar mass of U-235 is approximately 235 g mol^{-1} and 1.0 kg of pure U-235 contains $\dfrac{1.0}{0.235} \times N_A = 2.56 \times 10^{24}$ atoms.

Assuming that every fission reaction releases 184.8 MeV, the total energy available from 1.0 kg of uranium is $2.56 \times 10^{24} \times 184.8 = 4.73 \times 10^{26}$ MeV.

This is equivalent to $4.73 \times 10^{26} \times 10^6 \times 1.60 \times 10^{-19} = 7.6 \times 10^{13}$ J.

The neutrons emitted in the fission can go on to induce other fissions. When the size and geometry of the uranium bulk material are correct, a chain reaction can begin (Figure 3). This process was first suggested by Leo Szilard, who also recognized that careful geometry and engineering design are necessary to sustain the reaction in a nuclear reactor.

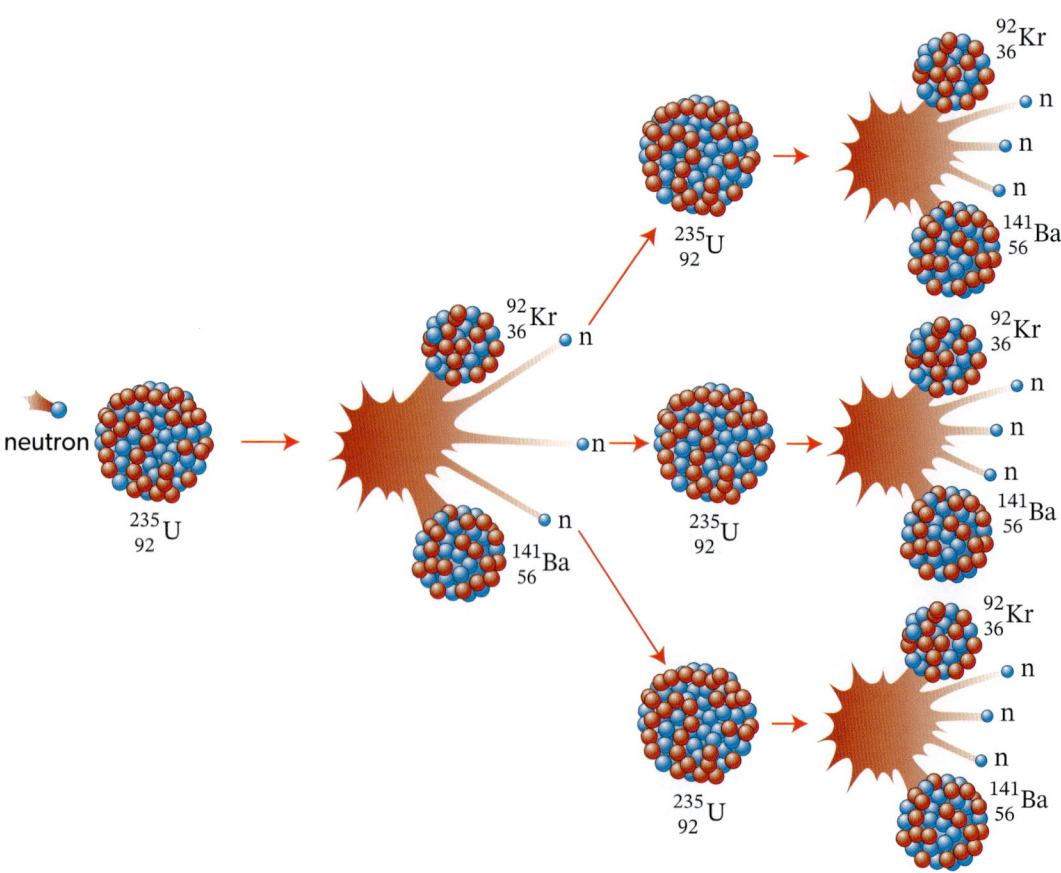

▲ **Figure 3** A chain reaction where three neutrons emitted in one fission go on to cause further fissions. It would be unlikely for each fission to produce the same fission products

This description of a working nuclear reactor focuses on the pressurized water reactor (PWR), which is a common type throughout the world. Other designs are also used commercially and for research.

The reactor is designed to control and transfer the energy released in the fission process and to convert it via a series of stages to a useful form, usually electrical energy. Figure 4 shows a schematic diagram of a PWR nuclear reactor.

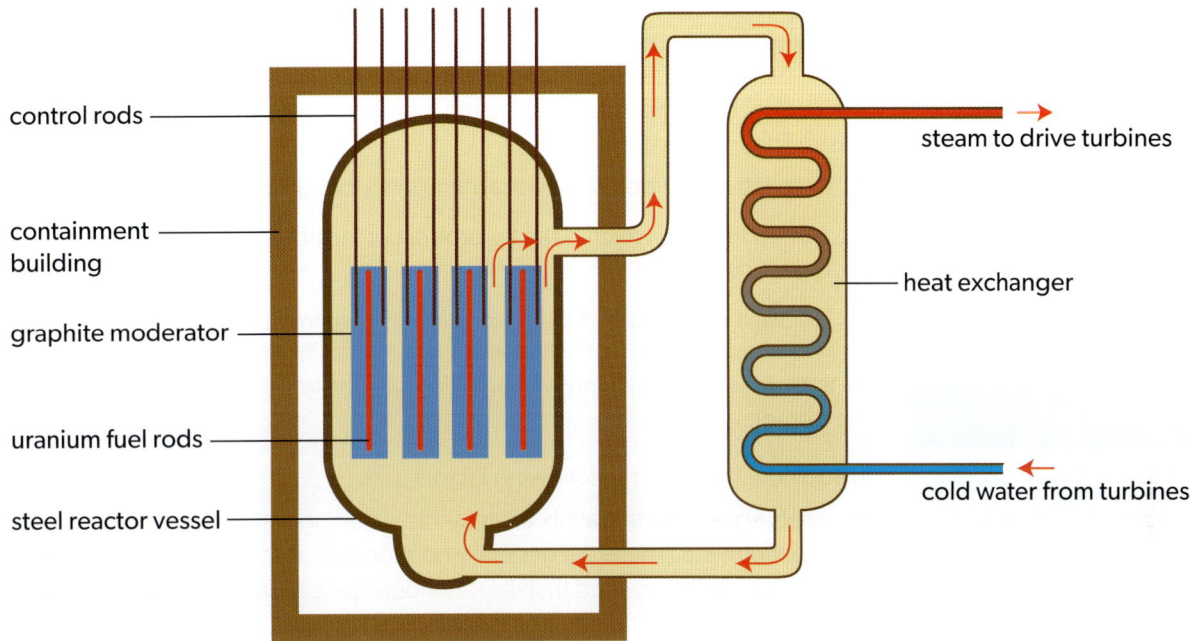

▲ **Figure 4** A schematic diagram of a pressurized water reactor

Details of the process include the following.

- Uranium ore is too poor in U-235 and is U-238 rich.
- Neutrons are travelling very fast after emission (typically around $0.1c$). They are most effective in causing further fissions when moving at about $2\,\text{km}\,\text{s}^{-1}$ — speeds similar to those of atoms in matter that is at about room temperature. Neutrons at these speeds are known as thermal neutrons. Also, U-238 nuclei are good absorbers of high-speed neutrons, so the neutrons must be slowed down in a moderator. The reactor is designed so that fast neutrons leave the fuel rods after emission, and re-enter the fuel rods only after they have been decelerated.
- It is vital to be able to control the output of the reactor and to be able to shut it down completely in an emergency. Control rods perform these functions. These rods, made of a good neutron absorber such as barium, can be raised or lowered into the reactor chamber (also called the pile) to achieve this. In some designs, small barium spheres or barium dust can be released in an emergency to coat all the reactor surfaces to stop the fission reactions very quickly.
- A heat exchanger is used to remove energy from the reactor to the outside world. This means that active substances remain inside the reactor pile.
- The turbines must be connected to generators, which convert the rotational kinetic energy of the turbine to electrical energy.

> The uranium ore must be enriched to boost the percentage of U-235 and reduce the percentage of U-238.

> Good **moderators** must not absorb neutrons and they should have a mass close to that of the neutron. They should also be inert in the extreme conditions of a nuclear reactor. Carbon as graphite and heavy water (deuterium oxide) are typical moderators.

> The heat exchange medium is often high-pressure steam, which goes directly to turbines. Other media are possible, such as liquid sodium.

E Nuclear and quantum physics

You should be able to use the physics in Topics A.2 and A.3 to show that the change in kinetic energy ΔE_k after one collision between a neutron of mass m and initial kinetic energy E_k and an initially stationary moderator atom of mass M is $\Delta E_k = \dfrac{4Mm}{(m+M)^2} E_k$ in a head-on collision.

You should be able to predict the number of head-on collisions required to reduce the energy (and hence speed) of a neutron by the required amount.

Assessment tip

Many topics in the IB Diploma Programme physics course link to the physics of the nuclear reactor. These include the Theme E topics concerned with fission and the activity and types of radioactivity of the fission products.

The conversion of energy to the water in the heat exchanger and the transfer of thermal energy involve the physics of Topic B.1.

The conversion of energy to the final electrical form is covered in Topic D.4.

The dynamics of the moderator and the turbine/generator involve much of Theme A.

You should be prepared to meet these ideas in a composite topic such as nuclear fission.

There are many engineering issues involved in the design of a commercial reactor.

- Thick steel is required to withstand the high temperatures and stresses of the reactor interior.
- A thick concrete casing plays the dual role of absorbing emitted particles and providing strength to the structure. This is known as the shielding.
- Safety mechanisms must be reliable.
- Robots are used to remove fuel rods and repair the internal (active) parts of the mechanism.

The nuclear reactor produces small but highly active quantities of radioactive waste. This includes:

- radioactive nuclides with proton numbers around 35 and 55 (see Figure 2)
- plutonium-239, which is the by-product of the absorption of one neutron by U-238
- U-236, a by-product of the absorption of a neutron by U-235.

All these must be dealt with when making safe the waste that is removed from the reactor.

In addition, there are the following factors.

- The fuel rods contain a mix of elements including krypton, barium, strontium, etc. These are generally neutron-rich, decay by β^- emission and are relatively short-lived (half-lives of hours to decades). The rods are stored under water in cooled ponds for 5–10 years before reprocessing to extract, and then reuse, the uranium-235 that did not fission.
- The fuel rods become distorted when many uranium atoms have split into two separate species. The rods must be removed and replaced before they become stuck in their channels in the reactor.
- The plutonium-239 reduces the efficiency of the U-235 fission overall and must be removed.
- There is low-level (low activity) waste produced in the reactor and as a by-product of the industrial activity in the nuclear plant. This must also be dealt with.
- Most nuclear reactors have design lifetimes of 20–50 years. After this the reactor core must be carefully decommissioned to remove the radioactive material and protect the site, parts of which may remain radioactive for up to a century.

Sample student answer

One possible fission reaction of uranium-235 is:

$$^{235}_{92}U + ^{1}_{0}n \rightarrow ^{140}_{54}Xe + ^{94}_{38}Sr + 2\,^{1}_{0}n$$

Data for this question

Mass of one atom of U-235 = 235u

Binding energy per nucleon for U-235 = 7.59 MeV

Binding energy per nucleon for Xe-140 = 8.29 MeV

Binding energy per nucleon for Sr-94 = 8.59 MeV

a) (i) State what is meant by the binding energy of a nucleus. [1]

E.4 Fission

This answer could have achieved 0/1 marks:

> The binding energy is the energy required to separate the protons from the neutrons in a nucleus.

▼ The answer needs to be clear that all the nucleons are separated from each other. "To separate the protons and neutrons completely" would have scored the mark.

(ii) Outline why quantities such as nuclear binding energy are often expressed in non-SI units. [1]

This answer could have achieved 1/1 marks:

> If you use joules then the numbers have very small powers of ten.

▲ The answer is not expressed well. It would have been better to talk about the "values when expressed in SI units". But it is clear what the candidate means.

(iii) Show that the energy released in the reaction is about 180 MeV. [1]

This answer could have achieved 0/1 marks:

> Binding energy from the U-235 = 235 × 7.59 = 1784 MeV
> Binding energy from the Sr = 140 × 8.59 = 1203 MeV
> Binding energy from the Xe = 94 × 8.29 = 779 MeV
> Energy released = 1982 − 1784 = 198 MeV

▼ The answer is incorrect because the student has used 140 for the Sr and 94 for the Xe instead of the other way round. If your answer is very different from the "show that" value, always check carefully to see where your error is. It is easy to transpose numbers in this type of calculation.

A nuclear power station uses U-235 as fuel. Assume that every fission reaction of U-235 transfers 180 MeV of energy.

b) (i) Estimate, in joules, the energy produced when a mass of 1.0 kg of U-235 fissions. [2]

This answer could have achieved 1/2 marks:

> Mass of one atom of U-235 = 235 × 1.66 × 10^{-27} = 3.9 × 10^{-25} kg
> ∴ Number of atoms in one kg = $\frac{1}{3.9 \times 10^{-25}}$ = 2.6 × 10^{24} atoms
> Energy from one fission = 180 × 1.6 × 10^{-19} = 2.9 × 10^{-17} J
> So the energy from one kg is 73.9 MJ

▼ The answer is 10^6 too small because the student has omitted the M in MeV. There is only one mark for arriving at the correct number of atoms in 1.0 kg. Another issue here is that the working oscillates between 2 and 3 s.f. It is best to make all calculations with all the s.f. until the last stage.

(ii) The power station has a useful power output of 1.2 GW and an efficiency of 36%. The station runs continuously. Determine the mass of U-235 that undergoes fission in each 24-hour period. [2]

This answer could have achieved 2/2 marks:

> In one day the energy produced is 1.2 × 10^9 × 24 × 3600 = 1.0 × 10^{14}
> ∴ the energy from the fission must be ÷ $\frac{36}{100}$ = 2.9 × 10^{14}
> ∴ the mass is $\frac{2.9 \times 10^{14}}{73.9 \times 10^6}$ = 3.9 × 10^6 kg

▲ The answer is now 10^6 too large because the student has carried forward the error in the previous part. However, this is accepted. The second line is confused and the working is not completely clear here. Because the answer follows from the data that the student has used, there is full credit for it.

Waste produced by the nuclear reactor contains 1.0 kg of strontium-94. This nuclide is radioactive and undergoes β^- decay to form a daughter nuclide **X**. The waste is removed from the reactor at time $t = 0$. The graph shows the activity A of the strontium-94 with time t.

E Nuclear and quantum physics

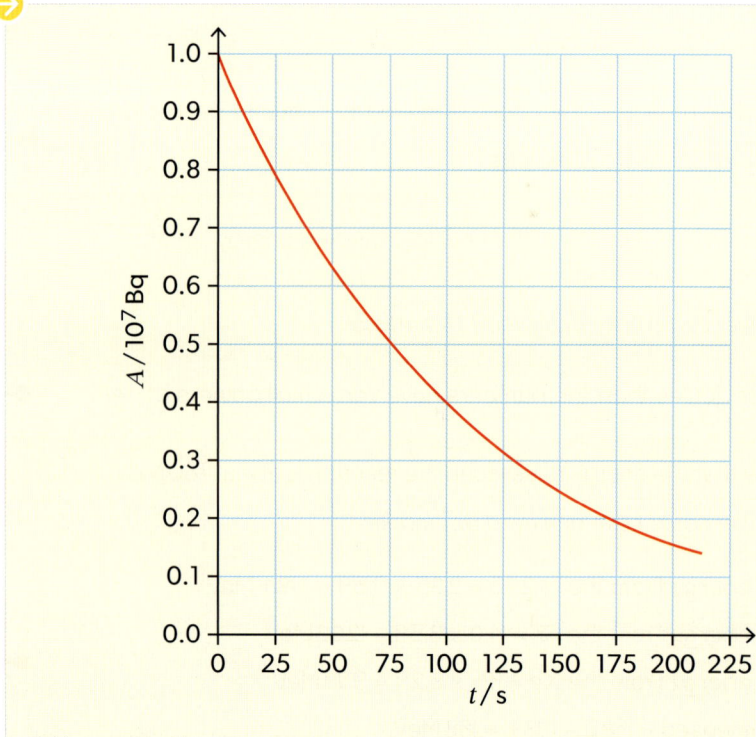

c) (i) State the equation for the strontium-94 decay.

$${}^{94}_{38}\text{Sr} \rightarrow \text{X}$$ [2]

This answer could have achieved 1/2 marks:

$${}^{94}_{38}\text{Sr} \rightarrow {}^{94}_{39}\text{X} + {}^{0}_{-1}e^- + v_e$$

▼ The student has forgotten to specify that the third particle is an electron antineutrino. There should be a bar over the symbol.

(ii) Determine the mass of strontium-94 that decays in the first 5 minutes. [3]

This answer could have achieved 1/3 marks:

1.0 → 0.5 takes 75 seconds so the graph shows that the half-life of the strontium is $1\frac{1}{4}$ minutes.

5 minutes is 4 lots of $1\frac{1}{4}$ minutes so there are 4 half-lives and the mass must be $\frac{1.0}{4} = 0.250 = 0.25$ kg.

▼ Each half-life, the mass remaining is halved. In four half-lives the mass remaining is $\frac{1.0}{2^4} = 6.3 \times 10^{-2}$ kg. The student has not read the question carefully. It asks for the amount of strontium that has disappeared, not the amount remaining. The answer is 1.00 − 0.06 = 0.94 kg.

Practice problems

Problem 1

A nucleus of uranium-235 undergoes neutron-induced fission according to this reaction:

$${}^{235}_{92}\text{U} + {}^{1}_{0}\text{n} \rightarrow {}^{137}_{52}\text{Te} + {}^{97}_{40}\text{Zr} + x {}^{1}_{0}\text{n}$$

The binding energies per nucleon are given in the table.

Nuclide	Binding energy / A
${}^{235}_{92}\text{U}$	7.591 MeV
${}^{137}_{52}\text{Te}$	8.280 MeV
${}^{97}_{40}\text{Zr}$	8.604 MeV

a) (i) Calculate the number of neutrons released in the reaction.

(ii) Outline the role of these neutrons in fission energy production.

b) Calculate, in MeV, the energy released in the reaction.

c) A nuclear power station outputs 1.5 GW of electrical power at an overall efficiency of 0.40. Use your answer in part (b) to estimate the time for 1.0 kg of U-235 to undergo fission in this power station.

Problem 2

A practical nuclear reactor uses enriched uranium as fuel. The uranium is formed into fuel rods.

a) Outline why the fuel rods are surrounded by a substance that is capable of reducing the kinetic energy of neutrons.

b) Explain how the power output from the reactor is regulated.

c) Suggest why the fuel rods are removed from the reactor well before all the uranium has undergone fission.

Problem 3

a) Outline why the products of nuclear fission require long-term storage after being removed from the reactor.

b) Describe some of the safety risks associated with the storage of the products of nuclear fission.

E.5 Fusion and stars

You must know:

- that nuclear fusion is the energy source for a star
- the conditions for fusion to occur in a star
- the mechanisms for stability in a star
- how stellar mass affects the evolution of a star
- what is meant by stellar parallax.

You should be able to:

- solve problems involving the energy transfer in fusion reactions
- solve problems that involve the Hertzsprung–Russell (HR) diagram and its main regions
- solve problems about the main properties of stars on the HR diagram
- solve problems involving stellar parallax
- solve problems involving the determination of stellar radii.

Stars form from dust and gas clouds that can be stable for millions of years until they are disturbed by some event such as an interaction with another cloud. The cloud can then become unstable and collapse. As it does so, its temperature rises and nuclear fusion can begin.

When collapse begins for a small cloud, pressure waves in the gas (travelling at the speed of sound) can cross the collapsing region quickly and restore stability. When the cloud is large, it takes too long for the waves to reach the unstable region. Gravity compresses the gas, which then heats up, before the sound wave has time to restore stability.

Nuclear fusion is the energy source for a star. In Sun-like stars, the predominant fusion process is the proton–proton cycle (Figure 1).

E Nuclear and quantum physics

There are three stages of the **proton–proton cycle**.

- Stage I: overall, two protons fuse to produce a deuterium, hydrogen-2 (^{2}H), nucleus with the release of a positron particle (an anti-electron) and an electron neutrino. Once the protons have fused, one of them decays by a β⁺ reaction to form a neutron (which remains bound to the proton forming the deuterium) plus the positron and the neutrino.

- Stage II: a deuterium nucleus now fuses with another proton to form a helium-3 nucleus. A gamma-ray photon is also emitted during this interaction.

- Stage III: two helium-3 nuclei formed from the previous stage fuse to give a helium-4 nucleus with the release of two protons.

- Each stage leads to a release of energy.

Overall, four hydrogen nuclei (i.e. four protons) fuse to give one helium-4 nucleus.

Assessment tip

Note that, to produce one helium-4 nucleus, there must be:

- two stage I fusions
- two stage II fusions
- one stage III fusion.

The total overall energy release in one complete cycle ($4\,^1_1\text{H}^+ + 2\text{e}^- \rightarrow\, ^4_2\text{He}^{2+} + 2\nu_e$) is 26.73 MeV as computed from mass differences (Topic E.3), but some of this is lost to the emitted neutrinos.

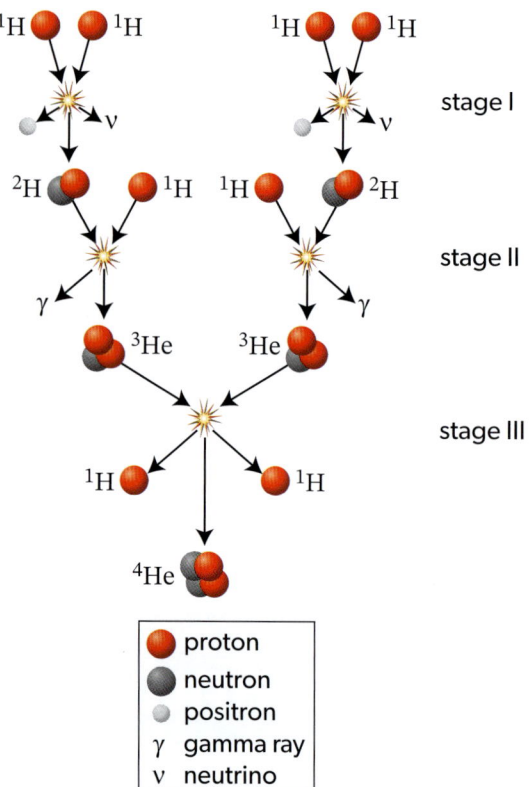

▲ **Figure 1** The three stages of the proton–proton fusion cycle that provides energy for stars

These three distinct stages do not proceed at the same rate.

- Each stage I fusion releases 1.4 MeV of energy, including the subsequent annihilation of the positron (anti-electron) with a nearby electron to give two gamma photons. This stage requires the two protons to remain bound long enough for the β⁺ decay to occur. This is unlikely and the average proton in the Sun will wait a few billion years for this reaction to happen.

- Each stage II fusion has an energy release of 5.5 MeV. As the protons are abundant, this happens quickly, within about one second of the production of the stage I deuterium.

- The single stage III reaction releases 12.9 MeV and it takes about 400 years for the reaction to occur between a pair of tritium (hydrogen-3) nuclei.

Example 1

The third stage of the proton–proton cycle is $^3_2\text{He} + ^3_2\text{He} \rightarrow\, ^4_2\text{He} + 2\,^1_1\text{H}$.

The following data are given.

Nuclide	Binding energy / A
^3_2He	2.573 MeV
^4_2He	7.074 MeV

Determine the energy released in this reaction.

Solution

Total binding energy of the two helium-3 nuclei = $2 \times 3 \times 2.573 = 15.438$ MeV

Binding energy of the alpha particle (helium-4 nucleus) = 4×7.074
$= 28.296$ MeV

The two protons are released as free particles and do not contribute to the binding energy of the products.

The binding energy of the alpha particle is greater than the combined binding energy of the reacting nuclei, and the difference is transferred to the kinetic energy of the products.

Energy released = $28.296 - 15.438 = 12.86$ MeV

For much of its lifetime (i.e. while it is on the main sequence—see Figure 2), a star is stable and in equilibrium. Two balanced forces maintain this equilibrium.

- There is a **gravitational attraction** inwards between the interior of the star and layers at a greater radius from the centre.
- There is a force acting outwards that arises from the **thermal and radiation pressures** generated inside the star.
- The thermal pressure is due to the very high temperatures in the interior of the star. This means that the constituents are moving at very high speed and transferring momentum during collisions.
- The radiation pressure originates from the presence of gamma photons that are released during the fusion processes. These photons can also transfer momentum (changing wavelength as they do so). It is possible to show that the radiation pressure is proportional to T^4, where T is the interior temperature of the star.

These ideas lead to two requirements for the interior of a star: it should have:

- a **high density** (so that the fusing protons are initially close together and their collision rate is high enough for fusion reactions to be efficient)
- a **high temperature** (so that the protons have very large energies in order for them to overcome the electrostatic repulsion between them).

Most stars emit a continuous spectrum from their high-temperature internal regions. This radiation passes through cooler, low-density gas in the star's outer regions, which absorbs wavelengths before re-emission. Absorption lines are characteristic of the chemical elements in the cooler gas. The hottest stars often fail to show absorption lines because the hydrogen gas is completely ionized, so has no electrons to be promoted out of the ground state and absorb photons from the star's interior.

Spectral analysis of a star is difficult because:

- there can be many elements present and the absorption lines are superimposed on each other
- Doppler broadening of the lines occurs because the atoms move
- stars often rotate—one limb of the star approaches an observer while the opposite limb moves away, which also causes Doppler shift.

In stars that are at significantly greater temperatures than those of our Sun, other reactions can occur in the proton–proton cycle. These are not discussed in the IB Diploma Programme physics course.

In stars that are significantly more massive and older than the Sun, further cycles such as the carbon–nitrogen–oxygen (CNO) cycle occur. This cycle is one of those responsible for the processes of nucleosynthesis, in which elements up to the proton number of iron and nickel are produced. Other processes such as neutron capture can create elements of even greater proton number than nickel.

Stellar-fusion processes are driven by differences in binding energy per nucleon, as shown in the plot of the magnitude of binding energy per nucleon against nucleon number (Figure 3 in Topic E.3). Iron and nickel are at the peak of the plot and represent the most stable nuclides. Use this essential idea in Topic E.5 as an opportunity to review and consolidate your understanding of Topic E.3.

The production of emission and absorption spectra are covered in Topic E.1.

Topics B.1 and B.2 cover other important areas linked to astronomy, including the Stefan–Boltzmann law and Wien's law that are intensively used by professional astronomers.

Doppler shift is the subject of Topic C.5.

E Nuclear and quantum physics

> The term luminosity L is introduced in Topic B.1. It is principally an astronomical term meaning the total power output of a black-body radiator—in this case, the star. This links to the Stefan–Boltzmann law, also outlined in Topic B.1, with $L = \sigma A T^4$, where A is the surface area of the star and T is the absolute temperature of the surface of the star, treated as a black body.

The mass of the star plays an important role in the equilibrium.

- The greater the mass, the greater the gravitational attraction and so the outward (thermal and radiation) pressure must also be greater to achieve equilibrium.
- This implies that, when the mass is larger, the core temperature must be larger also.
- The larger the internal temperature, the more probable are the fusion events that lead to energy transfer and the rate of fusion will be increased.
- Thus, the most massive stars have the highest luminosities and consume their fuel quickly, so their overall lifetimes are shorter than stars with less mass.

The conclusion that the most massive stars are the shortest lived is counterintuitive. Considerations of the power radiated by a star lead to the understanding that $\dfrac{\tau}{\tau_\odot} = \left(\dfrac{M_\odot}{M}\right)^{2.5}$ where τ and M are the lifetime and mass of a star, and $\tau_\odot$ and $M_\odot$ are the lifetime and mass of the Sun. This shows that a star 10 times the mass of the Sun will have an expected lifetime that is only 0.3% of that of the Sun. (This mass–lifetime relationship is not tested in the IB Diploma physics programme.)

Patterns of stellar behaviour and evolution are incorporated into the Hertzsprung–Russell (HR) diagram, which is a useful tool for the comparison of stars.

Figure 2 shows the main features of an HR plot of luminosity against temperature, with stars grouped according to type.

Assessment tip

You need to be aware of some particular features of the **Hertzsprung–Russell** diagram. Note, for example, the unusual axes.

- The temperature scale is reversed from the usual direction—it runs from high to low.
- Both axes are logarithmic. In Figure 2, the relative intensity axis goes up in factors of 10 and the temperature axis halves every division.

Examiners will expect accuracy in your sketches of HR diagrams. Make sure that the Sun's position is correct (it has a relative intensity—a relative luminosity—of 1 and a temperature of about 5700 K). You do not need to draw the stars in the main sequence as a series of points—a band will do to indicate the position from about (20 000, 10^4) down to (2500, 10^{-4}).

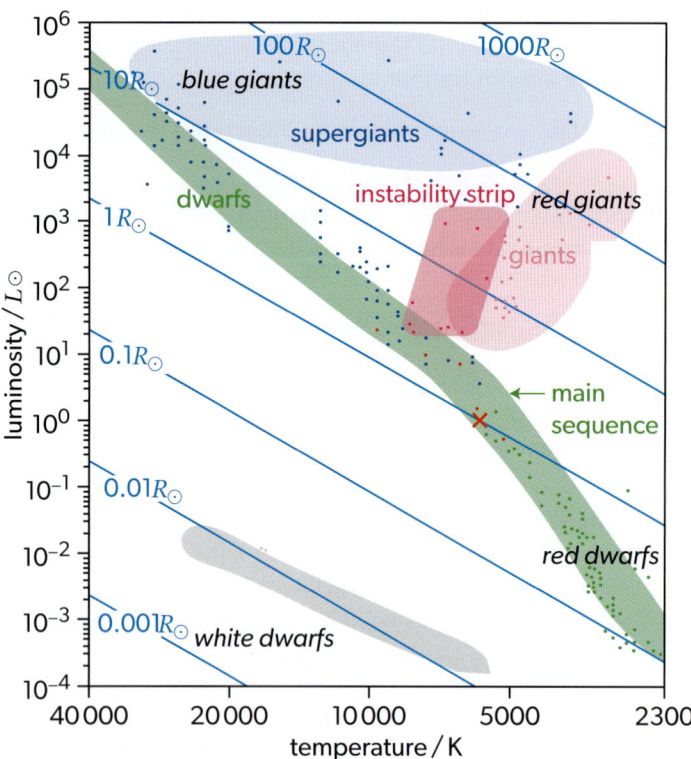

▲ Figure 2 The Hertzsprung–Russell diagram. The diagonal straight blue lines are the lines of constant radius and are marked with the radius in terms of the size of the Sun. The green strip is the main sequence. × is the position of the Sun

The **main sequence** consists of stars producing energy by fusing hydrogen and other light nuclei. About 90% of all stars are on the main sequence. They move along it throughout their lifetimes as their luminosity and temperature change. At the bottom right of the HR diagram are small, cool red stars (sometimes known as red dwarf stars). At the top left are large, hot blue stars (sometimes known as blue giant stars). The present position of the Sun is shown with an ×.

Red giant stars have lower temperatures but higher luminosities than the Sun—their surface areas and diameters are much larger than the Sun.

Supergiant stars are rare, very bright and much larger than red giants. A typical supergiant emits 10^5 times the power of the Sun. About 1% of stars are red giants or supergiants.

White dwarf stars are very dense and constitute the remains of old stars. They have low luminosity and small surface area. They are cooling down and will take billions of years to do so. About 9% of all stars are white dwarfs.

The **instability strip** is a region where variable stars are found. A variable star is one whose brightness (as seen from Earth) changes with time. This variation can be due to the normal evolution of the star or it can be a cyclic variation in brightness caused by processes inside the star. It is the second of these processes that leads to a star being positioned on the instability strip. Inside such a star, changes in the outer atmosphere give a variation in pressure that causes the diameter of the star to increase and decrease.

Lines of constant radius are a set of diagonal lines for constant stellar radius. The constant lines go from the upper left to the lower right of the diagram and indicate stars of the same physical size.

Nature of science

When astrophysicists want to model the interior of a star there is little direct data that they can use. The best they can do is to examine the visible surface of the Sun. The researchers make assumptions about the nature of the stellar medium from observations of many, distant stars. They make hypotheses about the plasma that is the internal stellar material. These hypotheses are likely to use a (modified) kinetic theory of gases and the known terrestrial details of nucleons and electrons.

Stars spend different times on the main sequence. As noted earlier, stars with large mass and high temperature burn their fuel more quickly than small, cool stars.

A **protostar** is essentially a dust/gas cloud that gains mass by gravitational accretion (the particles are attracted and therefore accelerated to each other). As this happens, the protostar's temperature increases sufficiently for fusion to begin. Now that it is a star, it joins the main sequence by moving over from the right-hand side of the diagram. It will remain on the main sequence for as long as it has hydrogen to convert to helium.

When most of the core hydrogen is used, the star moves off the main sequence. Outward radiation pressure is no longer equal to the inward gravitational forces and the star shrinks. Another temperature increase occurs because of this shrinking—this allows the remaining hydrogen in the outer layers to fuse and expand so that the size of the star increases again.

The star has now become either a red giant or a supergiant and it moves from the main sequence to the relevant part of the HR diagram.

However, the core continues to shrink and heat up so that heavier elements (such as carbon and oxygen) can form by fusion. Fusion continues in the most massive stars so that they produce elements up to (the most stable) iron and nickel. From this point on, the evolution of the star depends on its mass.

Table 1 compares stars up to four solar masses and those greater than four solar masses.

E Nuclear and quantum physics

Stars up to 4 solar masses	Stars greater than 4 solar masses
• The core temperature in these stars during the red giant phase or the supergiant phase is not large enough for fusion beyond carbon to occur. As the helium becomes exhausted, the core shrinks while still radiating. • Outer layers of the star are blown away as a planetary nebula. • Eventually, the core will have reduced to about the size of Earth and will contain carbon and oxygen ions and free electrons. • Further shrinkage is prevented because electrons cannot possess identical quantum states (the Pauli exclusion principle). The electrons provide a repulsion that counters the gravitational attraction that is attempting to collapse the star. • The star is now a white dwarf with a high density ($\approx 10^9$ kg m^{-3}) and gradually cools.	• In the red-giant or supergiant phase of these stars, the core is still large and at a high temperature so that nuclei fuse to create elements heavier than carbon. • The star ends its red-giant phase as a layered structure with elements of decreasing proton number from the centre to the outside. • Gravitational attraction is opposed by electron degeneracy pressure, but it is impossible for the star to stabilize. • With a core larger than a particular limit (1.4 solar masses), electrons and protons combine to produce neutrons and neutrinos. The star then collapses and the neutrons rush together to approach as closely as possible. The outer layers collapse inwards too, but when they meet the core, they bounce outwards again forming a **supernova**. The effects of this are to blow the outer layers away leaving what remains of the core as a **neutron star**. • Neutron degeneracy pressure opposes any gravitational collapse. • When the mass of the neutron star is sufficiently large (between 1.5 and 3.0 solar masses), then the star will collapse gravitationally, to form a **black hole**.

▲ **Table 1** Comparison between stars up to four solar masses and those greater than four solar masses

Black holes form when large neutron stars collapse. Nothing can escape from a black hole, including photons (hence the name). Matter is attracted by, and spirals into, a black hole so that the mass of the black hole increases with time.

> Observations that may confirm the existence of **black holes** include:
> - radiation emitted (X-rays), because as the matter spirals in it heats up
> - the emission of giant jets of matter by some galaxies—it is suggested that these are caused by rotating black holes
> - the modification of the trajectories of a star near a black hole by the gravitational field of the black hole
> - the gravitational waves that are emitted when two black holes in a double system spiral towards each other and merge.

Measuring astronomical distance

Distances in astronomy are very large and involve large powers of ten when measured in metres. Non-SI units are frequently used to avoid this complication. They include the light year, the astronomical unit and the parsec.

Parallax measurements are used to determine distances to the nearest stars. As the Earth moves across a diameter of its orbit over a six-month period, the positions of the nearest stars move relative to the "background" of fixed distant stars (Figure 3).

> The **light year** (ly) is the distance travelled by light in one year.
> 1 ly = 9.46×10^{15} m.
>
> The **astronomical unit** (AU) is the average distance between the Earth and the Sun. 1 AU = 1.50×10^{11} m.
>
> The **parsec** (pc) is defined using parallax angle. A star that is 1 pc from Earth will subtend a parallax angle of 1 arc-second. 1 pc = 3.26 ly.

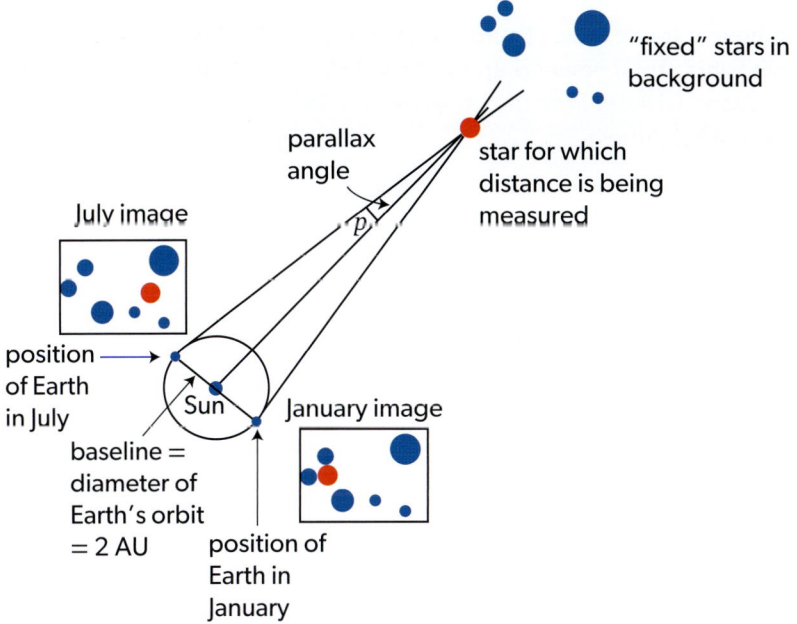

▲ **Figure 3** The basis of stellar parallax measurements

The distance across the baseline is two astronomical units (2 AU) so the parallax angle (half of the six-month variation) p is related to the distance d to the star by $d = \frac{1}{p}$, where d is in parsec and p is in arc-seconds. One arc-second is $\frac{1}{3600}$ of one degree (1°) and the arc-second symbol is ", so one milliarc-second can be written as 0.001".

Stellar parallax measurements made from the surface of Earth allow distance estimates up to about 100 pc because turbulence in the atmosphere limits the smallest angle that can be measured. When an orbiting satellite outside the atmosphere is used, the distance measured by parallax goes up to 10 000 light years (ly).

Example 2

A star has a parallax angle from Earth of 0.419 arc-seconds.

a) Outline what this parallax angle means.

b) Calculate, in light years, the distance to the star.

c) State why the stellar parallax method can only be used for stars less than a few hundred parsecs away.

Solution

a) This is half of the angle subtended by the star at Earth over a six-month period when Earth is at two extremes of its orbit.

b) For the parallax angle p:

Distance to star $d = \frac{1}{p} = \frac{1}{0.419} = 2.39 \text{ pc} \equiv 7.78 \text{ ly}$

c) For larger distances, the parallax angle becomes small and the distortions introduced by the atmosphere produce large fractional errors in the result.

Knowledge of the star–Earth distance and the apparent brightness of the star allow the radius of the star to be deduced.

A spherical star of radius R and surface temperature T will have a luminosity L (total power output) of $\sigma \times 4\pi R^2 \times T^4$. When the star is a distance d away from the observer, $L = 4\pi b d^2$, where b is the apparent brightness.

- A stellar parallax measurement gives d.
- The value of b can be measured—this gives L.
- The temperature T can be measured using the peak wavelength in the black-body spectrum.
- The same measurements can be determined for the Sun:

$$\frac{L}{L_\odot} = \frac{R^2 T^4}{R_\odot^2 T_\odot^4},$$ where $\odot$ indicates a measurement relating to the Sun;

hence $\dfrac{R}{R_\odot} = \dfrac{T_\odot^2}{T^2}\sqrt{\dfrac{L}{L_\odot}}$.

Example 3

The apparent brightness of star **X** is $4.6 \times 10^{-8}\,\text{W}\,\text{m}^{-2}$. **X** has a luminosity that is 420 times that of the Sun.

Determine, in parsec, the distance of **X** from the Sun.

Luminosity of Sun = $3.8 \times 10^{26}\,\text{W}$

Solution

Luminosity of **X** = $(3.8 \times 10^{26} \times 420)$

Rearranging $b = \dfrac{L}{4\pi d^2}$ gives:

$d = \sqrt{\dfrac{L}{4\pi b}} = \sqrt{\dfrac{3.8 \times 10^{26} \times 420}{4\pi \times 4.6 \times 10^{-8}}} = 5.3 \times 10^{17}\,\text{m}$

Use $1\,\text{ly} \equiv 9.46 \times 10^{15}\,\text{m}$ to convert from metres to light years: $\dfrac{5.3 \times 10^{17}}{9.46 \times 10^{15}} = 56\,\text{ly}$.

Use $1\,\text{pc} \equiv 3.26\,\text{ly}$ to convert from light years to parsecs: $\dfrac{56}{3.26} = 17\,\text{pc}$.

Example 4

The luminosity of Antares is 98 000 times that of the Sun. The surface temperature of Antares is 3400 K and that of the Sun is 5800 K.

a) Deduce $\dfrac{R}{R_\odot}$, where R is the radius of Antares and $R_\odot$ is the radius of the Sun.

b) Suggest the likely stellar type of Antares.

Solution

a) From $L = 4\pi\sigma R^2 T^4$:

luminosity ratio of Antares to the Sun = $\dfrac{L}{L_\odot} = \left(\dfrac{R}{R_\odot}\right)^2 \left(\dfrac{T}{T_\odot}\right)^4$

$\dfrac{R}{R_\odot} = \sqrt{\dfrac{L}{L_\odot} \times \left(\dfrac{T_\odot}{T}\right)^4} = \sqrt{98\,000 \times \left(\dfrac{5800}{3400}\right)^4} = 910$

The radius of Antares is 900 times that of the Sun.

b) Antares has a lower surface temperature than the Sun, but its radius and luminosity far exceed those of the Sun.

Therefore, Antares is a red supergiant star.

Sample student answer

Alpha Centauri is a star system that contains two stars that orbit each other. Both stars are on the main sequence.

	Alpha Centauri A	Alpha Centauri B
Luminosity	$1.5L_\odot$	$0.5L_\odot$
Surface temperature / K	5800	5300

a) Explain what is meant by the term *main sequence*. [2]

This answer could have achieved 0/2 marks:

> The main sequence is the line that joins white dwarfs and red giants on a HR diagram.

▼ The main sequence is a region of stars that predominantly fuse hydrogen to give helium. Although stars move along the main sequence during their lifetimes as their conditions change, the white dwarf and red giant regions are well away from the main sequence and these star types do not fuse hydrogen as their principal energy source.

b) (i) Calculate $\dfrac{b_A}{b_B} = \dfrac{\text{apparent brightness of Alpha Centauri A}}{\text{apparent brightness of Alpha Centauri B}}$ [2]

This answer could have achieved 0/2 marks:

> $b = \dfrac{L}{4\pi d^2}$ $\dfrac{1.5L_\odot \times 5800}{0.5L_\odot \times 5300} = 3 \times \dfrac{58}{53}$
>
> $\dfrac{b_A}{b_B} = 3.28\,\text{W m}^{-2}$

▼ The student should have used the relationship $b = \dfrac{L}{4\pi d^2}$ for the two stars. The distance d is the same for both, so $b \propto L$. The temperature is not required at this point in the question.

(ii) The luminosity of the Sun is 3.8×10^{25} W. Calculate the radius of alpha Centauri A. [2]

This answer could have achieved 0/2 marks:

> $b_A = 3.28 \times b_B$ $3.28 \times b_B = \dfrac{3.8 \times 10^{25}}{4\pi d^2}$

▼ The student is confusing the apparent brightness–luminosity relationship with the equation $L = 4\pi\sigma R^2 T^4$, which leads directly to R.

Practice problems

Problem 1
Deuterium (^2_1H) fuses with a proton (^1_1H) into a helium-3 nucleus (^3_2He) according to the reaction

$^2_1\text{H} + ^1_1\text{H} \to ^3_2\text{He} + \gamma$

The following data are available.

Atomic mass of ^2_1H = 2.014102 u

Atomic mass of ^1_1H = 1.007825 u

Atomic mass of ^3_2He = 3.016029 u

Calculate, in MeV, the energy released in the reaction.

Data for problems 2–4:
Luminosity of Sun = 3.8×10^{26} W

Surface temperature of Sun = 5800 K

Problem 2
Sirius A is a main sequence star of radius $1.7R_\odot$. The black-body radiation curve of Sirius A has a maximum at 290 nm.

a) Calculate the:
   (i) surface temperature of Sirius A
   (ii) luminosity of Sirius A. State the answer in terms of the solar luminosity $L_\odot$.

Sirius B has a surface temperature 25 000 K and luminosity $0.056L_\odot$.

b) (i) Calculate $\dfrac{\text{radius of Sirius B}}{R_\odot}$.
   (ii) Suggest the stellar type of Sirius B.

Problem 3
The luminosity of star X is 160 000 times that of the Sun and it has an apparent brightness of 8.0×10^{-9} W m^{-2}.

a) (i) Calculate, in parsecs, the distance of X from Earth.
   (ii) Explain why estimates of the distance of X from Earth using the stellar parallax method are affected by a significant uncertainty.

The surface temperature of X is 8500 K.

b) (i) Determine the radius of X. State the answer in terms of the solar radius $R_\odot$.
   (ii) State the star type of X.
   (iii) Suggest what elements are produced by fusion reactions in star X.

Another star Y has a surface temperature that is the same as X and is on the main sequence. The mass of Y is $2M_\odot$.

c) Suggest the subsequent evolution of star Y.

Problem 4
The parallax angle of star X is 0.027 arc-second and its apparent brightness is 6.0×10^{-9} W m^{-2}.

a) Calculate:
   (i) the distance, in metres, to X.
   (ii) the luminosity of X relative to the solar luminosity $L_\odot$.

The black-body radiation curve of X has a maximum at 580 nm.

b) (i) Calculate the surface temperature of X.
   (ii) Determine the radius of X.

c) Suggest the star type of X.

Tools for physics

You should be able to:
- ✔ use appropriate units for measurements and calculations, including fundamental and derived SI units
- ✔ check the consistency of equations by unit analysis
- ✔ express the results of measurements and calculations using scientific notation and SI prefixes
- ✔ apply the rules for significant figures in calculations
- ✔ make estimations and express answers to an order of magnitude when appropriate
- ✔ understand the types of errors in measurements and discuss their impact on the results
- ✔ express uncertainty in measurements as absolute, fractional and percentage uncertainty, and propagate uncertainties through calculations
- ✔ create and interpret graphs that represent relationships between physical quantities, including determining gradients, intercepts and areas under curves
- ✔ perform vector operations graphically and algebraically, including adding vectors and resolving vectors into components.

Units

Scientists need a shared language for communication between themselves and with the wider public. Part of this language involves an agreement on the units used to specify data. For example, if you are told that your journey to school has a value of 5000 then you need to know whether this is measured in metres (originally a European measure) or in fet (an old Icelandic length measure).

The agreed set of units and rules is known as the Système Internationale d'Unités (almost always abbreviated as SI). In this system, seven fundamental units are defined and all other units are derived from these. You are required to use six of the seven fundamental units—the seventh is the unit of luminous intensity, the candela, which is not used in the IB Diploma Programme physics course.

The six **fundamental units** used in the IB Diploma Programme physics course:

Measure	Unit	Abbreviation
mass	kilogram	kg
length	metre	m
time	second	s
quantity of matter	mole	mol
temperature	kelvin	K
current	ampere	A

There are many other derived units used in the course and the expression of these in fundamental units is usually given in this book when you meet the derived unit for the first time. Examples of these derived units include joule, volt, watt and pascal.

There are also some units used in the course that are not SI. Examples include $MeV\, c^{-2}$, light year (ly) and parsec (pc), which you find in Topic E.5. These have special meanings in some parts of the subject and are used by scientists in those fields. Their meaning is explained when you meet them in this book.

Often, the use of a **derived unit** avoids a long string of fundamental units at the end of a number, so $1\,V$ (volt) $= 1\,J\,C^{-1}$ $= 1\,kg\,m^2\,s^{-3}\,A^{-1}$.

Tools for physics

This format of $n.nn \times 10^n$ is known as **scientific notation** and should be used whenever possible. It can also be combined with the SI prefixes that are permitted.

Assessment tip

In physics, unless you are providing a final answer as a ratio or as a fractional difference, you must always quote the correct unit with your answer. Marks can be lost in an examination when a unit is missing or is incorrect. You should always link your answer value to its unit (together with the prefix where appropriate).

Assessment tip

Checking the consistency of units can be a powerful technique for answering multiple-choice questions in Paper 1a. When choices are presented in symbol (algebraic) form, then some, or all, of the incorrect alternatives can often be eliminated by checking the units.

Assessment tip

Many marks are lost through careless use of units in every IB Diploma Programme physics examination. Suppose a question begins "Calculate, in kg, the mass of...", if you do not quote a unit for your answer then the examiner will assume that you meant kilogram. If you worked the answer out in grams and did not say so, then you will lose marks.

Assessment tip

SI prefixes (metric multipliers) are added in front of a unit to modify its value, so 1012 s can be written as 1.012 ks. The full list of prefixes that you can use is included in the *Physics data booklet* and you can refer to it during examinations.

The SI also specifies how data in science should be written. Numbers in physics can be very large or very small. Expressing the diameter of an atom as 0.000 000 000 12 m is unhelpful — 1.2×10^{-10} m is much clearer.

Approaches to learning

It is possible to use units — or more accurately, dimensions — to check equations. This example uses an equation that is known to be correct: the time period for a mass–spring system (Topic C.1).

The prediction is that $T = 2\pi\sqrt{\dfrac{m}{k}}$. This dimensional analysis cannot say anything about the constant (2π) but it can check that the equation is consistent.

The single dimension of the left-hand side of the equation is **time**. The right-hand side contains a **mass** term and a **spring constant** term. So, in unit terms, the equation suggests: $[\text{time}] \equiv \sqrt{\dfrac{[\text{mass}]}{[\text{spring constant}]}}$

Turning these into SI units makes the relationship $[s] \equiv \sqrt{\dfrac{[\text{kg}]}{[\text{N m}^{-1}]}}$.

The newton is the unit of force or $[\text{mass}] \times [\text{acceleration}]$, so the right-hand side can be rewritten and simplified:

$$\sqrt{\dfrac{[\text{kg}]}{[\text{kg m s}^{-2}][\text{m}^{-1}]}} \equiv \sqrt{\dfrac{[\cancel{\text{kg}}]}{[\cancel{\text{kg}}\,\cancel{\text{m}}\,\text{s}^{-2}][\cancel{\text{m}^{-1}}]}} \equiv \sqrt{\dfrac{1}{[s^{-2}]}} \equiv \sqrt{[s^2]} \equiv [s]$$

This is the same unit as the left-hand side, so the variable part of the equation is confirmed.

Use the idea of dimensions to help you to consolidate your understanding of physics.

Prefix	Symbol	Factor	Decimal number
deca	da	10^1	10
hecto	h	10^2	100
kilo	k	10^3	1000
mega	M	10^6	1 000 000
giga	G	10^9	1 000 000 000
tera	T	10^{12}	1 000 000 000 000
peta	P	10^{15}	1 000 000 000 000 000
deci	d	10^{-1}	0.1
centi	c	10^{-2}	0.01
milli	m	10^{-3}	0.001
micro	μ	10^{-6}	0.000 001
nano	n	10^{-9}	0.000 000 001
pico	p	10^{-12}	0.000 000 000 001
femto	f	10^{-15}	0.000 000 000 000 001

There are some rules here too.

- Only one prefix is allowed per unit, so it would be incorrect to write 2.5 μkg instead of 2.5 mg.
- You can put one prefix per fundamental unit, so 0.33 Mm ks^{-1} would be acceptable for 330 m s^{-1} (the speed of sound in air) but nowhere near as meaningful.

Significant figures (s.f.) can lead to confusion. It is important to distinguish between significant figures and decimal places (d.p.). For example:
- 2.38 kg has 3 s.f. and 2 d.p.
- 911.2 kg has 4 s.f. and 1 d.p.

> The rule for the number of significant figures in a calculated answer is clear:
>
> Specify the answer to the **same number of s.f.** as the quantity in the question with the *smallest* **number of s.f.**

Example 1

A snail travels a distance of 33.5 cm in 5.2 minutes.

Calculate the speed of the snail.

State the answer to an appropriate number of significant figures.

Solution

The answer, to 7 s.f., is 1.073718×10^{-3} m s^{-1}.

It is incorrect to quote the answer to this precision as the time is only given to 2 s.f. (the fact that 5.2 minutes is 312 s is not relevant). The appropriate answer is 1.1×10^{-3} m s^{-1} (or if you prefer 1.1 mm s^{-1}).

Assessment tip

In Example 1, rounding up is needed. You should do this for every calculation—but only at the very end of the calculation. Rounding answers mid-solution leads to inaccuracies that may take you out of the allowed tolerance (allowed range) for the answer. This is particularly true in Paper 1b. Keep all possible s.f. in your calculator until the end and only make a decision about the s.f. in the last line. In Example 1, an examiner would be very happy to see:

"… = $1.073\,718 \times 10^{-3}$ m s^{-1} so the speed of the snail is 1.1×10^{-3} m s^{-1} (to 2 s.f.)."

as your working is then completely clear.

Estimations

Sometimes estimations are required in physics. This may be because:
- an educated guess is needed for all or some of the quantities in a calculation
- there is an assumption involved in a calculation.

Often it is appropriate to express your answer to an order of magnitude, which means that it is rounded to the nearest power of ten. The best way to express any order of magnitude answers is as 10^n, where n is an integer.

Example 2

Estimate the number of air molecules in a room.

Solution

The calculation is left for you, but use the following steps.

- Estimate the volume of a room by making an educated guess at its dimensions, in metres.
- The density of air is about 1.3 kg m^{-3}—call it 1 kg m^{-3} to make the numbers easy later.
    - *From the volume and density you can estimate the mass of gas in the room.*
- The mass of 1 mol of oxygen molecules is 32 g and 1 mol of nitrogen is 28 g—call the answer 30 g for both gases combined.
    - *This enables you to estimate the number of moles.*
- Each mole contains 6×10^{23} molecules.
    - *Now you can work out the answer.*

Assessment tip

When you see the command term "Estimate" in the examination, this may be because:
- you either lack some or all data for your calculation so that an educated guess is needed
- there is a significant element of assumption required in the question
- you are expected to make a gross simplification of the physics at some point.

It will be clear which of these apply.

In estimation questions, such as Example 2, make it clear what numbers you are providing for each step and how they fit into the overall calculation.

Orders of magnitude

You may see order of magnitude answers in Paper 1a (the multiple-choice paper) written as a single integer. When the response is, say, 7, this means 10^7.

It is also permissible to talk about "a difference of two orders of magnitude"—this means a ratio of 100 (or 10^2) between the two quantities.

Tools for physics

> **Random errors** are unpredictable changes in data collected in an experiment. Examples include any fluctuations in a measuring instrument or changes in the environmental conditions where the experiment is being carried out.
>
> **Systematic errors** are often produced within measuring instruments. Suppose that an ammeter always shows a reading of +0.1 A when there is no current between the meter terminals. This means that every reading made using the meter will read 0.1 A too high. The effect of a systematic error may produce a non-zero intercept on a graph when a line through the origin is expected.

Errors and uncertainties

All measurement is prone to error. Two basic types of error are implicit in the data you collect: random errors and systematic errors.

Random errors lead to an uncertainty in a value. One way to assess their impact on a measurement is to repeat the measurement several times and then use half of the range of the outlying values as an estimate of the absolute uncertainty.

When two or more measurements have to be combined by addition, multiplication or raising one or more quantities to a power, then their uncertainties have to be combined correctly. This is called the propagation of uncertainty.

> Uncertainty in measurement is expressed in three ways.
>
> - **Absolute uncertainty**: the numerical uncertainty associated with a quantity.
>   For example, when a length of quoted value 5.00 m has an actual value somewhere between 4.95 m and 5.05 m, the absolute uncertainty is ± 0.05 m. The length will be expressed as (5.00 ± 0.05) m.
> - **Fractional uncertainty** = $\dfrac{\text{absolute uncertainty in quantity}}{\text{numerical value of quantity}}$.
>   A fractional uncertainty has no unit.
> - **Percentage uncertainty** = fractional uncertainty × 100 and is expressed as a percentage. There is no unit.

Example 3

Five readings of the length of a small table are made. The data collected are:

0.972 m   0.975 m   0.979 m   0.981 m   0.984 m

a) Calculate the average length of the table.

b) Estimate, for the length of the table, its:

   (i) absolute uncertainty

   (ii) fractional uncertainty

   (iii) percentage uncertainty.

Solution

a) The average length is:
$$\frac{(0.972+0.975+0.979+0.981+0.984)}{5} = 0.978(2)\,\text{m}$$

b) (i) The largest and smallest values of the data are 0.972 and 0.984, which differ by 0.012 m. Half of this value is 0.006 m and this is taken to be the absolute uncertainty.

   The length should be expressed as (0.978 ± 0.006) m.

   (This absolute error is an estimate. Another estimate is the standard deviation of the set of measurements, which, in this case, is 0.004 m. 0.006 m is thus an overestimate.)

   (ii) The fractional uncertainty is $\dfrac{0.006}{0.9782} = 0.006(13) = 0.006$.

   This is a ratio of lengths and has no unit.

   (iii) The percentage uncertainty is 0.006 × 100 = 0.6%.

Combining uncertainties

You will often need to combine quantities mathematically: for example, a pair of lengths, both with uncertainty, may need to be added to give a total length. This **derived quantity** will also have an uncertainty.

Suppose we have two quantities p and q. These have absolute uncertainties Δp and Δq, respectively.

- When the quantities are **added** or **subtracted** the **absolute uncertainties are added.**

  The absolute uncertainty in r, which is the sum $p + q$, is $\Delta r = \Delta p + \Delta q$.

  The absolute uncertainty in s, which is the difference $p - q$, is $\Delta s = \Delta p + \Delta q$.

- When the quantities are **multiplied** or **divided** the **fractional uncertainties are added.**

  The fractional uncertainty in t, where $t = p \times q$, is $\dfrac{\Delta t}{t} = \dfrac{\Delta p}{p} + \dfrac{\Delta q}{q}$.

  The fractional uncertainty in u, where $u = \dfrac{p}{q}$, is $\dfrac{\Delta u}{u} = \dfrac{\Delta p}{p} + \dfrac{\Delta q}{q}$.

- When a quantity is **raised to a power** n, the **fractional uncertainty is multiplied by** n.

  The fractional uncertainty in v, where $v = p^n$, is $\dfrac{\Delta v}{v} = n \times \dfrac{\Delta p}{p}$. (This follows because, for example, $v = p^2 \equiv p \times p$ and so $\dfrac{\Delta v}{v} = n \times \dfrac{\Delta p}{p} = \dfrac{\Delta p}{p} + \dfrac{\Delta p}{p}$ using the addition of fractional uncertainties rule). When a power is negative, this rule still applies, but the fractional uncertainty should be multiplied by the absolute value of the exponent.

> **Assessment tip**
>
> Absolute uncertainties are normally rounded to one significant figure. However, when the leading digit is 1, the uncertainty can be given to two significant figures.
>
> Make sure that you match the precision of the data and the precision of the uncertainty. For example, when the uncertainty is 0.2 m, the data value should be quoted as 4.8 m rather than 4.80 m.

Example 4

A block of wood has sides of length (180 ± 5) cm and (60 ± 3) cm.

a) Calculate the perimeter of the block.

b) Calculate the area of the block.

A cube of the wood has sides of length (0.20 ± 0.01) m.

c) Calculate the volume of the cube including its uncertainty expressed as a percentage.

Solution

a) $180 + 180 + 60 + 60 = 480$ m. The absolute uncertainty is $5 + 5 + 3 + 3 = 16$ cm.

   The perimeter is (480 ± 16) cm or (4.8 ± 0.2) m.

b) The area is $1.8 \times 0.60 = 1.08 \, \text{m}^2$.

   The two fractional uncertainties of the measurements are $\dfrac{0.05}{1.8} = 0.028$ and $\dfrac{0.03}{0.6} = 0.050$.

   The sum of these is 0.078 and this is the fractional uncertainty of the answer.

   The absolute uncertainty in the area is $0.078 \times 1.08 = 0.084 \, \text{m}^2$.

   The answer should be expressed as $(1.08 \pm 0.08) \, \text{m}^2$ or $(1.1 \pm 0.1) \, \text{m}^2$.

c) The volume is (length)^3 and is $0.20^3 = 8.0 \times 10^{-3} \, \text{m}^3$.

   The fractional uncertainty in length is $\dfrac{0.01}{0.20} = 0.05$ so the fractional uncertainty in the volume is 0.15.

   The percentage uncertainty in the volume is 15%.

   The volume of the cube is $(8.0 \times 10^{-3} \, \text{m}^3 \pm 15\%)$.

Tools for physics

It is possible that data points, all with an associated error, are presented on a graph. Therefore, there are errors associated with the gradient and any intercept on the graph.

The way to treat these errors is described on page 235.

Sample student answer

The refractive index of a glass microscope slide is measured.

A travelling microscope is used to determine the position x_1 of a mark on a sheet of paper. The slide is placed over the mark and the position x_2 of the image of the mark when viewed through the slide is determined. Finally, the microscope is used to determine the position x_3 of the top of the slide.

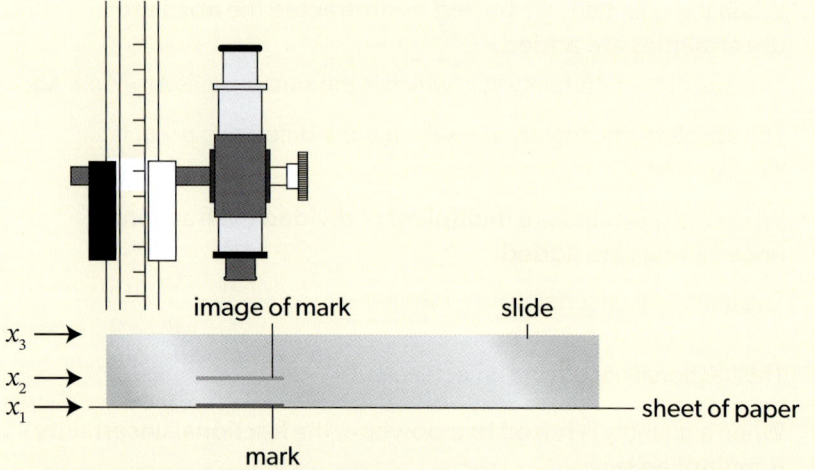

The table shows the average results of a large number of repeated measurements.

	Average position of mark / mm
x_1	0.20 ± 0.02
x_2	0.59 ± 0.02
x_3	1.35 ± 0.02

a) The refractive index of the glass from which the slide is made is given by: $\dfrac{x_3 - x_1}{x_3 - x_2}$.

Determine:

i) the refractive index of the glass to the correct number of significant figures, ignoring any uncertainty. [1]

This answer could have achieved 1/1 marks:

$$\dfrac{1.35 - 0.2}{1.35 - 0.59} = \dfrac{1.15}{0.76} = 1.51$$

▲ After an initial error which is condoned (the value of x_1 is 0.20), the remainder of the working is carried through correctly.

ii) Determine the uncertainty of the value calculated in part (i). [3]

This answer could have achieved 2/3 marks:

$$\dfrac{\Delta n}{n} = \dfrac{\Delta x}{x_3 - x_1} + \dfrac{\Delta x}{x_3 - x_2} = \dfrac{0.02}{1.15} + \dfrac{0.02}{0.76}$$

$\Delta n = 1.5(0.0437)$ where 1.5 is the refractive index n.

$\Delta n \approx \pm 0.0655 = \pm 0.1 = \Delta n$

▼ There are subtractions to find the differences in the image position but the uncertainties need to be obtained correctly. As this is a subtraction, the absolute uncertainties need to be added. The absolute errors in $x_3 - x_1$ and $x_3 - x_2$ are 0.02 + 0.02 not 0.02 alone as the solution indicates.

b) After the experiment, it is found that the travelling microscope is badly adjusted so the measurement of each value of x is too large by 0.05 mm.

232

Outline the effect that this error will have on the calculated value of the refractive index of the glass. [2]

This answer could have achieved 2/2 marks:

It will have no effect because this error will be cancelled out once $n = \dfrac{x_3 - x_1}{x_3 - x_2}$ is calculated.

▲ The point is that the refractive index is calculated by *differences*. Each measurement in $\dfrac{x_3 - x_1}{x_3 - x_2}$ is changed by the same amount and the overall ratio does not change. This could have been expressed more clearly in the answer.

Using graphs

Communication using a graph is essential in many sciences—graphs allow relationships between data points to be grasped visually and quickly.

It helps if you know the best graph to plot to display your data. First, consider how best to show your data as a straight line (in the form $y = mx + c$, where y and x are the data pairs, m is the gradient of the line and c is the intercept on the y-axis). This may involve algebraic manipulation.

Example 5

A simple pendulum is suspended from the ceiling of height D above the floor. There is no access to the point of its suspension from the ceiling. The period T of oscillation of the pendulum is measured as a function of the vertical distance h from the floor to the pendulum bob.

Suggest a suitable graph to display the data.

Solution

The length of the pendulum is $D - h$.

The equation for the period (see Topic C.1) is $T = 2\pi\sqrt{\dfrac{l}{g}} = 2\pi\sqrt{\dfrac{D-h}{g}}$.

A plot of T against h will not be a straight line.

However, squaring both sides and rearranging gives $T^2 = -\dfrac{4\pi^2}{g}h + \dfrac{4\pi^2}{g}D$.

This is now in the form $y = mx + c$.

So a plot of T^2 against h will give a straight line of gradient $-\dfrac{4\pi^2}{g}$ with an intercept on the y-axis of $\dfrac{4\pi^2}{g}D$.

Once you have collected data and chosen a suitable graph, the graph can be plotted.

Here are some guidelines for drawing graphs.
- Plotted graphs need labels and units on both axes—for example, "speed / m s^{-1}". The "/" does not mean "divide"; it means "measured in". You should write units in the form m s^{-1} rather than m/s.
- Scales on the axes should be straightforward using 1 : 2, 1 : 5 or 1 : 10. The ratios 1 : 3, 1 : 2, 1 : 6, 1 : 7 and 1 : 9 should *never* be used. Only use 1 : 4 when absolutely necessary.
- Always fill as much of the printed grid as possible. The minimum range for your plots should be half of the grid area—when you can double the scale, do so. This may mean using a false origin—that is, one that does not begin at (0, 0).
- Data points should be marked clearly using × or +. A single dot can be easily missed. If you must use a dot, put it in a circle ⊙.

Tools for physics

> **Assessment tip**
>
> A common question is "Draw the best-fit line". This does not mean draw the best-fit straight line, as the best-fit line could be curved with the data points distributed evenly about it.

> **Assessment tip**
>
> You should be completely familiar with the operation of your calculator so that you can carry out calculations without any slips during your routine work and especially in examinations. One particular point is to be careful to check whether you are in degree or radian mode for calculations involving trigonometry.

- All markings on a graph should be in sharp, black pencil (so that you can erase mistakes). As a guideline, if your pencil line is as thick as the thickest lines on the grid, your pencil is not sharp enough!
- Draw all straight lines with a ruler (preferably transparent).
- Use freehand for curves. Practise the curve several times first, drawing with your hand inside the curve and then draw the line in one continuous movement.
- Try to get the same number of points on both sides of the line, minimizing the total distance of all points from the line. Do not force your line to go through the origin unless you have a very good reason to distort the position of your line in this way.

Modern calculators have internal programs that can determine the slope and intercept of a line given the data points. Many calculators can do this—if yours can, then use this facility. Another quick way to find the best straight line is to divide your data into two groups: upper and lower (Figure 1). Find the separate x and y averages of both groups and plot these two points. Join them to give your straight line. The gradient can also be easily obtained from the two mean values.

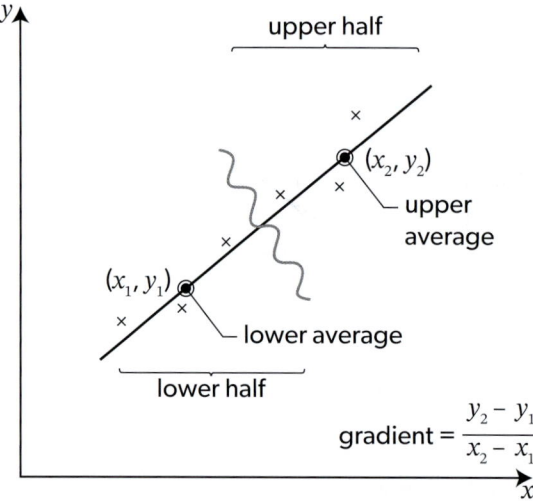

▲ Figure 1 Divide your data plots into two groups to find the approximate best-fit line

> **Assessment tip**
>
> Draw the line for which you need to calculate a gradient as long as possible on the grid. Then use at least half of this line for the determination. The longer the line, the smaller the fractional uncertainty in your gradient result.

> **Assessment tip**
>
> Where there is a false origin, never extend a grid by drawing to obtain an intercept from the grid—the axis lines do not show the true origin. Calculate the intercept using similar triangles or by substitution into $y = mx + c$.

Once your graph is plotted, much more information is available to you. Remember that, besides being a way of visualizing your data, a graph is an averaging technique from which you can obtain more information.

- The **gradient of a straight line** can be found by calculating $\dfrac{\text{change in } y\text{-coordinates}}{\text{change in } x\text{-coordinates}}$.
- The **gradient at a point on a curve** can be found by drawing the tangent at the point concerned and calculating the gradient of this new straight line. (To draw this tangent in the laboratory or for your IA, use a small plane mirror and align it so that, when looking into it, the curve and its reflection appear continuous. The mirror is then at 90° to the curve. Drawing along the mirror will give you the normal at this point—a protractor gives the tangent line directly from this.)
- The **area under a graph** can be found algebraically or, when the line is straight, it can be made into a series of triangles and rectangles. When it is a curve, the area can be found by counting squares and determining the area value of each square.
- The **intercept** on either axis can be found by directly reading it off or it can be calculated using trigonometry when there is a false origin. This will almost certainly involve extrapolation of your plotted line to the intercept.

Every data point on your graph has an error associated with it. When the absolute error in the point is much less than the size of the grid, you need do nothing further. However, when the absolute uncertainty is larger than the size of the smallest square, then you need to consider the use of error bars.

When every point on a graph has an error bar, it is possible to construct the maximum and minimum gradients. The construction is shown in Figure 2. The two lines must lie on the outer edges of the relevant error bars. Once they have been drawn, the two gradients can be calculated in the usual way. The absolute uncertainty in the gradient is found using $\dfrac{\text{maximum gradient} - \text{minimum gradient}}{2}$.

This is similar to the calculation of an absolute error from the overall range of data for a particular datum point. The absolute uncertainty in the gradient can then be used in further error calculations.

Similar ideas will apply to the intercepts on the axes. In Figure 2, the intercept on the y-axis is $(108 \pm 5)\,\text{m s}^{-1}$.

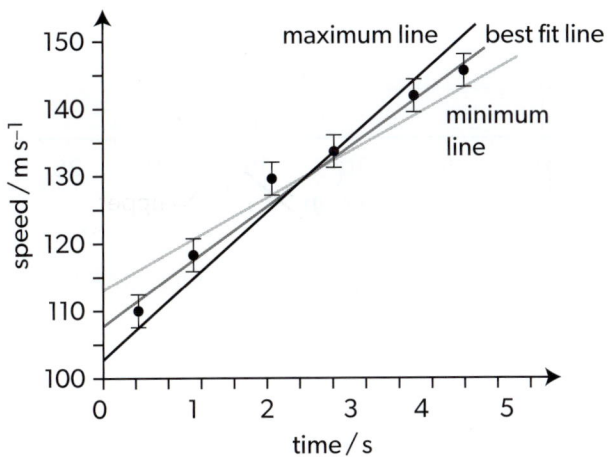

▲ Figure 2 Maximum and minimum gradients on a graph just touch the tops and bottoms of the error bars

Spreadsheet software, such as Excel, has graph-plotting tools that can quickly and accurately draw points and error bars, and carry out graphical calculations. It is worthwhile teaching yourself how to use this software early in your IB Diploma Programme physics course: it will save time and effort throughout your studies—and well beyond them.

Remember that no graphing software is available to you during written examinations, so make sure that you can draw a graph neatly by hand and that you have the skills to *calculate* gradients and intercepts quickly and accurately using your calculator.

> An **error bar** is an I-shape or H-shape centred on the datum point. The width or height of the bar indicates the error in the measurement. The nature of the experiment will determine whether each point on the graph has the same or different absolute errors.

Approaches to learning

Sometimes in an experiment you will find that a datum point is well away from the line or curve. This is a strong argument for plotting a rough graph on a grid while the apparatus is still set up and available. Points that are obviously away from the trend can be re-checked and then, if necessary, re-plotted.

Assessment tip

Always look closely at the origin of graphs printed in examination booklets. When there is a false origin, take this into account. When the origin is shown as (0, 0), this may be important in a multiple-choice or other question. When no marking is shown at the point where the x-axis and y-axis cross, this is a signal that you should only pay attention to the shape of the curve because this may be a false origin.

Tools for physics

Sample student answer

A student suggests that the relationship between I and x is given by

$$\frac{1}{\sqrt{I}} = Kx + KC$$

where K and C are constants.

Data for I and x are used to plot a graph of the variation of $\frac{1}{\sqrt{I}}$ with x.

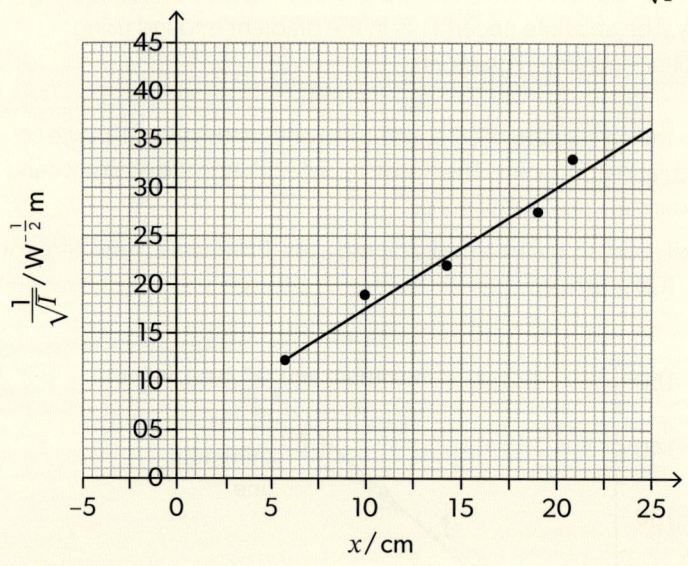

a) Estimate C. [2]

This answer could have achieved 0/2 marks:

> y intercept is (0, 5)
>
> C = 5 m

▼ The y-intercept is correct. However, the candidate was asked for the value of C. When $\frac{1}{\sqrt{I}} = 0$, $x = -C$ and this is the intercept on the x-axis.

b) Determine P to the correct number of significant figures including its unit, where $P = \frac{4\pi}{K^2}$. [4]

This answer could have achieved 4/4 marks:

> The gradient $K = \frac{30 - 5}{20 - 0} = 1.3 \, W^{-\frac{1}{2}}$ by considering (0, 5) and (20, 30).
>
> $K = 2\sqrt{\frac{\pi}{P}}$ so $\frac{K}{2} = \sqrt{\frac{\pi}{P}}$
>
> $\frac{K^2}{4} = \frac{\pi}{P} \Rightarrow P = \frac{4\pi}{K^2}$
>
> Therefore, $P = \frac{4\pi}{K^2} = 8.0 \, W \, m^{-2} \, cm^2$
>
> $= 8.0 \times 10^{-4} \, W \, m^{-2} \, m^2$
>
> $= 8.0 \times 10^{-4} \, W$

▲ There are a number of good points about this answer: it is clear and well presented. The candidate makes the source of the data for the gradient clear, and the data points concerned are far apart and on the line. The calculations are correct and the unit and number of significant figures are correct too.

c) Explain the disadvantages that a graph of I versus $\frac{1}{x^2}$ has for the analysis in parts (a) and (b). [2]

This answer could have achieved 1/2 marks:

> Since the slope will be $\frac{1}{K^2}$, it will be too small to measure from the graph to the right number of significant figures.
>
> y intercept will be too small → hard to measure c.

▼ There is some credit given for the point about the gradient. However, this is not the whole story: the equation becomes $I = \frac{1}{K^2(x+C)^2}$ so, unless C is known, the graph cannot be plotted to give a straight line.

236

Vectors and scalars

Quantities you meet in IB Diploma Programme physics course are either scalars or vectors. (There is a third type of physical quantity that is not used in the course.)

A vector can be represented by a line with an arrow. When drawn to scale, the length of the line represents the magnitude and the direction is as drawn.

Both scalars and vectors can be added and subtracted. Scalar quantities add just as any other number in mathematics. With vectors, however, you need to take the direction into account.

Figure 3 shows vector addition by drawing. The vectors must be drawn to the same scale and the direction angles must also be drawn accurately. A further construction produces the parallelogram with the red solid and dashed lines. Then the magnitude of the new vector $v = v_1 + v_2$ is given by the length of the diagonal (blue) vector with the direction as shown.

> **Scalars** are quantities that have magnitude (size) but no direction. They generally have a unit associated with them.
>
> **Vectors** are quantities that have both magnitude and a physical direction. A unit is associated with the number part of the vector.

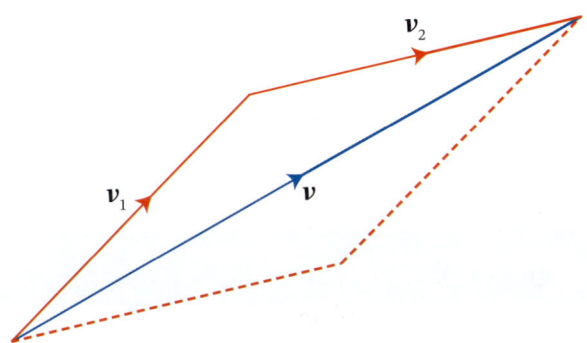

▲ **Figure 3** Adding two vectors v_1 and v_2 to give the sum v

Vectors can also be added algebraically. The most common situation you meet in the IB Diploma Programme physics course is when the vectors are at 90° to each other (Figure 4). As before, addition by drawing gives the resultant (red) vector which is the sum of $v_1 + v_2$. Algebraically, the use of trigonometry gives the magnitude of the resultant (added) vector as $\sqrt{v_1^2 + v_2^2}$ and its direction as $\theta = \tan^{-1}\left(\dfrac{v_2}{v_1}\right)$.

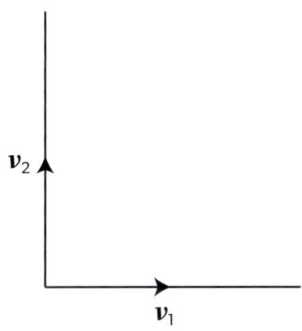

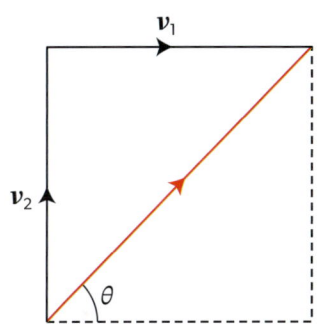

▲ **Figure 4** Adding two vectors that are at right angles

Example 6

A girl walks 500 m due north and then 1200 m due east. Calculate her position relative to her starting point.

Solution

This is similar to the situation in Figure 4 where the first vector has a magnitude of 500 m and the second has a magnitude of 1200 m.

The magnitude of the resultant is $\sqrt{500^2 + 1200^2} = 1300$ m.

θ is $\tan^{-1}\left(\dfrac{500}{1200}\right) = 22.6°$.

Another skill required is that of breaking a vector down into two components at right angles to each other—this is known as resolving the vector. A right angle is chosen because the two resolved components will be independent of each other. Figure 5 shows the process.

The vector F points diagonally upwards at angle θ to the horizontal. The length of F is the hypotenuse of the right-angled triangle. The other sides have lengths $F\cos\theta$ and $F\sin\theta$.

You can now add or subtract any non-parallel vectors algebraically. Figure 6 shows the notation used for this.

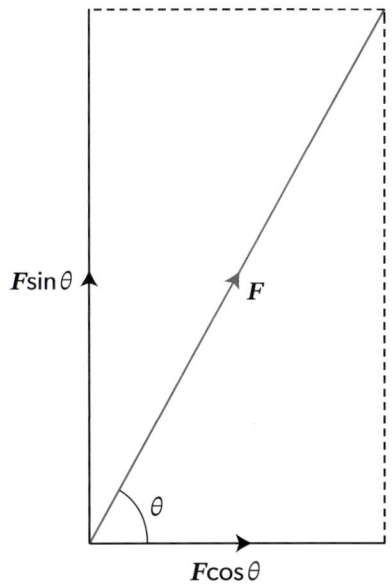

▲ **Figure 5** Resolving a vector into two perpendicular components

Tools for physics

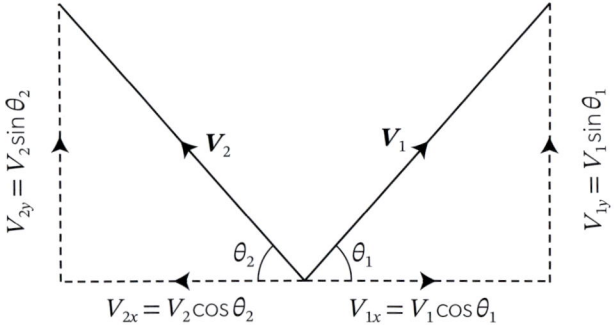

▲ Figure 6 The notation used for showing how vectors can be added algebraically

Horizontally, the addition gives $V_x = V_{1x} + V_{2x}$, which is $V_1 \cos\theta_1 - V_2 \cos\theta_2$. Vertically, the addition gives $V_y = V_{1y} + V_{2y}$, which is $V_1 \sin\theta_1 + V_2 \sin\theta_2$. These new vector components can be added to give the magnitude of new vector $V = \sqrt{V_x^2 + V_y^2}$ with an angle to the horizontal of $\theta = \tan^{-1}\left(\dfrac{V_y}{V_x}\right)$.

To subtract two vectors, simply form the negative vector of the one being subtracted (by reversing its original direction but leaving the length unchanged) and add this to the other vector.

Example 7

An object moves with a velocity 40 m s^{-1} at an angle N30°E. Determine the component of the velocity in the direction:

a) due east

b) due north.

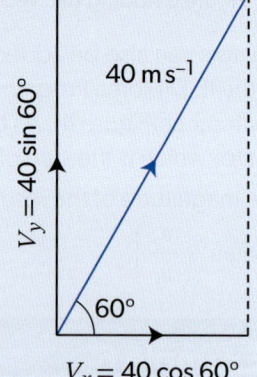

Solution

a) The angle between the vector and east is 60°.
So the component due east
= 40 cos 60° = 20 m s^{-1}.

b) Due north, the component is 40 cos 30°
= 40 sin 60° = 34.6 m s^{-1}.

Example 8

A girl cycles 1500 m due north, 800 m due east and 1000 m in a south-easterly direction. Calculate her overall displacement.

Solution

A drawing of the journey is shown.

Total horizontal component of displacement
= 800 + 1000 cos 45° = 1510 m

Total vertical component = 1500 − 1000 cos 45°
= 790 m.

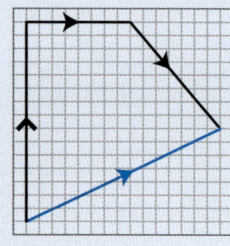

The displacement is $\sqrt{1510^2 + 790^2} = 1700$ m at $\tan^{-1}\left(\dfrac{790}{1510}\right) = 28°$.

Internal Assessment

During your course you carry out an internal assessment (IA) project. This project takes around ten hours and accounts for 20% of your total mark in the IB Diploma Programme physics assessment.

The type of project that you do is flexible. It might be:
- experimental data collection from laboratory work
- database investigation and analysis
- spreadsheet processing with data you have collected
- the investigation of an experimental simulation.

Collaboration between students is allowed in the early stages of the IA when data are being collected. However, you must carry out the data analysis independently. Your teacher can explain the rules surrounding collaboration by students.

To ensure that you earn as many of the available marks as possible, you should pay particular attention to the four assessment criteria, given in Table 1.

Assessment criterion	Weighting
Research design	25%
Data analysis	25%
Conclusion	25%
Evaluation	25%

▲ Table 1 The four assessment criteria for the IA

To see how these work in practice, you may be allowed to see examples of IA work produced in your school in previous years. If so, look carefully at the quality of work that gained high marks and, in contrast, identify why students who gained low marks scored poorly. Remember that you are seeing the final report after two weeks or so of work and some of the developmental thinking may be missing. However, a good report should reflect the refinements in both thinking and methodology that occurred as the project progressed.

The assessment criteria reflect the inquiry cycle for science (Figure 1).

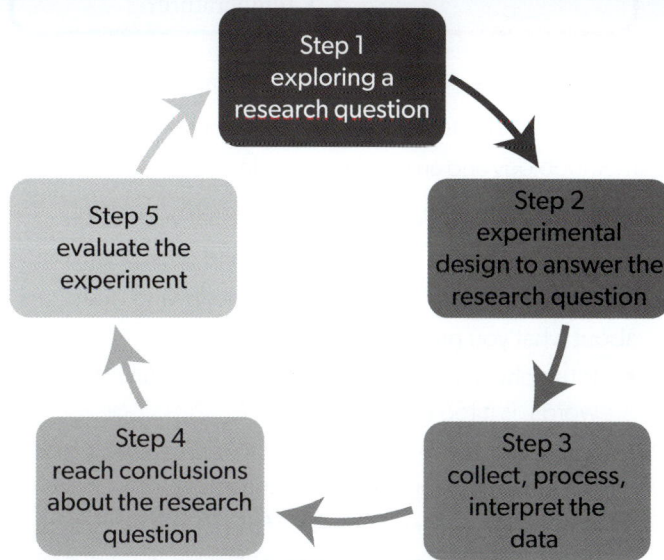

▲ Figure 1 The scientific inquiry cycle in five steps

Step 1: Exploring your research question

You need to decide on a research question. It must be focused and identify all the issues linked to it. The physics concepts and techniques should arise naturally from the research question and not be forced into it.

You should have a high level of engagement with your IA. You may show creativity in your thinking, or express scientific concepts in a personal and engaged way. A project based on a personal interest may be more likely to succeed. For example, if you play a woodwind instrument, you might investigate some aspect of the physics of this instrument. The key is to think creatively.

Sometimes, the aspects of your research will be developed in laboratory work. Alternatively, there may be suitable data on the internet or published in books. Look at what others have done, but do not use anyone else's work (words or data) without giving full credit to the author. Where possible, try to carry out preliminary experiments before you start your IA properly. This preliminary work should appear with evaluative comments in your final report.

When you have chosen your topic, make sure that your question is well defined.

> ▼ What happens when spheres drop through water?

> ▲ How does the terminal speed of a sphere dropping through water vary with temperature?

Of these two research questions, the one on the right is far better. This is because it leads immediately to a methodology and an experimental design.

Your question should clearly identify what you will be investigating, with the independent and dependent variables expressed. Ask yourself the following questions about what you propose to do.

- Is the physics of an appropriate standard (in other words, is it too easy or too hard)?
- Can you achieve your plans in the time available and still leave time for a thorough analysis?
- Is there scope for refinement of the IA as it progresses?
- Are your proposed experiments likely to lead to firm conclusions?
- Will you be as engaged with the project and its outcomes on the last day as on the first?

If you can answer "yes" to all of these, then your research task is suitable.

Planning an IA involves much more than just choosing the apparatus and the experimental methodology. You should make predictions from the outset and modify these in the light of your work.

> **Assessment tip**
>
> An **independent variable** is one that is changed and manipulated by you.
>
> A **dependent variable** is the one that you will measure or observe.
>
> **Control variables** are the ones that are held constant during the measurement.
>
> **Continuous variables** are ones that can take an infinite number of values (such as an ammeter reading).
>
> **Discrete variables** can only take fixed values (such as the surface on which a block slides: for example, glass, sandpaper or concrete).

Step 2: Design to answer your research question

If you choose an experimental topic, ensure that your research question will allow the collection of appropriate data. Consider the following.

- What are your independent variable(s) and your dependent variable(s)?
- Will the topic allow you to access a continuous dependent variable with a sensible independent variable? If you are forced to have one or more discrete variables, will this provide enough variety for 10 hours' work?
- Can your variables be made to vary over a wide range so that there is a significant difference between the lowest and highest values?
- What will you keep constant and how will you achieve this?
- Are the variables continuous or discrete?
- How will you vary the independent variable and measure it? How will you measure the dependent variable?
- Is the experimental method safe in all respects? Can safety be improved?
- What do you predict as the outcome? Is this informed by a sensible theory?
- What graphs will help you to know that the prediction is correct/incorrect? A negative result is just as significant as one that confirms your prediction. An IA does not "fail" because the research question has not been confirmed.

> **Assessment tip**
>
> For an experimental IA:
> - make a prediction and know what graphs/display techniques will demonstrate it
> - make measurements safely and display them clearly and effectively
> - manipulate data accurately and use them to confirm or reject your prediction and allow this to lead you to a new, refined prediction.

If you choose database analysis, spreadsheet processing or the investigation of an experimental simulation, then you should use similar questions to arrive at your choice of research question.

When following a spreadsheet or database pathway, you should ensure that it will provide the breadth of material that you need.

- For example, can you confirm that the variables used by the author were independent, and that the control variables are clearly defined?
- Are there sufficient data for your use?

When you choose the simulation route, are you using a simulation that you can modify easily? Are you constructing the simulation using a published writing package or is the simulation provided by another author? Will you write your own code for the simulation or construct your own interactive worksheet? Are your skills sufficient for this? As before, will the simulation provide the breadth of material and data that you need for ten hours' of work?

Step 3: Collecting the data

Here are some guidelines that you should consider in the data-collection phase of your project.
- Keep accurate daily notes of your work, including all references as you find them. Many working scientists use hard-backed notebooks, but if you keep an electronic record on a computer or tablet, make sure that you back up your work at the end of *every* session.
- Spend some time during each work session reviewing the work you have done. Remember that IB learners are reflective. Carry out the analysis of the day's data, plot any graphs from these data, and plan out the objectives and targets for the following session. It may be appropriate to do these analyses at home as laboratory time during a project is precious.

Step 4: Reaching conclusions

When reaching conclusions about your data:
- compare the outcome with the original prediction
- when appropriate, make use of a spreadsheet program for analysing the data and offering a possible equation to model them.

It is important to undertake a full analysis of the errors and uncertainties in your data. Remember that errors may be random or systematic (or include both in the same measurement). Look carefully at the apparatus and the data to decide on the magnitude of these errors. Make realistic estimates of them where possible.

Data from any experiment or investigation need to be processed, analysed and interpreted. The data-analysis criterion focuses on this aspect. The research question must lead to a detailed and valid answer. There must be enough data to justify your conclusion and for you to form a view of the experimental uncertainties in your data.

> **Assessment tip**
>
> In essence, the analysis must be directed towards answering the research question.

Graphs with a curved trend show the basics of how one variable varies with the change in another. Straight-line graphs give more information than this, including the exact nature of the relationship. Manipulate your data, if at all possible, to produce a straight-line trend. Look beyond the obvious and see if there are any hidden trends.

> **Assessment tip**
>
> Here are some possible transformations that allow you to obtain a straight-line trend.
>
> | Predicted relationship | Plot as | | Equation | Gradient | Intercept on y-axis | Notes |
> |---|---|---|---|---|---|---|
> | | y-axis | x-axis | | | | |
> | $y = kx + c$ | y | x | $y = kx + c$ | k | c | |
> | $y = kx^2$ | $\sqrt{y}$ | x | $\sqrt{y} = x\sqrt{k}$ | $\sqrt{k}$ | origin | Plotting this way reduces the impact of errors in x |
> | $y = kx^2 + c$ | y | x^2 | $y = kx^2 + c$ | k | c | |
> | $xy = c$ | y | $\frac{1}{x}$ | $y = \frac{c}{x}$ | c | origin | |
> | $y = x^n$ | $\ln y$ | $\ln x$ | $\ln y = n \ln x$ | n | origin | Only applicable when $x > 0$ |
> | $y = Ae^{kx}$ | $\ln y$ | x | $\ln y = \ln A + kx$ | k | $\ln A$ | The constant A must be positive |

Step 5: Evaluating the experiment

It is important to evaluate an experiment, whether you are a student or a professional scientist. Consider the quality of the work (experimental or analytical) and any issues that arose as you completed it. If possible, compare your results with accepted values or the published work of other scientists. If you recognize problems in your work, discuss these and suggest how they could be overcome (if you or someone else was undertaking the work again). Scientists should seek to identify any shortcomings in their work—the ability to do this is a strength not a weakness.

The internal assessment in DP science is about your ability to carry out a short project with competence from beginning to end, rather than your ability to invent new science—that is unlikely to happen. You *must* provide your teacher with evidence of your progress in the project.

To aid your evaluation, consider the following strategies.
- Write down everything day by day (you can edit it later).
- Consider producing a rough preliminary report draft about three-quarters of the way through the work. This will allow you to identify simple improvements when you still have time to implement them.
- Give details of preliminary work and show how your ideas developed.
- Make sure that your comments are aligned with your data and graph plots.
- Consider whether your safety precautions could be enhanced.
- Draw attention to anomalies in the results and, if possible, the reasons for them.
- Discuss the extent to which your results support your conclusion, which should be strongly linked to your research question.

Finally, be critical of your own work. There is no such thing as perfect science. A good scientist is self-critical.

▼ Poor evaluations use:
- simple statements that just repeat the trend or state that it is hard to draw a conclusion
- descriptions of practical issues that had to be overcome, rather than consideration of fundamental issues of the experimental method.

▲ Good evaluations:
- focus on the extent to which the research question can be answered
- discuss the limitations of the experimental method
- give realistic suggestions for improving the method
- identify any preliminary work that could have been carried out.

Assessment tip

Your evaluation must focus strongly on the research question and the extent to which you feel you have been able to answer it.

Scientific writing should be concise and effective. Aim for the highest standard of presentation that you can achieve. The main part of the text should be about 6–12 pages with a limit of 3000 words. Large quantities of data can be placed in an appendix—do not include large amounts of data in the main text as they will disrupt the flow.

A report should contain:
- a statement to put the work in context
- a statement of the research question
- an outline of the physics underlying the IA
- details of any preliminary work
- a complete account of the work you did—use an appropriate format for this, which may depend on the nature of your data collection
- full results (these may need to be in an appendix) and a complete analysis of the results that is linked to the physics quoted earlier in the report
- a conclusion that presents the findings clearly and unambiguously together with your evaluation and reflections on the whole investigation.

Assessment tip

Diagrams are helpful—they can and should be used where appropriate. They should be fully annotated.

Internal Assessment

> **Assessment tip**
>
> The 3000 word count for your IA report does not include:
> - charts and diagrams
> - data tables
> - equations, formulas and calculations
> - citations/references
> - the bibliography
> - headers.

You are allowed, and encouraged, to use secondary sources in your report, but you must make sure that you correctly reference any work that is not your own. It is academic theft to take someone else's work and pass it off as your own. Although the IB does not recommend a particular reference style, one common way to cite scientific material is the Harvard method.

In your main text, add a citation using the following format: (Homer and Piętka, 2024, p. 243). This reference cites this particular page in this book. Web references are similar: give the name of the author and the date if possible (for example, (Garza, 2014)).

Any text or image that is produced by artificial intelligence (AI) tools is *not* considered to be your own work. If you include any AI-generated content in your report, you must acknowledge this in the body of the text and then reference it correctly in the bibliography. The reference must include the date on which the content was created together with the prompt that you gave the AI tool.

Always consult your teacher for the latest IB Diploma guidance on using AI in your assessed work.

At the end of your report, you must list all the references in full, usually in alphabetical order. For web references, include the date when you accessed the reference.

> **REFERENCES**
>
> Garza, Celina. *Academic honesty – from principles to practice*. Oct. 2014. Viewed 15 March 2024. <https://www.ibo.org/contentassets/71f2f66b529f48a8a61223070887373a/academic-honesty.-principles-into-practice---celina-garza.pdf>
>
> Homer, D. R. and Piętka, M. 2024, *IB Prepared Physics*, 2nd edn. Oxford University Press.

Checklist

You can use this checklist to ensure that you have addressed all the marking criteria.

Research design	Tick (✓)
Have I **justified** my research question and the topic in general?	
Is the context of my research described in **specific** and **appropriate** ways?	
Is my data-collection method **relevant to the project** and will it yield sufficient data?	
Have I provided enough information so that someone could repeat my work?	
Have I identified and clearly stated the **independent, dependent and control variables**?	
Have I included these variables in the research question?	
Data analysis	
Is the way I recorded and processed the data described **clearly** and **precisely**?	
Have I made my **identification** and **treatment** of errors and uncertainties clear?	
Is my data processing **accurate** and **appropriate** to the research question?	
Have I identified a **trend line** in the processed data?	
Conclusion	
Is my conclusion **relevant** to the research question?	
Is my conclusion **consistent** with the data analysis that I made?	
Does my conclusion **explicitly address** any trends present in the processed data?	
Evaluation	
Have I identified the relative impact of any **weaknesses and limitations** in my investigation?	
Have I described and explained **improvements** to the investigation that are relevant to the weaknesses and limitations?	

Practice Exam Papers

At this point, you will have re-familiarized yourself with the content from the topics and options of the syllabus of the IB Diploma Programme physics course. Additionally, you will have picked up some key techniques and skills to refine your exam approach. It is now time to put these skills to the test; in this section, you will find practice examination papers 1 and 2, with the same structure as the external assessment you will complete at the end of the course.

Answers to these papers are available at **www.oxfordsecondary.com/ib-prepared-2e**.

Paper 1

SL: 1 hour 30 minutes [Paper 1A and Paper 1B]

HL: 2 hours [Paper 1A and Paper 1B]

Instructions to candidates

- For each question, choose the answer you consider to be the best and indicate your choice on the answer sheet (provided at **www.oxfordsecondary.com/ib-prepared-2e**).
- A calculator is required for this paper.
- A clean copy of the **Physics data booklet** is required for this paper.
- The maximum mark for SL papers 1A and 1B is **[45 marks]**.
- The maximum mark for HL papers 1A and 1B is **[60 marks]**.

Section A

*SL candidates: answer questions 1–25 **only**.*

*HL candidates: answer **all** questions.*

1. The acceleration of an object varies linearly from $10\,\text{m s}^{-2}$ to zero in a time of $5.0\,\text{s}$.

    What is the change in speed of the object?

    A. $2.0\,\text{m s}^{-1}$
    B. $5.0\,\text{m s}^{-1}$
    C. $25\,\text{m s}^{-1}$
    D. $50\,\text{m s}^{-1}$

2. A lift is used to raise 25 boxes through a vertical height of $20\,\text{m}$ in $50\,\text{s}$. Each box has a mass of $50\,\text{kg}$.

    What is the output power of the lift?

    A. $0.49\,\text{kW}$
    B. $4.9\,\text{kW}$
    C. $25\,\text{kW}$
    D. $250\,\text{kW}$

3. A ball falls freely from rest and gains a momentum p in time t.

    What is the relationship between p and t?

    A. $p = \text{constant} \times t^{-0.5}$
    B. $p = \text{constant} \times t^{0.5}$
    C. $p = \text{constant} \times t$
    D. $p = \text{constant} \times t^2$

4. Two springs P and Q obey Hooke's law. When a force F acts on spring P, P extends by a distance x and stores an energy E.

    Spring P has a spring constant k.

    When a force $\dfrac{F}{2}$ acts on spring Q, Q extends by a distance $2x$.

    What is the spring constant of Q and what is the energy stored in Q when it extends by $2x$?

	Spring constant of Q	Energy stored in Q
A.	$\dfrac{k}{4}$	E
B.	$4k$	$\dfrac{E}{2}$
C.	$\dfrac{k}{4}$	$\dfrac{E}{2}$
D.	$4k$	E

5. A block of mass m and initial velocity v collides elastically with another identical block of initial velocity $-v$.

   What are the total momentum and the total kinetic energy of the blocks after the collision?

	Total momentum	Total kinetic energy
A.	0	0
B.	0	mv^2
C.	$2mv$	0
D.	$2mv$	mv^2

6. The useful power output of an engine is 30 kW. The engine is 20% efficient and consumes a volume of fuel equal to 5.0×10^{-6} m³ every second.

   What is the energy density of the fuel?

   A. 1.5×10^7 J m⁻³
   B. 3.0×10^7 J m⁻³
   C. 1.5×10^{10} J m⁻³
   D. 3.0×10^{10} J m⁻³

7. A heater transfers a power P to a liquid for a time t. During this time, the temperature of the liquid changes by $\Delta\theta$. The specific heat capacity of the liquid is c.

   What is the mass of the liquid?

   A. $\dfrac{Pt}{c\Delta\theta}$
   B. $\dfrac{Pc\Delta\theta}{t}$
   C. $\dfrac{Pc}{t\Delta\theta}$
   D. $\dfrac{P\Delta\theta}{ct}$

8. A black body emits electromagnetic radiation that has a peak frequency of f_{peak} and a maximum intensity of I_{peak}.

   The temperature of the black body is increased.

   What are the peak frequency and the maximum intensity after the temperature increase?

	Peak frequency	Intensity
A.	$>f_{peak}$	$<I_{peak}$
B.	$<f_{peak}$	$>I_{peak}$
C.	$>f_{peak}$	$>I_{peak}$
D.	$<f_{peak}$	$<I_{peak}$

9. The table gives data for two gas samples P and Q. Both P and Q can be assumed to be ideal.

Ideal gas	Molecular mass	Temperature / K
P	m	T
Q	$0.5m$	$4T$

   The average (translational speed)² of the molecules of P is v^2.

   What is the average (translational speed)² of the molecules of Q?

   A. $\dfrac{v^2}{2}$
   B. $2v^2$
   C. $8v^2$
   D. $64v^2$

10. Three identical lamps are connected to a cell so that all the components are in parallel. The emf of the cell is 6.0 V and it has a negligible internal resistance. The lamps are designed to operate when the potential difference across them is 6.0 V.

    One lamp breaks and stops conducting.

    What happens to the other two lamps?

    A. The lamps burn out immediately.
    B. The current in them decreases.
    C. The current in them stays the same.
    D. The current in them increases.

11. Four resistors, each of 3.0 Ω resistance, are connected in a network.

    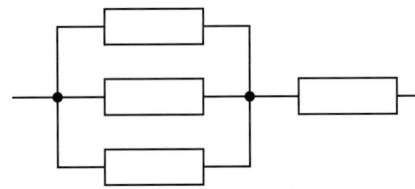

    What is the total resistance of this network?

    A. 0.75 Ω
    B. 3.0 Ω
    C. 4.0 Ω
    D. 12 Ω

12. A battery with an emf of 12 V and an internal resistance r is connected across a resistor of resistance R.

    The potential difference across the resistor is 10 V.

    What is the value of r in terms of R?

    A. $\dfrac{R}{6}$
    B. $\dfrac{R}{5}$
    C. $5R$
    D. $6R$

13. A mass–spring system X has a mass m and a spring constant k. It oscillates with simple harmonic motion with a time period T.

    Another mass–spring system Y has a mass $4m$ and a spring constant $\dfrac{k}{4}$.

    What is the time period of Y?

    A. $\dfrac{T}{4}$
    B. $\dfrac{T}{2}$
    C. T
    D. $4T$

14. A travelling wave moves on a string. The graph shows how the displacement of the string varies with distance along it.

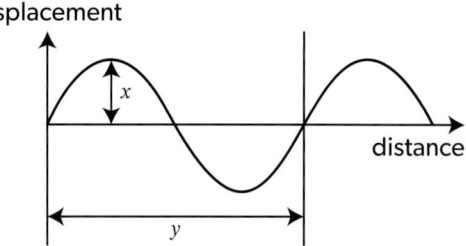

What quantities do x and y represent?

	x	y
A.	half-amplitude	period
B.	half-amplitude	wavelength
C.	amplitude	period
D.	amplitude	wavelength

15. The refractive index for light passing from air into medium A is 1.4.

    The refractive index for light passing from air into medium B is 1.7.

    Which statement is correct?
    A. Light travelling from A into B will refract away from the normal.
    B. Light travels slower in A than in B.
    C. The speed of light in A is about $0.59c$.
    D. The critical angle for light travelling from B into A is about 55°.

16. A string is stretched between fixed points P and S. P and S are 1.00 m apart. Q and R are positions on the string.

    PQ = 0.20 m   PR = 0.60 m

    When the string is sounding the first harmonic, Q and R oscillate in phase.

    What is the next harmonic for which Q and R oscillate in phase?
    A. second   C. fourth
    B. third   D. fifth

17. Planet X is a uniform sphere of radius R. The gravitational field strength at the surface of X is g.

    What is the mean density of X?
    A. $\dfrac{3g}{4\pi RG}$ C. $\dfrac{3G}{4\pi gR}$
    B. $\dfrac{3RG}{4\pi g}$ D. $\dfrac{3Rg}{4\pi G}$

18. An electron moves through a distance of 0.10 m parallel to the field lines of a uniform electric field of strength $2.0\,kN\,C^{-1}$.

    What is the work done on the electron?
    A. 0   C. 3.2×10^{-17} J
    B. 1.6×10^{-17} J   D. 1.6×10^{-20} J

19. Two chambers are shown from above. Magnetic fields act in the chambers. The magnetic field strengths are different in each. The magnetic field directions are vertical in both chambers.

    A charged particle enters the first chamber moving on a horizontal plane and is deflected in a circular path of radius r_1. The particle then enters the second chamber and follows a circular path of radius $\dfrac{r_1}{2}$. The particle leaves the second chamber with speed v.

    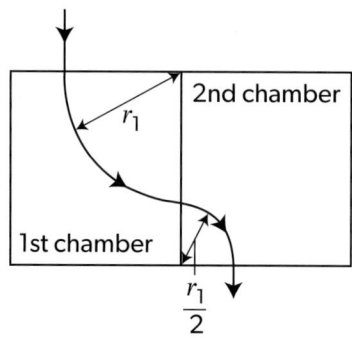

    What was the initial speed of the particle as it entered the first chamber?
    A. $\dfrac{v}{2}$ C. $2v$
    B. v D. $4v$

20. Electrons in a horizontal beam are moving due north.

    What is the magnetic field direction vertically above the beam due to the motion of the electrons?
    A. to the west
    B. to the east
    C. vertically upwards
    D. vertically downwards

21. A radioactive nuclide has a half-life of $T_{\frac{1}{2}}$. A pure sample of the nuclide has an initial uncorrected count rate of 7.6 counts s^{-1}. The background count rate in the laboratory is 0.40 counts s^{-1}.

    What is the uncorrected count rate of the sample after $3T_{\frac{1}{2}}$?
    A. 0.90 counts s^{-1} C. 1.0 counts s^{-1}
    B. 0.95 counts s^{-1} D. 1.3 counts s^{-1}

22. Radioactive Pb-211 decays through a series of decays to form Pb-207. In this series, only alpha particles and beta-minus particles are emitted.

    What numbers of alpha particles and beta-minus particles are emitted?

	Number of alpha particles	Number of beta-minus particles
A.	1	1
B.	1	2
C.	2	1
D.	2	2

23. Which of these materials can be used as a moderator in a nuclear reactor?
    A. boron   C. uranium
    B. concrete   D. water

24. What is the equation for the overall hydrogen cycle in the Sun?

   A. $6_1^1H \rightarrow {}_2^4He + 2{}_1^1H + 2\beta^+ + 2\gamma + 2\nu_e$
   B. $6_1^1H \rightarrow {}_2^4He + 2{}_1^1H + 2\beta^+ + 2\gamma + 2\bar{\nu}_e$
   C. $6_1^1H \rightarrow {}_2^4He + 2{}_1^1H + 2\beta^- + 4\gamma + 2\nu_e$
   D. $6_1^1H \rightarrow {}_2^4He + 2{}_1^1H + 2\beta^+ + 4\gamma + 2\bar{\nu}_e$

25. The radius of star X is 100 times that of the Sun. The two stars have the same surface temperature.

   The luminosity of the Sun is $L_\odot$.

   What is the luminosity of X?

   A. $10^{-4} L_\odot$
   B. $10^{-2} L_\odot$
   C. $10^{2} L_\odot$
   D. $10^{4} L_\odot$

The following questions are for HL candidates only.

26. An object rotates from rest about a fixed axis. The moment of inertia about this axis is 5.0×10^{-3} kg m^2.

   The constant angular acceleration is 0.25 rad s^{-2}. The object rotates twice about the axis.

   What is the final angular momentum of the object?

   A. 0
   B. 2.5×10^{-2} kg m^2 s^{-1}
   C. 1.3×10^{-2} kg m^2 s^{-1}
   D. 3.1×10^{-2} kg m^2 s^{-1}

27. Sphere P rotates around an axis through its centre. P has an angular velocity ω and a rotational kinetic energy E.

   Sphere Q has the same density as P and also rotates about an axis through its centre. Q has half the radius of P and its angular velocity is 4ω.

   The moment of inertia of a sphere of radius r and mass m rotating about an axis through its centre is $\frac{2}{5}mr^2$.

   What is the rotational kinetic energy of Q?

   A. $\frac{E}{2}$
   B. E
   C. $2E$
   D. $4E$

28. A particle travels at a speed of $0.80c$ relative to an observer. In the reference frame of the observer, the particle travels a distance of 2.5 km between two points X and Y.

   What, in the reference frame of the particle, is the distance between X and Y?

   A. 1.5 km
   B. 2.0 km
   C. 2.5 km
   D. 4.2 km

29. Energy is transferred to a system by heating and work is done on the system.

   Which transfers must result in an increase in the internal energy of a system?

	Energy input to the system	Work done on the system
A.	negative	positive
B.	positive	negative
C.	positive	positive
D.	negative	negative

30. A simple pendulum is displaced and released from rest at time $t = 0$ to complete one oscillation.

   How does the kinetic energy E_k of the system vary with t?

   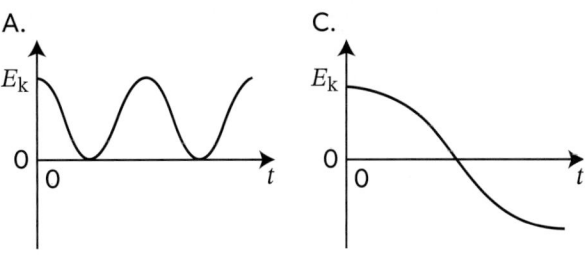
   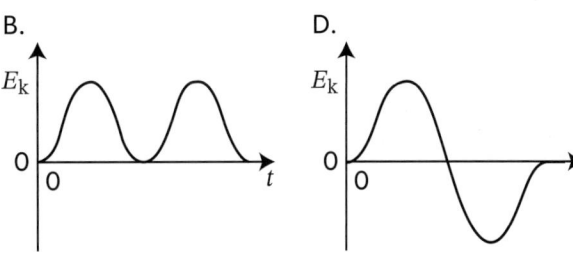

31. A diffraction pattern is produced when monochromatic light is incident on a single slit of width d.

   The width of the slit is changed to $2d$.

   What are the changes to the width of the central maximum and the maximum intensity of the diffraction pattern?

	Width of central maximum	Intensity of central maximum
A.	decreases	increases
B.	decreases	decreases
C.	increases	increases
D.	increases	decreases

32. A line in a spectrum has a wavelength of 500 nm when measured on Earth. This line is observed on Earth after emission from a galaxy that is moving towards Earth at a speed of $0.10c$.

   What is the observed wavelength?

   A. 50 nm
   B. 450 nm
   C. 500 nm
   D. 550 nm

33. The gravitational potential at the surface of spherical planet X is $-V$.

   Spherical planet Y has double the radius and double the density of X. Both planets have uniform density.

   What is the gravitational potential at the surface of Y?

   A. $-\dfrac{V}{4}$
   B. $-4V$
   C. $-8V$
   D. $-16V$

34. Lines of electric equipotential and electric field lines
    A. are straight when the field is radial.
    B. are curved when the field is radial.
    C. meet at 90° for a uniform field.
    D. meet at 90° for a field that is uniform or radial.

35. A solid conducting sphere of radius 0.40 m is charged. The electric field strength E varies with distance d from the surface of the sphere as shown.

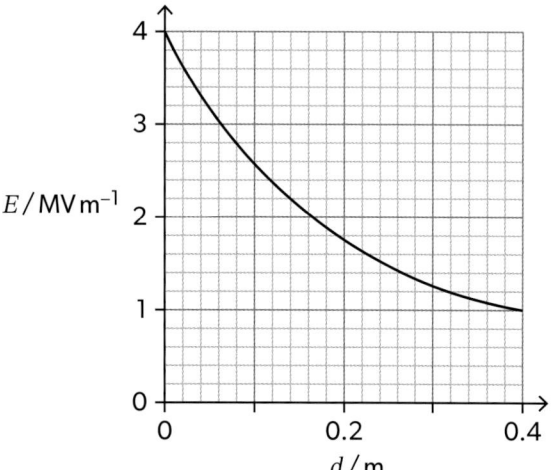

What is the potential difference between the centre of the sphere and a point 0.40 m above the surface of the sphere?
    A. 800 kV
    B. 1.6 MV
    C. 2.0 MV
    D. 6.0 MV

36. The plane of a coil PQRS is parallel to a uniform magnetic field.

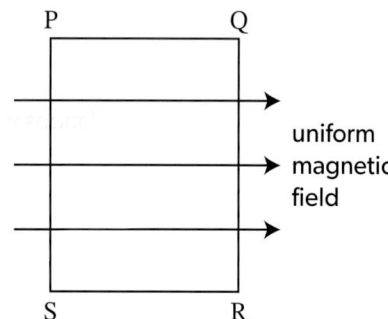

What change will cause an emf to be induced in the coil?
    A. translation in the direction PQ
    B. translation in the direction PS
    C. rotation about PQ
    D. rotation about PS

37. The diagram of the energy levels of an atom is drawn to scale.

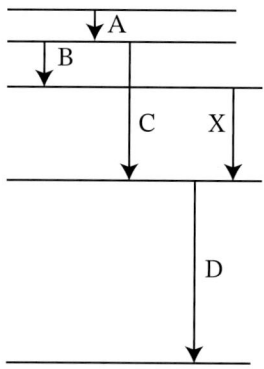

Transition X leads to the emission of a photon of wavelength λ.

Which transition leads to the emission of a photon of wavelength 2λ?

38. The radius of an oxygen-16 nucleus is R and its density is ρ.

What is the radius and density of the nucleus of hydrogen-2?

	Radius	Density
A.	$\dfrac{R}{8}$	ρ
B.	$\dfrac{R}{8}$	$\dfrac{\rho}{2}$
C.	$\dfrac{R}{2}$	ρ
D.	$\dfrac{R}{2}$	$\dfrac{\rho}{2}$

39. An electron is travelling at 3% of the speed of light.

What is the de Broglie wavelength of the electron?
    A. 2×10^{-14} m
    B. 8×10^{-14} m
    C. 2×10^{-11} m
    D. 8×10^{-11} m

40. A radioactive nuclide has an activity X_0 at time $t = 0$.

When:
- $t = 1$ hour the activity is X_1
- $t = 2$ hour the activity is X_2
- $t = 3$ hour the activity is X_3.

What is $\dfrac{X_0}{X_2}$?

    A. $\dfrac{X_0}{8}$
    B. $\dfrac{X_0}{2}$
    C. $\dfrac{X_2}{X_3}$
    D. $\dfrac{X_1}{X_3}$

Section B

*SL candidates: answer **all** questions.*

*HL candidates: answer **all** questions.*

1. An annulus is the region between two concentric circles.

   A thin sheet in the shape of an annulus is shown placed on a sheet of graph paper that has a one-millimetre grid.

   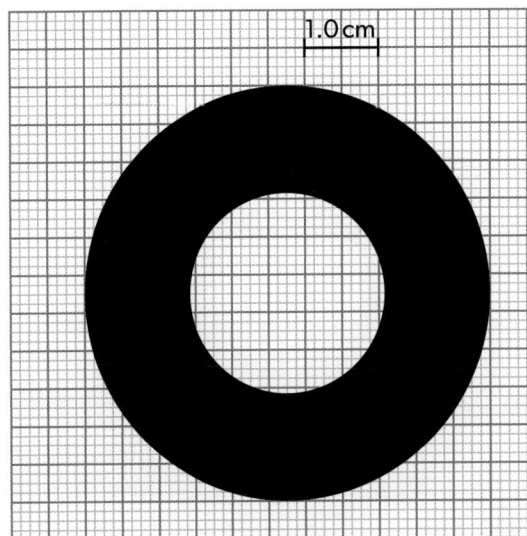

   (a) Calculate, giving its percentage uncertainty, the outer radius of the annulus. [3]

   (b) The inner radius of the annulus is (13 ± 1) mm. Calculate the area of the annulus. Give the absolute uncertainty of your answer. [3]

2. Two groups of students want to measure the height H of their laboratory.

   Both groups suspend a simple pendulum from the ceiling. The pendulum bob is a distance x from the laboratory floor.

   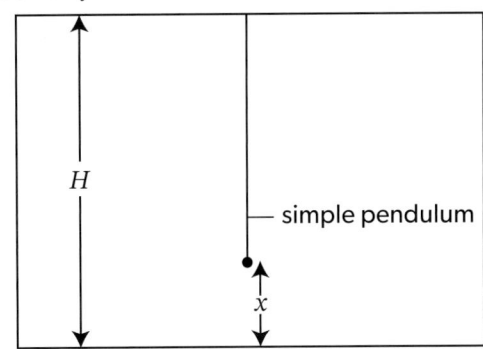

   One group of students measures the time period T of the simple pendulum for various values of x.

   (a) Show that $T^2 = kH - kx$, where $k = \dfrac{4\pi^2}{g}$. [1]

   These students plot a graph to show how T^2 varies with x.

   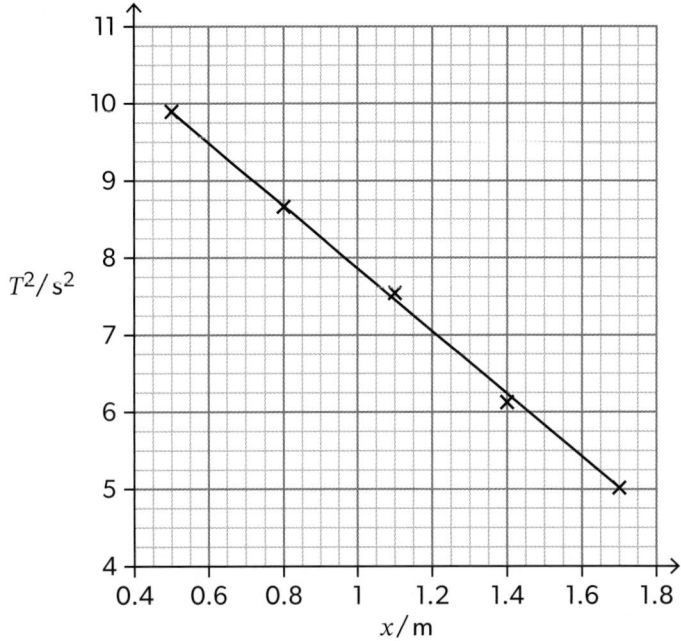

   (b) Determine, using the graph:
   (i) g [3]
   (ii) H. [3]

   (c) The students who plotted the graph measured each T by timing 20 oscillations of the pendulum.

   A second group of students measured T for only one value of x and used the result to calculate H. This group measured 100 oscillations of the pendulum. The students assumed a value for g to calculate H.

   The same total number of oscillations was used by both groups of students.

   Explain why the graphical analysis is a better approach than the single measurement of T. [2]

3. The variation of resistance R with current I is shown for an electrical device.

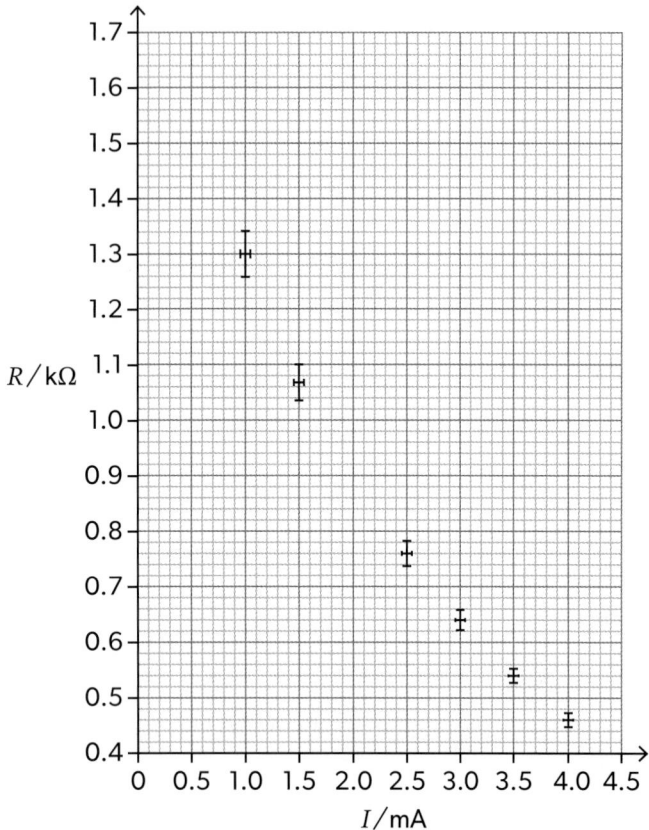

Error bars are given on the graph.
(a) Draw the best-fit line for these data. [1]
(b) Estimate R when:
   (i) $I = 2.0$ mA [1]
   (ii) $I = 0.50$ mA. [1]
(c) Explain which of these estimates is more reliable. [2]

Paper 2

SL: 1 hour 30 minutes
HL: 2 hours 30 minutes

Instructions to candidates
- For each question, choose the answer you consider to be the best and indicate your choice on the answer sheet (provided at **www.oxfordsecondary.com/ib-prepared-2e**).
- A calculator is required for this paper.
- A clean copy of the **Physics data booklet** is required for this paper.
- The maximum mark for SL paper 2 is **[50 marks]**.
- The maximum mark for HL paper 2 is **[90 marks]**.

SL candidates: answer questions 1–6 only.

HL candidates: answer questions 1–5 and 7–10 only.

Question 1
(a) Outline **two** ways in which a standing wave differs from a travelling wave. [2]

(b) A string of length 1.8 m is fixed at both ends. The string is made to vibrate in its third-harmonic mode.
   (i) Draw the standing wave produced. Label all the nodes present. [2]
   When the frequency is increased by 23 Hz, the fourth-harmonic standing wave forms on the string.
   (ii) Determine the speed of the wave on the string. State the answer to an appropriate number of significant figures. [3]

Question 2
An asteroid orbits the Sun in a circular orbit of radius 4.2×10^{11} m.
(a) Calculate, in years, the orbital period of the asteroid. [2]
(b) (i) State the fundamental SI units of gravitational field strength. [1]
   (ii) Determine the ratio $\dfrac{\text{gravitational field strength at the orbit of Earth}}{\text{gravitational field strength at the orbit of the asteroid}}$. [2]

Question 3
The following data are given about the star Pollux.

Parallax angle = 0.0965 arc second
Apparent brightness = 9.75×10^{-9} W m^{-2}
Peak wavelength of the black-body spectrum = 632 nm

(a) Calculate, in m, the distance to Pollux. [1]
(b) Calculate, in W, the luminosity of Pollux. [1]
(c) Determine the radius of Pollux. [3]
The luminosity of the Sun $L_\odot = 3.83 \times 10^{26}$ W.
(d) Plot, on the HR diagram, the position of Pollux. [1]

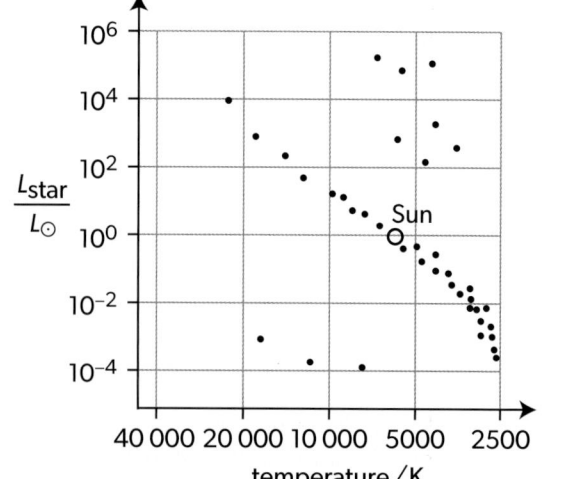

Question 3(e) is for HL candidates only.

(e) Deduce which has a greater core temperature: Pollux or the Sun. [2]

Question 4

A closed container of fixed volume 1.2×10^{-3} m³ holds 0.050 mol of an ideal gas. The initial temperature of the gas is 500 K.

(a) Calculate the initial pressure of the gas. [1]

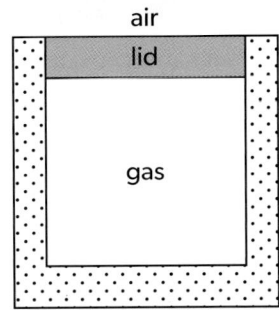

The container has insulated walls and a thermally conducting lid. The lid has a surface area of 8.5×10^{-3} m³, thickness 1.4 cm and is made of a material of thermal conductivity 0.16 W m⁻¹ K⁻¹.
The air above the lid has a constant temperature of 280 K.

---

*Question 4(b) is for SL candidates **only**.*

(b) Calculate the initial rate of thermal energy transfer through the lid. [1]

*Question 4(c) is for HL candidates **only**.*

(c) Consider a system consisting of the gas in the container and the surrounding air. Determine the initial rate at which the entropy of the system increases. [3]

(d) Explain, with reference to the kinetic theory of gases, why the pressure of the gas in the container decreases. [3]

When a thermal energy of 21 J is transferred from the gas, the temperature of the gas decreases by 10 K. The molar mass of the gas is 44 g mol⁻¹.

(e) Calculate the specific heat capacity of the gas. [2]

Question 5

(a) (i) State what is meant by the binding energy of a nucleus. [1]

(ii) The binding energy per nucleon can be graphed against the nucleon number A. State and explain which feature of this graph predicts that energy is released in nuclear fusion of nuclei with small A. [2]

Plutonium-240 ($^{240}_{94}$Pu) decays with an emission of an alpha particle into uranium (U).

(b) (i) Calculate the neutron number of the uranium nucleus produced in the decay. [1]

(ii) The following data are given about nuclear masses.

Nucleus	Mass / u
Plutonium	240.0022
Uranium	235.9951
Alpha particle	4.0015

Show that the energy released in the decay is about 5 MeV. [2]

(c) (i) Explain how Newton's third law of motion leads to the idea of conservation of momentum in alpha decay. [3]

(ii) Show that the kinetic energy of the alpha particle emitted in alpha decay of plutonium-240 is about 98% of the total energy released in the decay. [3]

(iii) Calculate the speed of the alpha particle. [2]

(d) The alpha particle now enters a region of a uniform electric field between a pair of parallel metal plates. The path of the alpha particle is initially parallel to, and midway between, the plates.

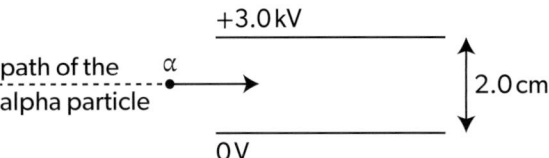

The potential difference between the plates is 3.0 kV and the plates are separated by 2.0 cm.

(i) Calculate the electric field strength in the region between the plates. [1]

(ii) F_e is the electric force on the alpha particle and F_g is the gravitational force. Determine the ratio $\frac{F_e}{F_g}$. [2]

(iii) Calculate the acceleration of the alpha particle. [1]

(iv) Outline the subsequent path of the alpha particle. [2]

---

*Question 6 is for SL candidates **only**.*

Question 6

An object is released, from rest, in air above the Earth's surface.

The graph shows the variation of the speed v of the object with time t. Air resistance is not negligible.

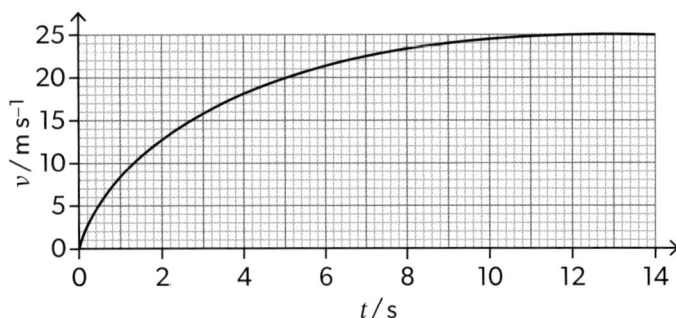

(a) Explain how the magnitude of the acceleration of the object varies with time. [2]

(b) Estimate the distance from its starting point at which the object reaches its terminal speed. [3]

SL paper ends here.

Questions 7–10 are for HL candidates only.

Question 7
(a) State what is meant by electric potential at a point. [2]

The electric potential at the surface of a charged conducting sphere is +6.0 kV. The radius of the sphere is $R = 2.0$ cm.

(b) Calculate the charge of the sphere. [2]

(c) Sketch, on the axes, a graph to show how the electric potential V varies with distance r from the centre of the sphere. [2]

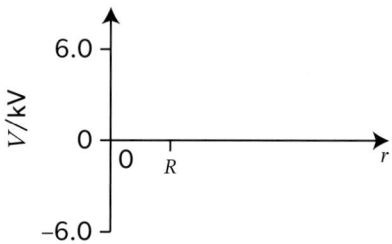

A point charge +2.6 nC is placed at a distance of 5.0 cm from the centre of the sphere and released.

(d) Calculate the work done by the electric field in moving the point charge to infinite separation from the sphere. [1]

Question 8
Monochromatic light of wavelength 532 nm is incident normally on a diffraction grating that has 600 lines per mm.

(a) Calculate the angle between the central maximum and the first-order maximum of the diffracted light. [2]

(b) Deduce the number of maxima present in the diffracted light. [2]

The light source is replaced by one that emits white light.

(c) Describe the diffraction pattern after the change. [2]

Question 9
No photoelectron emission is observed from a particular metal surface when monochromatic light incident on it is below a certain minimum frequency.

(a) Outline why the wave theory for light cannot explain this observation. [2]

Radiation of frequency 7.4×10^{14} Hz is incident on the metal surface. The maximum kinetic energy of the emitted photoelectrons is 1.0 eV.

(b) Determine the threshold frequency of the metal. [4]

The intensity of the radiation is increased.

(c) State and explain the effect of this change on the maximum kinetic energy of the photoelectrons. [2]

Question 10
A solid cylinder of radius R and mass M rolls without slipping down a ramp that makes an angle of 25° with the horizontal. Air resistance is negligible. The moment of inertia of the cylinder about the central axis is $\frac{1}{2}MR^2$.

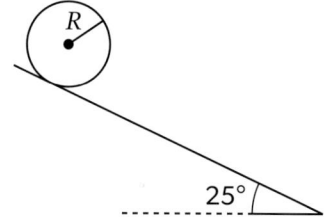

(a) (i) Identify the force that provides torque to the cylinder. [1]

(ii) Show that the linear acceleration of the cylinder is $\frac{2}{3}g \sin 25°$. [4]

The cylinder starts from rest and rolls down a distance of 1.5 m along the ramp.

(b) Calculate the final linear speed of the cylinder. [2]

The cylinder is replaced with a block of the same mass. The block slides down the ramp through the same distance and reaches the same final speed as the cylinder.

(c) (i) Calculate the coefficient of dynamic friction between the block and the ramp. [2]

(ii) Calculate $\dfrac{\text{kinetic energy of the block}}{\text{kinetic energy of the cylinder}}$ at the bottom of the ramp. [3]

The speed of the block is measured with a motion sensor that is placed at the top of the ramp so that the block moves away from the sensor. The motion sensor emits a sound wave of frequency 24 kHz towards the block and detects the frequency of the wave reflected off the block and returning to the sensor. The speed of sound in air is 340 m s^{-1}.

(d) Calculate the minimum frequency detected by the sensor. [2]

In another experiment, a ramp is formed by a pair of parallel conducting rails that make an angle of 25° with the horizontal. The rails are electrically connected at the bottom of the ramp. A conducting rod is released from rest and slides down the ramp without friction. The system is in a uniform vertical magnetic field B.

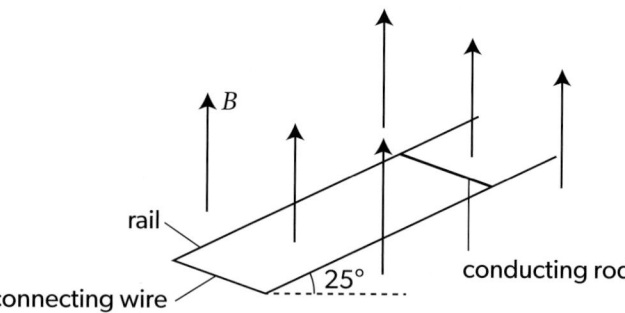

(e) (i) Explain why the rod achieves a constant speed. [3]

(ii) The constant speed of the rod is 0.45 m s^{-1}. The mass of the rod is 0.12 kg, its resistance is 2.4 Ω and the resistance of the rails is negligible. Determine the current induced in the rod. [3]

Index

Page numbers in *italics* refer to Practice Problem and Exam questions. Key words are in **bold**.

absolute refractive index 114
absorption spectra 183–4
ac generators 177–9
acceleration 1–6, 9, 11–13
   angular 32–3, 35–6
   charged particles 167
   due to free-fall (g) 2, 3, 4, *137*
   instantaneous 2–3, 99
   simple harmonic motion 97, 98, 99, 102–3, 105
   uniform 2–3
acceleration–displacement graphs 97
acceleration–time graphs 1–2, 99, 102–3
activity 205
adiabatic changes 78, 80–1
air (fluid) resistance 5–6, 10
albedo 64–5, *66–8*
alpha decay 197, 198–9, 204
alpha particles 180–2, 198
ammeters 88
amplitude 98, 123, 124, 128
angular acceleration 32–3, 35–6
angular displacement 19, 33
angular frequency 97, 98
angular impulse 35, 36
angular momentum 34–5
   change in 36–7
angular speed 19, 22
angular velocity 19, 32–3, 35–7
antineutrinos 197, 198
antinodes 123–4
apparent brightness 60, 61, 223–4
Archimedes' principle 10
assessment iv–v, *239, 243*
assumptions 4, *71–2*
astronomical distances 139, 222–5
astronomical units 139, 222
atomic mass 201
atomic structure 180–7, *187–8*
atoms 53, 58–9
Avogadro constant 70

background radiation 205–6
banking 20
beta-minus decay 197, 198–200
beta particles 198–200
beta-plus decay 197, 198, 207

binding energy 201–3, 218–19
black bodies 60, 61, 64
black body radiation 63–4
black holes 222
blueshift 132
Bohr model of hydrogen atom 184–5
boiling 53–4
Boltzmann constant 54
bound states 185
buoyancy 10, 11

carbon–nitrogen–oxygen (CNO) cycle 219
Carnot cycle 80–1
centre of gravity 138
centrifugal force 19
centripetal acceleration 19, 20
centripetal force 19, 20
chain reactions 212
charged particles 163–4, 167–71
chemical cells 93–4
circuits 87–94, *95–6*
circular motion 18–22
climate change 63, 65, 67
closed systems 77
coherent waves 115
coils 156, 173, 175–9
collisions 13, 14, 15–16, 17–18
combining uncertainties 231
command terms v–vi
Compton scattering 191–2
condensing 53–4
conduction 57–9, 87
conductors 89–90, 159, 160
   in a magnetic field 164–5, 173–9
conservation of angular momentum 35
conservation of energy 14, 24, 25
conservation of momentum 13, 14, 17
conservative forces 25
constructive interference 116
contact forces 9–11
convection 57, 59–60
conventional current 93
Coulomb constant 150, 181
Coulomb's law 150–1, 181
count rate 205
critical angle 114–15
critically damped oscillations 128

damping 128
de Broglie wavelength 186, 193
decay constant 208–9
decay equations 197

density 59, 72, 73
derived units 227
destructive interference 116
diffraction 117–20
diffraction gratings 119–20, 183, 193
diffraction patterns 118–19
direct current 93
displacement 1, 2, 6, 109
   angular 19, 33
   SHM oscillations 98, 99, 102–3, 104
   waves 115
displacement–distance graphs 108, 110
displacement–time graphs 1, 99, 102–3, 104, 108
distance 1, 139, 222–5
distance–time graphs 1–2, 102–3
distance travelled 1, 3, 6
Doppler effect 131–4, *135*, 219
double-source interference 115–16
drag force 5–6, 10–11, 144–5
drift speed 87
driving frequency 128

efficiency 27–30
Einstein, Albert
   photoelectric effect 189, 192
   special relativity 41–2
elastic collisions 14, 17–18
elastic potential energy 25, 99, 100
elastic restoring force 10
electric charge 87–8, 149–55, 157, 159–60
electric currents 87–8, 164–7
electric field lines 153–5, 158
electric field strength 151–2, 157–8, 159
electric fields 150–5, 157–61, *162*
   charged particles in 167–8, 170–1
electric force 152, 167, 170
electric potential 157–9
electric potential difference 88, 92, 93, 157
electric potential energy 157, 160
electrical power 89–90
electrical resistance 88–9
electromagnetic fields, motion in 163–71, *171–2*
electromagnetic radiation 60–1, 189–91
electromagnetic waves 60–1, 110, 115, 132
electromotive force (emf) 93, 173–9
electron antineutrinos 198
electron neutrinos 198, 199–200

electrons 174, 180, 185–7, 189–93
electrostatic forces 12, 19, 168
electrostatic induction 149–50
emission spectra 183, 184–5, 219
emissivity 64, 66–7
empirical equations 10
empirical laws 72, 138
energy 24–30, *30–1*
    binding energy 201–3, 218–19
    internal energy 55–6, 78, 81
    of photon 191–2
    radioactive decay 198–203
    released in fission 211
    released in fusion 218–19
    see also **kinetic energy**; **potential energy**
energy density 29
energy levels 183, 185–6, 199
energy pathways 24–5
energy states 66
energy stores 24–5, 29
energy transfers 24–30, 77–8, 80–3
    SHM oscillations 99–101, 104
    thermal 53–61, *61–2*
    waves 123
enhanced greenhouse effect 65, 66
entropy 81–3
equilibrium position 97, 98
equipotentials 141, 158
error bars 235
errors 230, 241
escape speed 144
estimations 229
exact simple harmonic motion 100
exam preparation vii, *244–52*
exponential change 208

Faraday's law 174–5
field lines 137, 141, 153–6, 165, 173
first harmonic 124, 125
fission 198, 202, 203, 211–16, *216–17*
Fleming's left-hand rule 165
fluids 10–11, 59
force–distance graphs 27
force–time graphs 16
forces 8–22, *23*, 25–6
    action–reaction pairs 12
    balanced/unbalanced 11
    on charged objects 150–2
    and **circular motion** 18–22
    contact forces 9–11
    electric 152, 167, 170
    electrostatic 12, 19, 168
    external 13
    free-body diagrams 8–9
    gravitational 12, 19, 136–7

magnetic 19, 163–7, 169, 170
and **momentum** 13–15, 16
on moving charged particles 163–4, 167–8, 169, 170
Fraunhofer lines 183–4
free-body diagrams 8–9
freefall 2–3
freezing 53–4
frequency 98, 123, 124–5, 128–9
    angular frequency 97, 98
    Doppler effect 131–4
frictional forces 9–10, 20
fundamental units 227
fusion 198, 202, 203, *225*
    in stars 217–19, 220, 221, 222

Galilean relativity 40–1, 42
Galilean transformations 41, 42
gamma decay 197, 199, 207
gas laws 69–71, *76*
gases 53–5, 69–75, *76*, 77–81
Geiger–Marsden–Rutherford experiment 180–1
general gas equation 70–1
geostationary orbit 143
geosynchronous orbit 143
graphs 233–6
    area under 1–2, 16, 27, 36–7, 234
    force–time graphs 16
    gradient 1–2, 3, 99, 234
    I–V characteristics 89–90
    intercept 234
    motion graphs 1–3, 6, 12
    pressure–volume (*PV*) graphs 77, 78, 79, 80, 81
    SHM graphs 97, 99, 102–3, 104–5
    torque–time graphs 36
    wave graphs 108, 110
gravitational attraction 219
gravitational field strength 136, 137–8, 142–3
gravitational fields 136–47, *147–8*
gravitational forces 12, 19, 136–7
gravitational potential 140–2
gravitational potential difference 140
gravitational potential energy 25, 99, 141, 142, 143
greenhouse effect 63–8, *68–9*
greenhouse gases 65–7
grey bodies 64
ground state energy level 185, 186
gyroscopes 34

half-life 205–6, 207, 209
harmonic series 124, 125
heat capacity 55–7

heat engines 80–1
Hertzsprung–Russell diagram 220–1

I–V characteristic graphs 89–90
ideal ammeters 88
ideal gases 55, 70–1, 73–5, 78–80
ideal voltmeters 88
impulse 16
incidence angle 113, 114
induction 149–50, 172–9, ***179***
inelastic collisions 14, 16
inertia, moments of 33–4, 35
inertial reference frames 40–1, 42–50
inexact simple harmonic motion 100
infinity 140, 141
infrared radiation 65
instability strip 221
intensity 60, 61, 63, 67–8, 112–13
interference 115–20
internal assessment (IA) 239–43
    checklist 243
    conclusions 241
    data collection 241
    experiment evaluation 242
    referencing 243
    research question 239–41
internal energy 55–6, 78, 81
invariant quantities 43
inverse Lorentz transformations 41–2
inverse square law 113, 157
ionizing radiation 196, 197
ions 87
isobaric changes 78
isolated systems 77
isothermal changes 78, 80–1
isotopes 196–7
isovolumetric changes 78, 84

Kelvin scale 54
Kepler's laws 138–9
kinematic equations 2–5, 6
kinetic energy 14, 15–16, 18, 25
    of a gas particle 73–4
    particles 54–5, 58–9, 181–2
    rockets and satellites 143–5
    rotational 35, 37, 38
    simple harmonic motion 99–100, 104
kinetic model of a gas 16, 71–5

latent heat 53, 55, 56–7
length contraction 43, 49
Lenz's law 174
light waves 117–18, 119–21
light years 222
linear speed 19

lines of constant radius 221
liquids 53–4, 56, 59, 70
longitudinal waves 107–9, 110
Lorentz transformations 41–3
luminosity 60, 61, 220, 224

magnetic field lines 155–6, 173
magnetic field strength 164, 173, 177
magnetic fields 155–7, *162*
    charged particles in 168–71
    coil rotating in 177–9
    force on a moving charge 163–4
    force on current-carrying conductor 164–7
    induction 172–9
magnetic flux 173, 174–5
magnetic flux density 173, 175
magnetic forces 19, 163–7, 169, 170
main sequence stars 221
mass 11, 13, 17, 138
    of Earth 137
    mass–spring systems 97–102, 103, 105
    stars 220, 221–2
mass defect 201–2
mechanical energy 25
melting 53–4, 56–7
method of mixtures 56
microstates 82–3
Millikan's experiment 154–5
Minkowski diagram *see* **spacetime diagrams**
modelling 54
molar mass 70
mole (mol) 70
molecules 53, 70–4
moments of inertia 33–4, 35
momentum 13–18, *23*
    see also **angular momentum**
monochromatic light 117–18, 119, 189
motion 1–6, *7*, 12, 18–22, 32–3
    in electromagnetic fields 163–71, *171–2*
    see also **Newton's laws of motion**
motion graphs 1–3, 6, 12
muon decay experiment 49–50

natural frequency 128, 129
neutrinos 197, 198, 199–200
neutron number 180, 203
neutron stars 222
neutrons 180, 196–7, 201, 203–4
    neutron-induced fission 211–14
newtons 11, 228
Newton's law of gravitation 136–7
Newton's laws of motion 11–13
    circular motion 18, 19, 20

rotational motion 34, 35
nodes 123–4, 126
non-conservative forces 25
normal force 9
nuclear binding energy 201–3
nuclear density 182
nuclear radius 182
nuclear reactors 212–14
nucleon number 180, 182, 202, 203
nucleus 180–2, 198–9, 201–3
nuclides 197, 202, 203–4, 205
    decay constant 208–9
    fission 211–12

ohmic conductors 89–90
Ohm's law 89
orbits 138–9, 143–5, 186
orders (of a spectrum) 119
orders of magnitude 229
oscillations 97–106, 124–5
    forced 128–9

parallax measurements 222–4
parsecs 222
particles 53, 70–5
pendulum bob 100
pendulums 97, 100–1
period 19
permittivity of free space 150
perpendicular fields 170–1
phase angle 98, 99
phase changes 53
phase difference 117, 123, 129
photoelectric effect 189–91, 192
photoelectric equation 189
photons 65–6, 183, 184, 189–92
Planck constant 183
position 1
postulates of special relativity 41
potential difference 88, 92, 93, 140, 152, 157
potential energy 25, 55, 140
    elastic 25, 99, 100
    electric 157, 160
    gravitational 25, 99, 141, 142, 143
    simple harmonic motion 99–100, 104, 105
potential gradient 141
potentiometers 91–2
power 27–30, *30–1*
    electrical 89–90
power stations 29, 213–14
pressure 69–71, 72–3, 74–5, 77–8
    longitudinal waves 109
    in stars 219
pressure–volume (*PV*) graphs 77, 78, 79, 80, 81
principal quantum number 185
projectile motion 3–6
proper length 43
proper time interval 43
proton number 180, 203
proton–proton cycle 218–19
protons 180, 185, 186, 196–7, 201, 203–4
protostars 221

quantum 183
quantum physics 189–94, *195*

radiation 63–5, 189–91
    background radiation 205–6
    ionizing 196, 197
    thermal radiation 57, 60–1
radiation pressure 219
radioactive decay 196–209, *210*
    decay constant 208–9
    energy 199–203
    half-life 205–6, 207, 209
    nuclear stability 202–4, 208–9
    uses 206–7
radioactive waste 214
random radioactive decay 197
rays 112–13
real gases 70, 71–2, 74–5
red giant stars 221
redshift 132
reference frames 40–1, 42–50
reflected angle 113–14
reflection 113–15, 123, 126
refraction 113–15
refraction angle 113
refractive index 114
relative permittivity 150
relativistic velocity addition equations 42
resistance 88–9, 90–2
resistivity 89
resistors 90–2, 93
resonance 65–6, 128–9, *130*
resultant forces 8–9, 10, 11
right-hand rule 156
rigid body mechanics 32–9, *39*
rockets 143–4
rolling objects 37–8
rotational kinetic energy 35, 37, 38
rotational motion 32–3
Rutherford scattering 180–1

Sankey diagrams 29
satellite orbits 138–9, 143–5
scalar quantities 1, 11

Index

scalars 237
scientific notation 228
scientific writing 242
self-induction 177
SI prefixes 228
SI units 227–8
significant figures 229
simple harmonic motion 97–106, *106–7*
simple pendulums 100–1
Snell's law 113
solar cells 93–4
solar constant 63
solids 53–4, 56, 58–9, 70
sound waves 109, 116, 131–2, 133
space-like intervals 43
spacetime diagrams 44–9
spacetime interval 43
special relativity 41–3, 50, *51–2*
specific charge 169
specific heat capacity 55–7
specific latent heat 55, 56–7
speed 1–6, 12, 19, 22
    escape speed 144
    and **momentum** 18
    terminal speed 5, 10–11
    wave speed 108–9, 110
speed–time graphs 1–3, 12
spontaneous radioactive decay 197
spring constant 10, 101, 228
standing waves 122–7, *130*
stars 217–25, ***225–6***
states of matter 53
Stefan–Boltzmann law 60
Stokes' law 10–11
strong nuclear interaction 201
supergiant stars 221, 222
supernova 222
superposition 115, 124
surface charge density 154
surface frictional force 9–10
suvat **equations** 2, 3, 4, 5

temperature 53–5, 78
    of Earth 63–9
    of a gas 70–5
    of stars 219, 220, 221
tension 20, 22
terminal speed 5, 10–11
theoretical models 72
theory 138
thermal conductivity 58–9
thermal efficiency 81
thermal-energy power stations 29
thermal energy transfers 53–61, ***61–2***
thermal neutrons 213
thermal pressure 219
thermal radiation 57, 60–1
thermodynamic systems 77
thermodynamics 77–85, ***85–6***
    first law 77–80
    second law 81–3
threshold frequency 189
time dilation 43, 49
time-like intervals 43
time period 98
torque 34–7
total internal reflection 114
transverse waves 107–9
triple point 54

ultraviolet radiation 65, 189
uncertainties 230–1
unified atomic mass unit 201
units 1, 11, 88, 227–9
    astronomical distances 139, 222
    energy 152
upthrust 10

variables 240, 241
vector quantities 1, 11, 13
vectors 237–8
velocity 1, 3–4, 6, 11
    angular 19, 32–3, 35–7
    constant 12
    instantaneous 99
    and **momentum** 13, 15–16, 18
    relativistic velocity addition 42
    simple harmonic motion 99, 102–3
velocity–time graphs 1, 6, 99, 102–3
viscosity 10, 11
viscous drag force 5–6, 10–11, 144–5
voltage 88
voltmeters 88
volume of a gas 70–1, 73–4, 77–8, 80–1, 83

wave–particle duality 192–3
wave speed 108–9, 110
wave theory 192
wavefronts 112, 131–2
wavelength 108–9, 123, 191–2
waves 107–10, *111*
    Doppler effect 131–4, *135*, 219
    standing waves 122–7, *130*
    wave phenomena 112–21, *121–2*
weight 9, 12, 22, 154
white dwarf stars 221
Wien's displacement law 60
work 81–3
work done 25–7, *30–1*
    by or on a gas 77–8, 80–1
work function 189
worldlines 44–5

X-rays 191–2

Young's double-slit interference 116